Karl Wilhelm Nageli

The Microscope in Theory and Practice

Karl Wilhelm Nageli

The Microscope in Theory and Practice

ISBN/EAN: 9783337075507

Printed in Europe, USA, Canada, Australia, Japan

Cover: Foto ©berggeist007 / pixelio.de

More available books at **www.hansebooks.com**

THE MICROSCOPE

IN

THEORY AND PRACTICE.

Translated from the German

OF

PROF. CARL NAEGELI

AND

PROF. S. SCHWENDENER.

WITH NUMEROUS ILLUSTRATIONS.

London:
SWAN SONNENSCHEIN, LOWREY & CO.,
PATERNOSTER SQUARE.
1887.

PREFACE.

This translation of Nægeli and Schwendener's well-known treatise "Das Mikroskop" was commenced by Mr. Frank Crisp, Secretary of the Royal Microscopical Society, immediately after the publication of the last (German) edition (1877), with the intention—as indicated by him in a communication to the Quekett Microscopical Club—of filling up a blank in English microscopical literature in regard to the scientific technical treatment of the theory of the Microscope, in which English text-books were so deficient.

The student refers in vain, even at the present date, to English works on the Microscope for explanations of the theory of the construction of objectives, eyepieces, &c., or for the discussion of the phenomena of diffraction and polarisation in their connexion with the Microscope, or for any scientific treatment of the question of interpreting microscopical images or the theory of microscopic observation. These subjects are dealt with systematically in German works only, and notably in that of Nægeli and Schwendener.

The translation was thus undertaken with a view to placing before English readers the then best known collective exposition or technical treatment of these points by German writers.

When the rough draft of the translation was completed, the first five sheets (80 pp.) were revised and put in type, but in consequence of prior claims upon his time in connexion with the Royal Microscopical Society, Mr. Crisp was compelled to relinquish the task of further revision and of passing the volume through the press, a labour which was undertaken by Mr. John Mayall, jun., one of the editors of the Society's Journal.

Just as the printing was completed a fire destroyed the premises of the printers, and the whole of the printed sheets of the volume were burnt except one set as far as p. 374, which the publishers had retained in their possession, together with a few of the woodcuts.

Under these circumstances the publishers had to consider the alternatives (1) of abandoning the issue of the volume; or (2) of incurring the additional expense of re-translating the portion of the work totally lost by the fire, replacing the missing woodcuts, and reprinting the whole; or (3) of reprinting as far as p. 374 only, omitting therefore Part VIII. (Microphysics), Part IX. (Microchemistry), and Part X. (Morphology). It was finally decided to adopt the last course,—hence the present issue.

Whilst it is much to be regretted that this translation should only now be issued, microscopists will no doubt appreciate the advantage of having a version in English of a work which has received high commendation from both English and foreign critics; and it is hoped this volume may be supplemented before long by an English version of the further researches in Microscopical Optics by Prof. E. Abbe, of Jena, which have extended so much our knowledge of the matters dealt with in Nægeli and Schwendener's work.

CONTENTS.

CHAP.		PAGE
	PREFACE	v—vi

PART I.

THEORY OF THE MICROSCOPE.

I.	INTRODUCTION	1—9
II.	ANALYTICAL DETERMINATION OF THE PATH OF THE RAYS IN REFRACTING SYSTEMS	9—23
II.	DETERMINATION OF THE DISTANCES OF CORRESPONDING IMAGE-POINTS FROM THE AXIS IN THE CASE OF ANY GIVEN INCLINATION OF THE RAYS	24—26
IV.	THE COMPONENT PARTS OF THE MICROSCOPE ...	27—50
	1.—THE OBJECTIVE ... ... 27—34	
	2.—THE EYE-PIECE :—	
	a. The Campani (or Huyghenian) Eye-piece 35—42	
	b. The Ramsden Eye-piece 42, 43	
	c. The Aplanatic and the Orthoscopic Eye-piece ... ... 43—45	
	d. The Erecting Eye-piece 45, 46	
	e. The Spectral Eye-piece 47, 48	
	3.—MEANS OF DIVIDING THE PENCILS OF RAYS 48—50	
V.	CHROMATIC AND SPHERICAL ABERRATION	50—65
	1.—CHROMATIC ABERRATION ... ... 50—53	
	2.—SPHERICAL ABERRATION ... 53—65	
VI.	INFLUENCE OF THE COVER-GLASS ...	65—68
VII.	THE FLATNESS OF THE FIELD OF VIEW ...	68—78
VIII.	THE CENTERING OF THE SYSTEMS OF LENSES ...	79—84
IX.	BRIGHTNESS OF THE FIELD OF THE MICROSCOPE ...	84—90

CONTENTS.

PART I.—*continued.*

CHAP.		PAGE
X.	THE OPTICAL POWER OF THE MICROSCOPE	90—101
	1.—Defining and Penetrating Power according to the earlier authors	91, 92
	2.—Importance of the Angle of Aperture	93—97
	3.—Fusion of the Interference and Dioptric Images	98, 99
	4.—Ratio of Aperture and Focal Length	99—101
XI.	DIFFRACTIONAL ACTION OF THE APERTURE OF THE LENSES	102, 103
XII.	ILLUMINATION	103—115
	1.—Illumination by Transmitted Light	104—111
	2.—Illumination by Reflected Light	111—115

PART II.
THE MECHANICAL ARRANGEMENT OF THE MICROSCOPE.

I.	GENERAL RULES FOR THE CONSTRUCTION OF STANDS	116—121
II.	THE STANDS OF MODERN OPTICIANS	122—125

PART III.
TESTING THE MICROSCOPE.

I.	TESTING OF THE OPTICAL POWER IN GENERAL	126—149
	1.—Absolute Power of Discrimination:—	
	Images of Wire-gauze as Test-objects	127—132
	Organic Test-objects	132—139
	Nobert's Test-plate	139, 140
	2.—Relative Power of Discrimination	140—147
	3.—Defining Power	148, 149
II.	TESTING THE SPHERICAL ABERRATION	149—158
III.	TESTING THE CHROMATIC ABERRATION	158—167
IV.	TESTING THE FLATNESS OF THE FIELD OF VIEW	168, 169
V.	TESTING THE CENTERING	170, 171
VI.	DETERMINATION OF THE ANGLE OF APERTURE	172—177
VII.	DETERMINATION OF THE MAGNIFYING POWER AND FOCAL LENGTH	178—183
	1.—Magnifying Power	178—181
	2.—Focal Length	181—183
VIII.	DETERMINATION OF THE CARDINAL POINTS	184—188

PART IV.

THEORY OF MICROSCOPIC OBSERVATION.

CHAP.		PAGE
	INTRODUCTION	189—191
I.	SPHERICAL AND CYLINDRICAL OBJECTS	191—217
	1.—Air-Bubbles in Water	191—202
	2.—Globules of Oil in Water	202—206
	3.—Hollow Spheres and Hollow Cylinders	206—217
II.	OBJECTS OF IRREGULAR FORM	217—226
	1.—Membranes with small Depressions or Holes	217—219
	2.—Membranes bounded by one Plane and one Undulating Surface	219, 220
	3.—Membranes bounded by Parallel Undulated Surfaces	220, 221
	4.—Alternate Solid and Aqueous Layers	221
	5.—Elevations and Depressions as opposed to Dense and Loose Layers	221, 222
	6.—Vision through Stereoscopic Binocular Microscopes	222—226
III.	INTERFERENCE PHENOMENA	226—247
	A. In the Microscope:—	
	1.—Delineation of the Fine Structure of Objects by Interference	226—236
	2.—Reflexion of Light by Small Spheres, Granules, Fine Threads, &c., and the Interference Phenomena thereby produced	236—239
	3.—Interference Lines caused by the withdrawal of a Source of Light of small extent	239, 240
	B. Interferences in the Plane of Adjustment:—	
	1.—Interference of Direct with Reflected Light	240—243
	2.—Interference of Refracted with Reflected Light	244
	3.—Interference of Refracted and Direct Light	245
	4.—Interference Colours of Thin Plates	245—247
IV.	OBLIQUE ILLUMINATION	247—251
V.	THE PHENOMENA OF MOTION	252, 253
VI.	DIFFERENCES OF LEVEL	254, 255

PART V.

THE SIMPLE MICROSCOPE AND THE LANTERN MICROSCOPE.

CHAP.   PAGE
- I. THE SIMPLE MICROSCOPE   256—267
  - A. GENERAL PRINCIPLES   256, 257
    1. —Aperture of the Effective Cones of Light   257—259
    2. —Brightness   259
    3. —Curvature of the Field of View   260
    4. —The Magnifying Power   260
    5. —The Extent of the Field of View   261
  - B. THE OPTICAL ARRANGEMENT   261—265
  - C. THE MECHANICAL ARRANGEMENT   266
- II. THE LANTERN MICROSCOPE   267

PART VI.

TECHNICAL MICROSCOPY.

- I. THE USE OF THE MICROSCOPE   268—274
  1. —Illumination   268—271
  2. —The Selection of the Magnifying Power   271
  3. —Employment of Cover-Glasses   271, 272
  4. —Preservation of the Instrument   273
  5. —Care of the Eyes   273, 274
  6. —The Work-table   274
- II. PREPARATION AND TREATMENT OF SPECIMENS   275—280
- III. THE PRESERVATION OF MICROSCOPIC SPECIMENS   280—287
- IV. THE MEASUREMENT OF MICROSCOPIC OBJECTS   288—296
- V. THE DRAWING OF MICROSCOPIC OBJECTS   296—303

PART VII.

THE PHENOMENA OF POLARISATION.

- I. ARRANGEMENT OF THE POLARISING MICROSCOPE   304—319
  1. —The Polariser   304—311
  2. —The Analyser   311—315
  3. —The Apparatus for Rotating the Objects   315—319

PART VII.—*continued*.

CHAP.		PAGE
II.	THE ACTION OF ANISOTROPIC CRYSTALLOID BODIES VIEWED SEPARATELY	319—341
	1.—The Ellipsoid of Elasticity ... ... 319—323	
	2.—The Phenomena of Polarisation in relation to the Ellipsoid of Elasticity ... ... 323—330	
	3.—Determination of the Axes of Elasticity 330—341	
III.	THE ACTION OF TWO SUPERPOSED CRYSTALLOID BODIES, WHOSE PLANES OF VIBRATION INTERSECT OBLIQUELY	341—349
IV.	THE ACTION OF CYLINDRICAL AND SPHERICAL OBJECTS COMPOSED OF CONCENTRICALLY GROUPED ANISOTROPIC ELEMENTS	349—366
	1.—Cylindrical Objects ... ... 349—362	
	2.—Spherical and Oval Objects:—	
	A. Objects with Equal Diameters ... ... 363—365	
	B. Objects with One Axis ... ... 365—366	
V.	ON SOME STRUCTURAL PECULIARITIES OF ORGANISED SUBSTANCES	366—368
VI.	COLLECTION OF EXAMPLES	368—373
	A. Cylindrical Objects ... ... 370—372	
	B. Spherical Objects 372	
	C. Disc-like Crystals of Amylo-dextrine ... 372, 373	
VII.	ON THE EMPLOYMENT OF NÖRRENBERG'S "POLARISING MICROSCOPE"	373, 374

PART I.

THEORY OF THE MICROSCOPE.

I.

INTRODUCTION.

THE compound dioptric Microscope consists, in principle, of two systems of convergent refracting lenses, of which the one directed to the object is termed the *Objective*, and the other, turned to the eye, the *Eye-piece*. The objective forms an inverted image of an object placed somewhat beyond its principal focus, and this image is viewed through the eye-piece just as through an ordinary magnifying lens.

The optical action of the Microscope may be readily comprehended from Fig. 1, which shows the process of the formation of the image. The rays of light proceeding from the object $a\ b$ unite after their passage through the objective A to form the real image $b'\ a'$, the position and magnitude of which depend solely upon the focal length of the objective and the distance of the object. If we denote these quantities by f and p respectively, and the distance of the image by p^*, then their mutual dependence is expressed by the well-known equation

$$\frac{1}{p} + \frac{1}{p^*} = \frac{1}{f}.$$

The image $b'\ a'$ acts as a real object to the eye-piece B, and as its distance is regulated so that it lies somewhat nearer to the eye-piece than its principal focus, a virtual image $b''\ a''$ is formed, which is visible to the eye when placed over B. If we denote by p_1 and

$p_1{}^*$ the distances of the images $b'\,a'$ and $b''\,a''$ from the eye-piece, and by f_1 the focal length of the latter, we obtain the equation

$$\frac{1}{p_1} - \frac{1}{p_1{}^*} = \frac{1}{f_1}.$$

If the distance between the eye-piece and the objective is constant, and their focal lengths f and f_1 are given, then $p_1{}^*$ is deter-

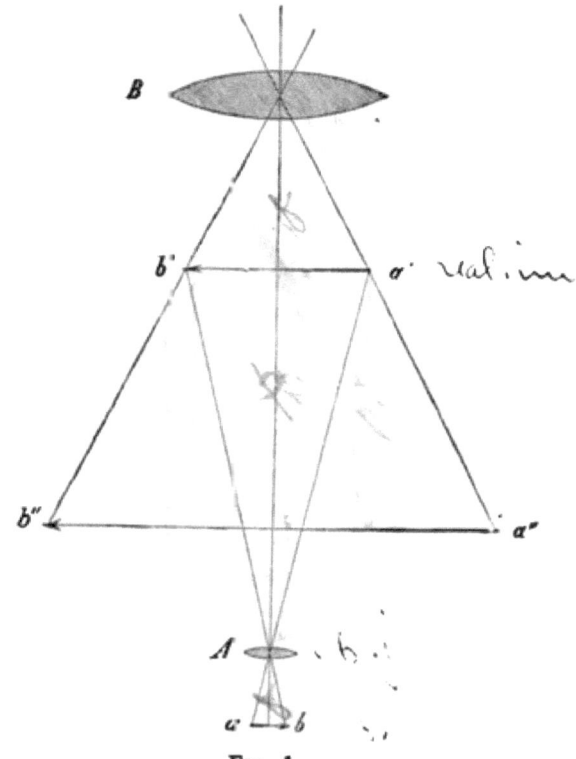

Fig. 1.

mined by p_1, p_1 by p^*, and p^* by p, and, consequently, the first quantity $p_1{}^*$ by the last p. It is always possible, therefore, so to regulate the distance of the object from the objective that the final virtual image shall be at the distance from the eye of distinct vision.

The linear dimensions of the image and object are in the ratio of their respective distances from the image-forming lens. In the case of the objective this relation is expressed by $\dfrac{p^*}{p}$, and in that of the eye-piece by $\frac{p_1^*}{p_1}$. The combination of the two in the Microscope gives therefore an amplification of $\frac{p^* p_1^*}{p \, p_1}$ times.

In order that the image seen by the eye may be distinct and bright, the spherical and chromatic aberration must be reduced to a minimum, and the apertures of the lenses must be as large as possible. These conditions are, however, in antagonism, as the aberrations of a single lens increase with its aperture. Distinctness can, therefore, be gained only at the cost of brightness, and conversely. This inconvenience reaches such a pitch in the older Microscopes, that when their images are at all satisfactory in point of distinctness they are very deficient in light, even with moderate magnifying powers.

It was only in more recent times that the happy idea was suggested of constructing the objective of two or three compound lenses approximately achromatic. It was soon found that by their skilful combination both kinds of aberration could be eliminated far more perfectly than had been at all feasible in the earlier arrangements by merely diminishing the aperture. In addition to this, the important advantage was gained, that, as the deflection of the rays which form the image was divided between several lenses, their curvatures might be considerably reduced, and thus a larger aperture obtained.

We shall, hereafter, examine more closely the principles of achromatism and aplanatism as far as they concern the construction of the Microscope, and will for the present assume that we have an ordinary objective composed of three compound lenses, just as it leaves the hands of the optician, and will trace the course of the rays of light from the object to their reunion in the image. Our explanation will be based upon the dioptric investigations of Gauss,[1] which expound the laws of the refraction of light in the case of a concentric system of any number of spherical surfaces, between which lie media with varying refractive indices. Though the suppositions which form the basis of these investigations are not applicable, in certain essential points, to the course of the rays in the Microscope, still the results furnish an excellent view of the conditions met with. In a subsequent chapter we shall give the

---

[1] C. F. Gauss: "Dioptrische Untersuchungen."—Abhandl. Göttinger Gesell., I., 1843.

actual theories of Gauss, with some abridgments and modifications, which will facilitate their comprehension by non-mathematical readers. It will be sufficient, here, if we summarize the most important of his conclusions, and demonstrate the points of difference and analogy which exist between an infinitely-thin lens and systems of lenses.

An infinitely-thin convex lens (that is, one whose thickness as compared with its radii of curvature may be disregarded) has the property of transmitting rays which are directed to its optical centre without refraction, and of uniting incident parallel rays in one point, the principal focus.

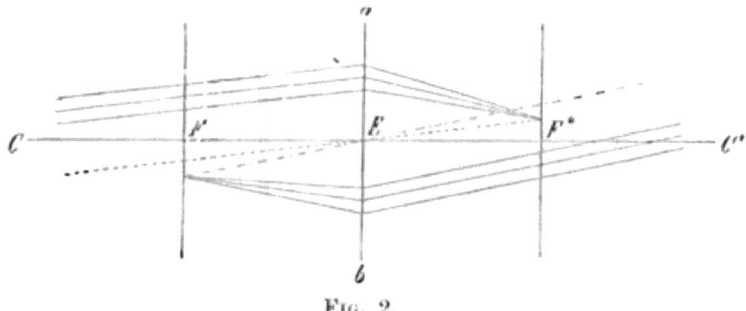

FIG. 2.

If $a\ b$ (Fig. 2) is the given lens, $C\ C'$ its axis (that is, the line in which the centres of curvature lie), E its optical centre, and $F\ F^*$ two points at a distance from the lens equal to its focal length, through which two planes are drawn at right angles to the axis; then the rays which start from a point in the plane F will emerge from the lens as parallel rays, and, conversely, rays which are parallel when they enter the lens will be united in one point in the plane F^*. To every incident cone of rays, whose apex lies in the plane F, there corresponds an emergent cylinder of rays, and to every incident cylinder of rays a corresponding cone of rays whose apex lies in the plane F^*. If the rays, in addition to being parallel to each other, are parallel to the axis of the lens, then their point of union is situated on the axis at the other side of the lens, that is, at F or F^*.

On this property, coupled with that possessed by spherical surfaces of so refracting rays proceeding from a point that they (or their prolongations) again intersect in a point, is based the familiar construction, by means of which the position and magnitude of the image of a given object can be easily determined. Let two lines $p\,r$ and $p\,q$ be drawn from the extremity p (Fig. 3) of the object $p\,t$, one of which cuts the axis in F, whilst the other proceeds parallel to it. From q and r, the points of intersection with the plane E, let two other lines $r\,p^*$ and $q\,p^*$ be drawn, whose directions are, as it were, interchanged with the former lines, $r\,p^*$ being parallel to the axis, and $q\,p^*$ intersecting it at F^*. It is evident that the two lines $p\,r$ and $p\,q$ represent two

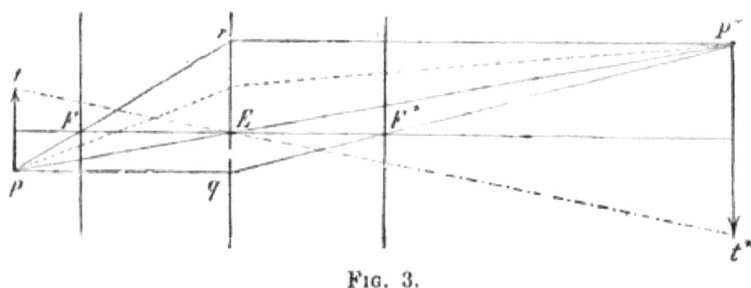

Fig. 3.

rays, which, after refraction, unite at p^*, and as all the other rays which proceed from p are refracted in a like manner to p^*, the image of the point p is formed at p^*. By a similar construction the image of the other extremity t, as well as of all other points between p and t, may be obtained. To simplify the matter still further we can replace one of the two lines by the *ray of direction* $p\,E\,p^*$, which passes through unrefracted.

If we would extend the terminology introduced by Gauss to this simple case of an infinitely-thin lens, the planes F and F^* would be termed the *focal planes*, and the plane E the *principal plane*. The distances of the focal planes from the principal plane represent the focal distances, and their points of intersection with the axis, the focal points.

It must, however, be borne in mind that the above construction, as well as the formulæ generally given in the text-books of Physics, by which the position of the image is calculated, are only accurate, even for infinitely-thin lenses, on the two suppositions that the rays from the object form very small angles with the axis, and that the effective portion of the refracting surface is only a small part of the whole spherical surface. These are limitations which we shall retain in the following examination of systems of lenses, so that the aberration due to spherical form will be disregarded.

When the refracting surfaces, which bring about the union of the rays (or of their prolongations) to form the image, are situated so far from each other that a neglect of these distances in comparison with the other constants is inadmissible, the rules by which the paths of the refracted rays are determined undergo an essential modification. This case may occur as well with single lenses of considerable thickness as with systems of lenses as employed in the Microscope, and in the treatise of Gauss, the most general case is supposed of there being no limitation of the distances between the refracting surfaces.

Such a system of refracting surfaces, with their centres of curvature lying in a straight line, has, in the first place, the property, in common with an infinitely-thin lens, of reuniting in a point rays emanating from a point, or, in other words, of refracting *homocentric* pencils of light in such a manner that after the refraction they still remain homocentric. Secondly, it may be shown that under all circumstances there exist two points on the common axis, F and F^*, which exactly coincide with the foci of a single lens, since the planes drawn through these points at right angles to the axis possess all the properties of focal planes. As, however, the latter may, under certain circumstances, lie within the limiting surfaces of the system, they can be characterized only by the definition already given above, which is universally applicable, viz. :—that to every incident homocentric pencil of rays whose centre lies in the anterior[1] focal plane, there corresponds an emergent parallel pencil; and, conversely, to each parallel pencil an emergent homocentric one whose centre lies in the posterior focal plane.

The distinguishing feature of a compound refracting system consists in this,—that two points, E and E^*, take the place of the optical centre, the former in reference to the incident rays, and the latter to the emergent rays. These points are called the *principal points* of the system, and the planes drawn through them at right angles to the axis are called the *principal planes*. Their importance will be at once evident, if we try to determine by construction the image of any object $a\,b$ (Fig. 4), as before in the case

---

[1] With regard to the direction of transmission.

of the single lens. A ray proceeding from b through the first principal point E undergoes, in a sense, no deviation, since the directions of transmission before and after refraction form the same angle; but the emergent ray, produced backwards, cuts the axis at the second principal point E^* instead of E. It may be treated, therefore, as an unrefracted ray which has been *displaced along the axis* to a distance equal to $E\ E^*$. A similar displacement must be taken into consideration in the case of every other ray, whatever its direction may be. The emergent rays always

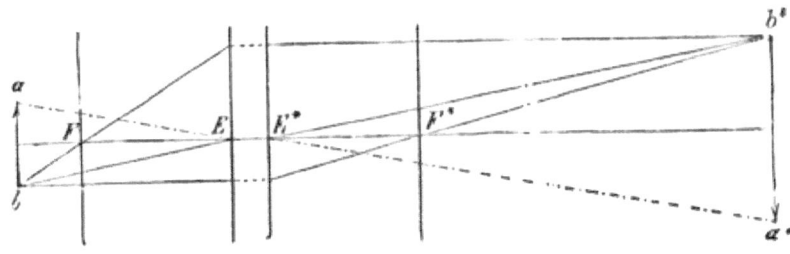

Fig. 4.

behave as if the refraction had taken place in the first principal plane, and then a displacement of the refracted ray had taken place along the axis to the distance $E\ E^*$, or, what amounts to the same thing, as if the displacement of the incident rays had taken place first, and then the refraction in the second principal plane had followed. We can, accordingly, determine without difficulty the direction of a ray which is incident parallel to the axis,—it must be produced to the second principal plane and then carried on through the focal point F^*. Conversely, a ray which cuts the axis in the first focal point F must be produced from the first principal plane in a direction parallel to the axis.

Consequently, from a given point b of the object (Fig. 4), not being on the axis, there may be drawn three distinctive rays, for which the corresponding emergent rays may be constructed just as readily as with the single lens. Two of them are sufficient to determine the point b^* of the image, for the others will, obviously, be refracted to this point also, as indicated in the figure.

It is evident, therefore, that the combined action of a refracting system is in every respect analogous, though not entirely equivalent, to the refraction of an infinitely-thin lens. A lens of equal focus placed at E will, in fact, depict an image of the object $a\ b$ (Fig. 4), which will be exactly similar to $b^*\ a^*$; in order to coincide with $b^*\ a^*$, however, the former image must be displaced along the axis parallel with itself to a distance equal to that between the two principal planes.

The focal lengths of a system of lenses are measured by the distances of the focal points from the respective principal points, consequently by the lines $F\ E$, $E^*\ F^*$. We have hitherto regarded them as equal to one another, because this equality actually exists when the terminal surfaces of the system are bounded by the same medium, as is usually the case in the Microscope. Where this condition is not fulfilled the two focal lengths vary as the refractive indices of the corresponding media. If, for instance, the incident ray passes into water, and the emergent ray into air, the anterior focal length is to the posterior as $1\tfrac{1}{3}$ to 1 ($1\tfrac{1}{3}$ being the approximate refractive index of water). On this supposition a ray directed to E is not only displaced to E^*, but is, in addition, refracted as though at E it passed out of water into air. The *ray of direction*, as the ray passing through unrefracted is also in this case usually called (in accordance with the terminology for *one* refraction), can, therefore, no longer be drawn through E and E^*, but must be drawn through points, of which the first is at a distance from its nearest corresponding focus equal to that of the posterior focal length, and the latter at a distance equal to the anterior focal length. These two points, which in the present case share with one another the function of the optical centre of a single lens, are called the *nodal points*. In the following examination of the Microscope, however, the introduction of these nodal points is superfluous, since the immersion of the objective in water, as well as the effect of the thin glass cover, can be as well taken into account afterwards.

If we represent the distance of the object from the first principal plane by p, the distance of the image from the second principal plane by p^*, and the focal length by f, we obtain the equation

$$\frac{1}{p} + \frac{1}{p^*} = \frac{1}{f},$$

which agrees with that above established for a single refraction. The quantities p and p^* are called the *conjugate focal lengths*; p^* is to be taken as negative, when image and object lie on the same side of the principal planes, which always happens when $p < f$. The ratio of p to p^* gives, in the same way as with the single lens, the linear magnifying power, so that its value, as found above, remains correct for all cases.

II.

ANALYTICAL DETERMINATION OF THE PATH OF THE RAYS IN REFRACTING SYSTEMS.

We have now to determine the points E and E^*, and F and F^* (which are called the *optical cardinal points*) for a given refracting system, in which the curvatures and distances of the refracting surfaces, as well as the refractive indices of the media, are known. This problem can, however, be reduced to a much simpler one in our special case, where we have to consider only lenses and combinations of lenses. It is sufficient to extend the theoretical consideration to *two* refracting surfaces, and to work out the calculation according to the formulæ already obtained for the separate lenses, or for the pairs of surfaces, of which the system is composed. Each two systems of cardinal points may then be combined in the same way as the effects of two refractions were combined for the determination of their cardinal points. The same, of course, holds good for the resulting systems, and the process of combination can be continued, until, at length, the total effect of the refracting surfaces is reduced to *one* system of principal and focal points.

Let the points N^0 and N' (Fig. 5) be the so-called vertices of a lens, that is, the points in which its limiting surfaces are intersected by the optic axis $O X$; r^0 and r' the respective radii of curvature, M^0, M' the corresponding centres, and n^0, n', n^* the indices of refraction of the media in succession from left to right: *in front of*, *within*, and *behind* the lens. If the first and the last medium is air, as is the case with most optical instruments, then n^0 and n^* may be taken as equal to 1. We will, however, leave these quantities provisionally undetermined for the sake of symmetry in the form of the equations.

If $S P$ is the line along which the incident ray is transmitted, its position in the plane of the paper may be determined by means of any given system of rectangular co-ordinates, to which, also, can be referred all other directions and points which come into consideration. If we take the straight line in which the centres of curvature lie as the abscissa-axis, and denote, for brevity, the abscissæ of the points N^0, M^0, N', M' by the same letters, so that

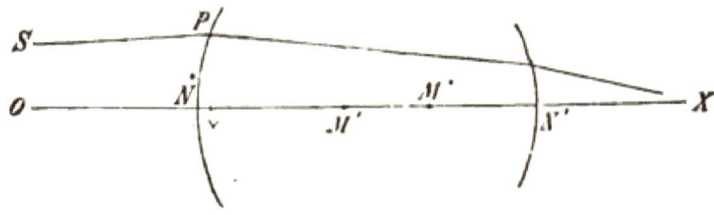

Fig. 5.

$r^0 = M^0 - N^0$, $r' = M' - N'$ (which values will therefore be positive for convex refracting surfaces and negative for concave), the equation for the incident ray assumes the form

$$y = \frac{\beta^0}{n^0}(x - N^0) + b^0. \quad (1)$$

For readers who are little conversant with mathematical modes of expression, we may add that y is the rectangular ordinate to the optical axis, x the corresponding abscissa (taken from any given point of origin), $\frac{\beta^0}{n^0}$ the tangent of the angle which the incident ray makes with the direction of the axis, and b^0 the ordinate of the point in which the ray meets a plane drawn through N^0 at right angles to the axis.

By the first refraction at P the ray takes another direction, which will be determined by the equation

$$y = \frac{\beta'}{n'}(x - N^0) + b',$$

in which, as is easily seen, the quantities $\frac{\beta'}{n'}$ and b' are dependent upon $\frac{\beta^0}{n^0}$ and b^0, as well as upon the curvature of the lens.

If the aperture of the lens is, as we assume, very small in proportion to the radii of curvature, the effective portion of the refracting surface will nearly coincide with a tangential plane drawn through the vertex. The plane is, therefore, intersected by the refracted ray (produced backwards) in a point whose distance b' from the axis is, within a very small quantity which we may disregard, equal to b^0. The equation for the refracted ray becomes, therefore,

$$y = \frac{\beta'}{n'}(x - N^0) + b^0. \tag{2}$$

It is evident that the ratio to the second vertex N' may be expressed in a similar way. For this purpose the term b^0 has only to be replaced by another, b^*, which determines the point of incidence ~~upon the second refracting surface, or, more precisely,~~ upon the tangential plane drawn through N'. We obtain, therefore,

$$y = \frac{\beta'}{n'}(x - N^*) + b^*. \tag{3}$$

By equating the expressions on the right-hand side of the equations (2) and (3) we get

$$b^* = b^0 + \frac{\beta'}{n'}(N^* - N^0). \tag{4}$$

In order, still, to determine the unknown quantity β' by means of β^0 we must draw through M^0 (Fig. 6) a perpendicular which cuts the refracted ray at Q' and the prolongation of the incident ray at

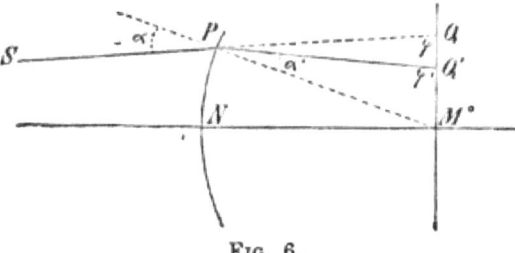

Fig. 6.

Q. Let the angles which this makes with PQ and PQ' be ϕ and ϕ', and the angles of incidence and refraction a and a'. In the triangles $P Q M^0$ and $P Q' M^0$, since $PM^0 = r^0$, the trigonometrical ratios are

$$M^0 Q = r^0 \frac{\sin a}{\sin \phi}, \quad M^0 Q' = r^0 \frac{\sin a'}{\sin \phi'}.$$

From this results

$$\frac{M^0 Q'}{M^0 Q} = \frac{\sin a' \sin \phi}{\sin a \sin \phi'},$$

or, since $\dfrac{\sin a'}{\sin a} = \dfrac{n^0}{n'}$ (according to the law of refraction),

$$M^0 Q' = \frac{n^0 \sin \phi}{n' \sin \phi'} M^0 Q. \tag{5}$$

From the equations (1) and (2) for the incident and refracted ray we get, however, for the ordinates of the points Q and Q', for which $x - N^0 = r^0$, the values

$$y \, (= M^0 Q) = b^0 + \frac{\beta^0 r^0}{n^0},$$

$$y \, (= M^0 Q') = b^0 + \frac{\beta' r^0}{n'}.$$

If we substitute these in equation (5), then

$$b^0 + \frac{\beta' r^0}{n'} = \frac{n^0 \sin \phi}{n' \sin \phi'} \left(b^0 + \frac{\beta^0 r^0}{n^0} \right),$$

or

$$\beta' = \frac{n^0 b^0 + \beta^0 r^0}{r^0 \sin \phi'} \sin \phi - \frac{n' b^0}{r^0}.$$

In this expression the first term on the right-hand side is strictly accurate; but, since ϕ and ϕ', according to our supposition, differ from a right angle only by small quantities, and, therefore, their sines differ from unity by quantities which, as compared with the latter, are very small, it follows (since the quotient must in addition to this be multiplied by the small quantities b^0 and β^0) that

$$\beta' = \beta^0 - \frac{n' - n^0}{r^0} b^0, \tag{6}$$

which is accurate for quantities up to the third approximation.

It is, therefore, clear, without further explanation, that if the equation for the path of the ray after the *second* refraction at the posterior surface of the lens is brought into the form

$$y = \frac{\beta^*}{n''} (x - N^*) + b^*, \tag{7}$$

the quantity β^* can be determined through β' in exactly the same way as the latter is determined through β^0. We have, therefore, only to change the accents in the expression just found for β'. It then becomes

$$\beta^* = \beta' - \frac{n'' - n'}{r'} b^*;$$

or, if we substitute for β' and b^* their values from (6) and (4),

$$\beta^* = \beta^0 - \frac{n'-n^0}{r^0}b^0 - \frac{n''-n'}{r'}\left[b^0 + \left(\beta^0 - \frac{n'-n^0}{r^0}b^0\right)\left(\frac{N^*-N^0}{n'}\right)\right].$$

If we take, for the sake of brevity,

$$\left.\begin{array}{c} u^0 = -\dfrac{n'-n^0}{r^0}, \quad u' = -\dfrac{n''-n'}{r'}, \\ t' = \dfrac{N^*-N^0}{n'} \end{array}\right\}; \quad (8)$$

then the above expression becomes

$$\beta^* = \beta^0 + u^0 b^0 + u' b^0 + u' t' \beta^0 + u^0 u' t' b^0,$$

or

$$\beta^* = (u^0 + u' + u^0 u' t') b^0 + (u' t' + 1) \beta^0.$$

Similarly, we obtain from equation (4), if for β' and $N^* - N^0$ we substitute their values,

$$b^* = (u^0 t' + 1) b^0 + t' \beta^0.$$

To further simplify these expressions let

$$\left.\begin{array}{c} u^0 t' + 1 = g \\ t' = h \\ u^0 + u' + u^0 u' t' = k \\ u' t' + 1 = l \end{array}\right\}; \quad (9)$$

consequently

$$\left.\begin{array}{c} \beta^* = k b^0 + l \beta^0 \\ b^* = g b^0 + h \beta^0 \end{array}\right\}. \quad (10)$$

Conversely, the quantities β^0 and b^0 can, of course, be determined through β^* and b^*, if the latter are known. It is only necessary, for this purpose, to substitute successively the values of b^0 and β^0, obtained from the one of the two equations, in the other equation, and in the reduction to bear in mind that $g l - h k = 1$. This latter ratio may be readily proved by multiplication of the corresponding expressions in (9). We obtain

$$\left.\begin{array}{c} \beta^0 = -k b^* + g \beta^* \\ b^0 = l b^* - h \beta^* \end{array}\right\}. \quad (11)$$

If P is a given point on the straight line (if necessary produced) which the incident ray describes, and ξ, η its co-ordinates, we have then, according to equation (1),

$$\eta = \frac{\beta^0}{n^0} (\xi - N^0) + b^0,$$

or, if we substitute for β^0 and b^0 their values from (11),

$$\eta = \frac{g\beta^* - kb^*}{n^0} (\xi - N^0) - h\beta^* + lb^*;$$

consequently

$$b^* = \frac{n^0\eta + [n^0 h - g(\xi - N^0)] \beta^*}{n^0 l - k(\xi - N^0)}.$$

If we put this value in the equation (7), which determines the direction of the ray after the second and last refraction, then the equation, if the terms multiplied by β^* are placed together, will be

$$y = \frac{\beta^*}{n^*}\left\{x - \left(N^* - \frac{n^0 h - g(\xi - N^0)}{n^0 l - k(\xi - N^0)} \cdot n^*\right)\right\} + \frac{n^0 \eta}{n^0 l - k(\xi - N^0)}, \quad (12)$$

and if we write by way of abbreviation

$$\left.\begin{aligned} N^* - \frac{n^0 h - g(\xi - N^0)}{n^0 l - k(\xi - N^0)} \cdot n^* &= \xi^* \\ \frac{n^0 \eta}{n^0 l - k(\xi - N^0)} &= \eta^* \end{aligned}\right\}, \quad (13)$$

the equation will then take the form

$$y = \eta^* + \frac{\beta^*}{n^*} (x - \xi^*).$$

Therefore, on the last path of the ray there must of necessity lie a point P^*, whose co-ordinates are ξ^* and η^*; for, if in the above equation we take $x = \xi^*$, then $y = \eta^*$. Since, then, ξ^* and η^* are determined solely through ξ and η, in combination with the optical constants entering into g, h, k, l, and are not dependent upon the quantities β^0 and b^0, it follows that every ray, which on its first path passes through the point P, must on its last path pass through the point P^*. In other words, to the incident rays whose lines of direction intersect in the point P there correspond emergent rays which (if necessary produced) intersect at P^*. The point P may, therefore, be regarded as the object, and P^* as its optical image; the object is, however, *real* only when P lies in the first medium, therefore if $\xi - N^0$ is negative. Similarly, the image can be *real* only when $\xi^* - N^*$ is positive, or P^* lies in the last medium. In the opposite cases object and image are *virtual* only, i.e., the rays do not actually proceed from P nor actually converge at P^*, P and P^* being merely the points of intersection of their prolongations.

If we draw through the points P and P^* planes at right angles to the axis, it is evident that to every point in the one plane there corresponds a co-ordinate image-point in the other; for, if ξ remains constant, ξ^* undergoes no alteration, being dependent in a given system upon ξ alone. The distances of the respective object- and image-points are to each other as the corresponding ordinates η and η^*; or, according to equation (13), as

$$1 : \frac{n^0}{n^0 l - k(\xi - N^0)}.$$

Every object of finite extent may be regarded as such a system of points, and it would give rise, therefore, to a continuous image whose linear dimensions are determined by the above-mentioned relation, the object being taken as unity. If the quantity $\frac{n^0}{n^0 l - k(\xi - N^0)}$ be represented by m, then this is the co-efficient of linear amplification. Its sign distinguishes whether the image is erect or inverted; if it is negative, and with it therefore $\frac{\eta^*}{\eta}$, it indicates that the object-point and the image-point lie on opposite sides of the axis.

The points P and P^* may, of course, assume all possible positions within the limits established in (13) as to their inter-dependence, since for every value of ξ the corresponding ξ^* can be calculated. Of these positions three in particular deserve special consideration, since they exhibit a more simple ratio between the incident and the emergent rays.

We will first bring the two points into such a position that they are at equal distances from the axis. Therefore $\eta = \eta^*$; or,

$$m = \frac{n^0}{n^0 l - k(\xi - N^0)} = 1;$$

whence we obtain $\xi - N^0 = -\frac{n^0(1-l)}{k}$;

consequently $\xi = N^0 - \frac{n^0(1-l)}{k}.$ \qquad (14)

Similarly, we obtain for ξ^*, if we put this value in the equation (13),

$$\xi^* = N^* + \frac{n^*(1-g)}{k}. \qquad (15)$$

If, therefore, we suppose that there are two points E and E^* on the optic axis, whose abscissæ (which we also denote by E and E^*) are equal to the values of ξ and ξ^*, which we have just found, and if we draw through them two planes at right angles to the axis, then the first will be met by the incident ray at a distance equal to that at which the second will be met by the emergent ray. The points in question are, therefore, no other than the *principal points* of the system, and the planes drawn through them are the *principal planes*.

It may also be easily proved that if $n^0 = n^*$ a ray directed to E will emerge without deviation. For, if, in the equation for the incident ray, that is, in

$$y = \frac{\beta^0}{n^0}(x - N^0) + b^0,$$

x is made equal to E, and for $(E - N^0)$ its value from (14) is substituted, then, since y will be 0, we get

$$y = b^0 - \beta^0 \left(\frac{1-l}{k}\right) = 0$$

consequently

$$b^0 = \beta^0 \left(\frac{1-l}{k}\right).$$

If we introduce this value in the expression for β^* (equation 10), then $\beta^* = \beta^0$, and, (since $n^* = n^0$), $\dfrac{\beta^*}{n^*} = \dfrac{\beta^0}{n^0}$; that is, the emergent ray forms the same angle with the axis as the incident ray. The ray directed to E is a *ray of direction*.

It is not necessary here to follow out further the consequences which result for the case of n^0 and n^* being unequal, and therefore $\dfrac{\beta^*}{n^*}$ and $\dfrac{\beta^0}{n^0}$ being to one another as n^0 to n^*.

As the *second* case, let the position of P and P^* be such that the image-point is at an infinite distance, and consequently

$\xi^* - N^* = \infty$. Therefore, evidently, $\eta^* = \infty$, and, as we assume that $\eta > 0$ has a finite value,

$$m = \frac{n^0}{n^0 l - k(\xi - N^0)} = \infty.$$

Since, in this expression, the numerator is a finite quantity, the denominator must $= 0$, and, therefore, $k(\xi - N^0) = n^0 l$, from which we get

$$\xi = N^0 + \frac{n^0 l}{k}. \qquad (16)$$

In like manner for the analogous *third* case, in which the object-point lies at an infinite distance, we get from equation (13) the corresponding value of ξ^*. If, for instance, we consider that the quantities $n^0 h$ and $n^0 l$, which occur in the numerator and denominator, disappear in comparison with the infinite quantity $(\xi - N^0)$, it is readily seen that

$$\xi^* = N^* - \frac{n^* g}{k}. \qquad (17)$$

These values of ξ and ξ^* evidently correspond to the abscissæ of the two focal points F and F^*, and the planes drawn through them at right angles to the axis are the focal planes. As soon as one of the two points lies in the corresponding focal plane, the other moves to an infinite distance.

The distances of the respective principal and focal points may be determined from their abscissæ, given in equations (14) to (17), by simple subtraction of the corresponding values of the abscissæ. We get

$$\left. \begin{array}{c} E - F = -\dfrac{n^0}{k} \\ F^* - E^* = -\dfrac{n^*}{k} \end{array} \right\}. \qquad (18)$$

The quantities $-\dfrac{n^0}{k}$ and $-\dfrac{n^*}{k}$, or the distances of the principal planes from the corresponding focal planes, are called the *focal lengths* of the system. They have under all circumstances—since n^0 and n^* are, from their nature, positive numbers—the same, and k the opposite, sign. If they are positive, and if, therefore, E lies behind F, and F^* behind E^*, the system is called a *convergent* one; it acts as a convergent lens and produces real images. In the opposite case the system is called *divergent*, because it acts as a divergent lens and produces only virtual images of real objects.

The distances of the object from the first, and of the image from the second, principal plane may be called their *conjugate focal lengths* in accordance with the terminology used in the case of single refractions. Their values may be easily calculated from what has been given above. The transposition of the equations (14) and (15), in which $\xi = E$ and $\xi^* = E^*$, gives

$$N^0 = E + \frac{n^0(1-l)}{k},$$

$$N^* = E^* - \frac{n^*(1-g)}{k}.$$

If these values are substituted in (13), where ξ, η and ξ^*, η^* represent the co-ordinates of the conjugate foci P and P^*, they become

$$\left. \begin{array}{l} \xi^* = E^* - \dfrac{n^*(E-\xi)}{n^0 + k(E-\xi)} \\ \eta^* = \dfrac{n^0 \eta}{n^0 + k(E-\xi)} \end{array} \right\} . \qquad (19)$$

From the former of these equations we get

$$\frac{n^0}{E-\xi} + \frac{n^*}{\xi^* - E^*} = -k;$$

or, if we denote $(E - \xi)$ by p, $(\xi^* - E^*)$ by p^*, and the focal lengths by f and f^*, and if we also add the values of $-k$ from (18),

$$\frac{n^0}{p} + \frac{n^*}{p^*} = -k = \frac{n^0}{f} = \frac{n^*}{f^*}. \qquad (20)$$

The analogy to the case of refraction at one surface is consequently apparent.

The magnifying power m, which expresses the ratio of the ordinates η^* and η, now assumes, in conformity with the expressions in (19) and (20), the following additional forms:

$$m = \frac{n^0}{n^0 + kp} = \frac{1}{1 + \frac{k}{n^0}p} = \frac{1}{1 - \frac{p}{f}} = \frac{f}{f - p} = \frac{n^* + kp^*}{n^*}$$

$$= 1 + \frac{k}{n^*}p^* = 1 - \frac{p^*}{f^*} = \frac{f^* - p^*}{f^*}.$$

For the completion of the analogies which exist between the optical action of a system and the case of refraction at *one* surface, the equivalent significance of the quantity k for the former and u^0 or u' for the latter must here be further emphasized. It may be shown that the relation expressed in (20), that is $k = -\dfrac{n^0}{f} = -\dfrac{n^*}{f^*}$, obtains in a manner precisely analogous for u^0 and u', and that if we represent the focal lengths of the first refracting surface by ϕ and ϕ' and those of the second by ϕ'' and ϕ''', then

$$u^0 = -\frac{n^0}{\phi} = -\frac{n'}{\phi'},$$
$$u' = -\frac{n'}{\phi''} = -\frac{n''}{\phi'''}.$$

For if, in the equation for the direction of the ray after the first refraction, we make $(x - N^0) = \phi'$, and, consequently, x equal to the abscissa of the focal point, and if we substitute for β' its value $\beta^0 + u^0 b^0$, then

$$y = \frac{\beta^0 + u^0 b^0}{n'} \phi' + b^0,$$

and, since this value must be 0 for rays incident parallel to the axis (for which $\beta^0 = 0$), there results

$$u^0 b^0 \frac{\phi'}{n'} + b^0 = 0;$$

consequently

$$u^0 = -\frac{n'}{\phi'}.$$

Similarly, we get from the equation for the incident ray, if for β^0 we put its value $\beta' - u^0 b^0$,

$$u^0 = -\frac{n^0}{\phi}.$$

It will be clear, without further explanation, that the ratios of u' to ϕ'' and ϕ''' may be deduced in a similar manner.

The quantities u^0, u', k are, consequently, all equal to $-\dfrac{n}{f}$, if f is the anterior or the posterior focal length, and n the index of refraction of the medium in which the ray moves, before the assumed refraction in the one case, and after it in the other. The deviation caused by the latter is, therefore, the same as that which would be brought about by an infinitely-thin lens bounded on both sides by air (for which $n = 1$) and whose focal length $= f$.

These ratios supply us with a ready means of combining any two systems of optical cardinal points just as easily as two refracting surfaces; for, by virtue of the equations already given, it is allowable to reduce all the refractions to the case of infinitely-thin lenses, and to take the quantities u^0 and u', which appear in the formulæ for β' and β'', as equal to the reciprocal focal lengths of those lenses, taken as negative.

If, for instance, E^0 and I^0 are the principal points of a lens, whose (calculated) focal length $= f^0$, and E', I' those of another lens, whose focal length is f', then, in accordance with the properties of principal points, a ray which passes through both will be refracted as if in I^0 and E' were situated infinitely-thin lenses with the same focal lengths. The distance of the points I^0 and E' has, therefore, the same signification as $\dfrac{N^* - N^0}{n'} = t'$ has in the case of a single lens. This also follows directly from the equations for the ray before and after refraction, referred to the principal points. For its direction before refraction we have

$$y = \beta^0 (x - E^0) + b^0;$$

after the *first* refraction

$$y = \beta' (x - I^0) + b^0,$$

or, referred to E',

$$y = \beta' (x - E') + b^*;$$

after the *second* refraction

$$y = \beta^* (x - I') + b^*.$$

From the middle equations, making the expressions on the right-hand side equal, results

$$b^* = b^0 + (E' - I^0) \beta',$$

whilst we found above [equation (4)]

$$b^* = b^0 + \frac{N^* - N^0}{n'} \beta'.$$

The analogy is, therefore, apparent. As, however, besides the refraction upon which b^* is alone dependent, there also comes into account the displacement of the rays from one principal plane to the other, then, with respect to the position of the emergent to the incident ray, the first principal point E^0 corresponds to the anterior lens-vertex, and the last I' to the posterior. If, therefore, in the preceding, E and E^*, in their position to N^0 and N^*, are determined by the equations

$$E = N^0 - \frac{1-l}{k}, \quad E^* = N^* + \frac{1-g}{k},$$

the principal points of the resulting system, denoted by **E** and **E***, will be given by

$$\mathbf{E} = E^0 - \frac{1-l}{k}, \quad \mathbf{E}^* = I' + \frac{1-g}{k}.$$

The same holds good also for the focal points. We have

$$\mathbf{F} = E^0 + \frac{l}{k}, \quad \mathbf{F}^* = I' - \frac{g}{k}.$$

Summary of the Results.

It is probable that many who are not able to follow readily the preceding analytical explanation will wish us to summarize the most important of its results.

The quantities which enter into the formulæ for the determination of the principal and focal points, and which evidently remain constant for a given system, are g, l, k; these are, in their turn, dependent upon two series of other quantities, viz., $n^0, n', n'' \ldots$ and $t', t'' \ldots$, in which $\frac{1}{n^0}$ denotes the focal length (taken as negative) of an infinitely-thin lens, which would produce a deviation equivalent to the first refraction, and n', n'' have the same signification for the second and third refraction, while $t', t'' \ldots$ denote the distances of the imaginary equivalent lenses. Hence it follows, that if, with any combination, two refractions only come into account, whether at surfaces or through imaginary lenses, the first series of these quantities are reduced to u^0 and u', and the second to t'. In the preceding mathematical consideration we have confined ourselves to this simple case.

If we denote by r^0 and r' the radii of curvature of the anterior and posterior lens-surfaces, and by f^0 and f' the focal lengths of lenses whose action is equivalent to the refraction at those surfaces, in the sense that to equal angles of incidence correspond equal angles of refraction, and if, further, the index of refraction of the lens-substance is represented by n, then, if the surrounding medium is air,

$$u^0 = -\frac{1}{f^0} = -\frac{n-1}{r^0},$$
$$u' = -\frac{1}{f'} = -\frac{1-n}{r'} = +\frac{n-1}{r'}.$$

It must also be observed that the radii of curvature r^0 and r' are to be taken as positive or negative according as the incident ray meets the convex or the concave side of the surface. If the surface is plane, then $r = \infty$ and, consequently, the corresponding $u = 0$. Similarly, the focal lengths, where these are brought into account, are to be taken as positive or negative according as the refraction causes the rays to converge or diverge,—that is, positive for converging and negative for diverging lenses and systems of lenses.

For the quantity t' we get, in the case of a single lens whose thickness $= d$,

$$t' = \frac{d}{n};$$

and, in the combination of two systems, if the principal points of the first are denoted by their abscissæ E^0 and I^0, and those of the second by E' and I',

$$t' = E' - I^0,$$

that is, t' is equal to the distance of the principal points which are turned towards each other.

The operations which are necessary for the determination of the optical cardinal points can now—since u^0, u', and t' may be regarded as given—be easily collected in a tabular form.

I.—For the optical constants:—

$$g = u^0 t' + 1$$
$$l = u' t' + 1$$
$$k = u^0 u' t' + v^0 + u' = u'g + u^0$$
$$h = t'$$

for which the equation

$$gl - hk = 1$$

serves as a check.

II.—For the cardinal points:—

a. Two refracting surfaces.	b. Two refracting systems.
$\mathbf{E} = N^0 - \dfrac{1-l}{k}$	$\mathbf{E} = E^0 - \dfrac{1-l}{k}$
$\mathbf{E^*} = N^* + \dfrac{1-g}{k}$	$\mathbf{E^*} = I^* + \dfrac{1-g}{k}$
$\mathbf{F} = N^0 + \dfrac{l}{k}$	$\mathbf{F} = E^0 + \dfrac{l}{k}$
$\mathbf{F^*} = N^* - \dfrac{g}{k}$	$\mathbf{F^*} = I^* - \dfrac{g}{k}$
Focal length $= -\dfrac{1}{k}$	Focal length $= -\dfrac{1}{k}$

The large letters here denote the abscissa-values of the corresponding points (reckoned from any point of emergence situate in the axis), that is, N^0, N^*, those of the first and second lens-vertices, E^0, I^*, and those of the first and last of the four principal points which enter into consideration in the combination of two systems, and $\mathbf{E}\ \mathbf{E^*}$, $\mathbf{F}\ \mathbf{F^*}$ those of the resulting principal and focal points.

If, for example, in a given case $\dfrac{l}{k} = -5$ mm., and $\dfrac{g}{k} = -4$ mm., then the first focal point $\mathbf{F}$ lies 5 mm. before the first lens-vertex, and the second focal point $\mathbf{F^*}$ 4 mm. behind the second. The other equations are similarly interpreted.

III.

DETERMINATION OF THE DISTANCES OF CORRESPONDING IMAGE-POINTS FROM THE AXIS IN THE CASE OF ANY GIVEN INCLINATION OF THE RAYS.

As already intimated, the assumptions upon which the formulæ of Gauss are founded are entirely at variance with the actual conditions of the formation of the image by means of systems of lenses of high power. In Microscopes we have to deal with very large angles of aperture of the image-forming pencils; the theory of Gauss assumes these angles to be so small that their cubes may be neglected.

Nevertheless, the conclusions which may be drawn by the aid of these formulæ afford a means of making an approximately correct, and, in many cases also, practically useful diagram of the path of the rays, and observation shows that the position of the cardinal points determined analytically does not differ very much from that found by measurement. For objects, therefore, whose marginal points are not at a great distance from the axis, the calculation of the corresponding magnitude of the image for any combination of refracting surfaces is fairly applicable, *i.e.*, the ratio between image-magnitude and object is about the same as that between the posterior and anterior focal lengths. A construction made after the analogy of Fig. 4 gives, therefore, for any system not merely the distances in the direction of the axis (abscissæ), but also the distances in the plane of the image (ordinates). Trigonometrically expressed, the latter, measured from the axis, are equal to the tangent of the angle at which the rays diverge from the direction of the axis. If we denote this angle by δ, and the posterior focal distance, as above, by p^*, the linear distance of the image-point is consequently equal to $p^* \times \tan \delta$.

A more accurate analytical determination of these distances, especially with reference to rays more obliquely inclined, leads, however, to the result, that in the above formula the sine of the angle δ must be taken instead of the tangent. With the usual methods of observation this does not make much difference; for the sine of 15°, for instance, is not as much as ·01 of the radius smaller than the tangent of that angle; whence it follows, among other things, that the small image of the diaphragm-aperture in the posterior focal plane of the objective appears, according to both formulæ, of nearly the same size, the incident pencil of light assumed to be 30° aperture. With an objective of 3 mm. focal length, the diameter, d, in the one case $= \dfrac{2 \times \tan 15°}{r} \cdot 3$, or 1·60 mm., and in the other $\dfrac{2 \times \sin 15°}{r} \cdot 3$, or 1·55 mm. The difference amounts, therefore, only to $\tfrac{1}{20}$ mm. But instances are found in practice, where the differences between sine and tangent are much greater, and where it is not permissible to disregard them. With reference to this we give the following theorem put forward by Abbe,[1] which is applicable to all angles that are met with in microscopic observation. It is—"*When an optical system is completely aplanatic for one of its focal points, every ray emerging from this point meets a plane drawn through the other focal point at a distance from the axis, the linear magnitude of which is equal to the product of the equivalent focal length of the system and the sine of the angle which this ray makes with the axis.*"

The proof of this proposition has not yet been published by Abbe. It will suffice, however, for our purpose to show that with any double-lens the tangent of the angle in question must, in fact, give somewhat too high values for the distance of the emergent rays from the axis,—that is, if we determine the cardinal points for such a lens, it will be found that the anterior principal plane will lie somewhat behind the geometrical centre of the crown-glass lens, and will, consequently, intersect the last refracting surface. Compare, for instance, the calculations given on the succeeding pages, or the construction of Fig. 7. If we suppose a ray proceeding from the focal point and passing without deviation to the point of intersection of the anterior principal plane with the last refracting surface, it necessarily meets this surface at a greater distance than if it were diverted inwards at the anterior surface of the lens, and then assumed, almost in the same vertical, a direction parallel to the axis. The differences that result from this may be shown,

[1] "Archiv für mikr. Anat." Bd. ix. p. 420.

by way of example, in a double-lens whose focal length = 8·773 mm. Let the distance of the focal point from the first lens-surface = 7·339, therefore, that of the anterior principal point of the same surface = 1·434 mm.; and let the refractive index of the flint-glass = 1·6 :[1] on these assumptions we get for the distance from the axis, y, of a ray emerging from the focal point and inclined at an angle of 14° after its passage through the lens:—

a. According to the trigonometrical calculation for the case of the point of emergence of the ray approximately coinciding with the point of intersection in the anterior principal plane, where, at the same time, the path described in the crown-glass need not be calculated separately,

$$y = 7·339 \cdot \tan 14° + 1·434 \cdot \tan 8° 40' = 2·048 \text{ mm}.$$

b. According to the formula y = focal length × tan 14°,

$$y = 8·773 \cdot \tan 14° = 2·187 \text{ mm}.$$

Our lens is not, however, aplanatic for rays which emanate from the focal point, but is noticeably over-corrected. The distance from the axis of the ray refracted by the flint-glass proves, therefore, somewhat too small; it may rise to about 2·07 mm. But, in substance, nothing is altered thereby; the deviation as compared with the tangent of the angle of inclination continues the same. It is, further, evident that the amount of this variation increases and decreases with the angle of inclination of the incident ray.

---

[1] Cf. the data for the second double-lens of the objective calculated in the succeeding pages.

IV.

THE COMPONENT PARTS OF THE MICROSCOPE.

1.—The Objective.

We will now turn to the determination of the cardinal points of an objective-system, consisting of three double-lenses, each of which is composed of one plano-concave flint lens and of an equi-convex crown lens. Let the refractive index of the flint be 1·6 and that of the crown 1·5. Under these suppositions we obtain for a *flint-glass* lens, if the thickness d and the radius of curvature r be left provisionally undetermined, according to the formula contained on p. 19 *et seq.*,

$$u^0 = 0, \; u' = \frac{3}{5r'}, \; t' = \frac{5}{8}d,$$

and, hence,

$$g = 1, \; k = u' = \frac{3}{5r'}, \; l = 1 + \frac{3d}{8r'};$$

consequently

$$E = N^0 + \frac{5}{8}d, \; E^* = N^*, \; f^0 = -\frac{5}{3}r'.$$

With a plano-concave lens, therefore, the second principal point coincides with the posterior lens-vertex, while the first is at a distance of $\frac{d}{n}$ (therefore, here, $= \frac{5}{8}d$) from the plane anterior surface. The focal length f^0 is identical with that of the curved surface; it is negative and $= -\frac{1}{u'} = \frac{r'}{n-1}$ (cf. p. 22).

Similarly, we obtain for a *crown-glass* lens, if the radius of curvature is denoted by r and the thickness by d,

$$u^0 = u' = -\frac{1}{2r}, \; t' = \frac{2}{3}d;$$

hence

therefore

$$E = N^0 - \frac{2dr}{d - 6r} = N^0 + \frac{2dr}{6r - d},$$

$$E^* = N^* + \frac{2dr}{d - 6r} = N^* - \frac{2dr}{6r - d},$$

$$f' = \frac{6r^2}{6r - d}.$$

The principal points of a bi-convex (or a bi-concave) lens with equal curvatures lie, therefore, symmetrically on both sides of the optical centre; the distances from the two lens-vertices are equal to each other. In a sphere, where $d = 2r$, the principal points lie in the centre. If d is greater than $2r$, as in cylindrical lenses, the second principal point will be in front of the first.

We will now give determinate values to the quantities r and d, in order to apply the formulæ we have obtained to the three double-lenses of our objective. For greater convenience, let f^0 and f' be, in each case, the focal lengths of the flint- and crown-glass lenses, d^0 and d' their thickness, and e^0, i^0, e', i' their principal points; further, let ϕ^0, ϕ', ϕ'' be the focal lengths of the first, second, and third double-lenses, and E^0, I^0, E', I', E'', I'' their principal points; and let $N^0, N^1, N^2 \ldots N^8$ be the successive lens-vertices. In the combination of the flint- and crown-glass lenses, therefore, t' ($= e' - i^0$) will be equal to the distance of the point e' from the surface of contact, $u^0 = -\frac{1}{f^0}$ and $u' = -\frac{1}{f'}$. As the calculation is easy, it will be sufficient to point out the operation to be performed for the first double-lens only, and, for the others, simply to compare the results obtained in a similar manner.

First double-lens.—Let $r = 1$, $d^0 = \frac{1}{2}$, $d' = 1$;

then

$$e^0 = N^0 + \frac{5}{8} d^0 = N^0 + \frac{5}{8} \times \frac{1}{2} = N^0 + \frac{5}{16}, \quad i^0 = N^1,$$

$$e' = N^1 + \frac{2dr}{6r - d} = N^1 + \frac{2 \times 1 \times 1}{6 \times 1 - 1} = N^1 + \frac{2}{5}, \quad i' = N^2 - \frac{2}{5},$$

$$f^0 = -\frac{1}{k} = -\frac{1}{\frac{3}{5} \times r} = -\frac{5}{3}, \quad f' = -\frac{1}{k} = -\frac{6r^2}{d - 6r} = \frac{6}{5}.$$

We obtain from the above, for the combination of both lenses,

$$u^0 = -\frac{1}{f^0} = \frac{3}{5}, u' = -\frac{1}{f'} = -\frac{5}{6}, t' = t' - i^0 = +\frac{2}{5};$$

further

$$g = u^0 t' + 1 = \frac{3}{5} \times \frac{2}{5} + 1 = +\frac{31}{25},$$

$$l = u't' + 1 = -\frac{5}{6} \times \frac{2}{5} + 1 = +\frac{2}{3},$$

$$k = u'g + u^0 = -\frac{5}{6} \times \frac{31}{25} + \frac{3}{5} = -\frac{13}{30}.$$

For the cardinal points of the double-lens, therefore,

$$E^0 = e^0 - \frac{1-l}{k} = e^0 - \frac{1-\frac{2}{3}}{-\frac{13}{30}} = e^0 + \frac{1}{3} \times \frac{13}{30} = e^0 + \frac{10}{13}$$

$$= N^0 + \frac{5}{16} + \frac{10}{13} = N^0 + 1\cdot08173,$$

$$I^0 = i'' + \frac{1-g}{k} = i'' + \frac{-\frac{6}{25}}{-\frac{13}{30}} = i'' + \frac{6}{25} \times \frac{30}{13} = i'' \times \frac{180}{325}$$

$$= N^2 - \frac{2}{5} + \frac{180}{325} = N^2 + \cdot153846,$$

$$\phi^0 = \frac{30}{13} = 2\cdot3077.$$

Second double-lens.—Let $r = 4, d^0 = \frac{2}{3}, d' = \frac{4}{3}$;

then

$$e^0 = N^3 + \frac{5}{12}, i^0 = N^4, e' = N^4 + \frac{72}{153},$$

$$i'' = N^5 - \frac{72}{153}, f^0 = -6\frac{2}{3}, f' = 4\frac{1}{5}.$$

From this

$$E' = e^0 + 1\cdot01803 = N^3 + 1\cdot43469$$
$$I' = i'' + \cdot6193 = N^5 + \cdot1487$$
$$\phi' = 8\cdot77347.$$

Third double-lens.—Let $r = 10$, $d^0 = \frac{3}{4}$, $d' = \frac{3}{2}$; then

$$e^0 = N^6 + \frac{15}{32}, \quad i^0 = N^7, \quad e' = N^7 + \frac{10}{19},$$

$$i'' = N^8 - \frac{10}{19}, \quad f^0 = -16\tfrac{2}{3}, \quad f'' = 10\tfrac{10}{39}.$$

Hence
$$E'' = e^0 + 1\cdot23456 = N^6 + 1\cdot70331$$
$$I'' = i'' + \cdot7732 = N^8 + \cdot2469$$
$$\phi'' = 24\cdot6913.$$

Since the second and third double-lenses are sometimes used without the first when lower magnifying powers are required, it will be to the purpose if we, first of all, combine these two, and then follow out somewhat further their united action as an objective. For this purpose, we must make the further assumption that N^6 coincides with I'; the lens-surfaces, which face each other, are therefore at a distance of $\cdot1487$ apart, so that t' (the distance of the principal points I' and E'' on that side) in this case $= 1\cdot70331$, while u^0 and u', as usual, are to be taken as equal to the reciprocals of the relative focal lengths, that is, $-\dfrac{1}{\phi'}$ and $-\dfrac{1}{\phi''}$. For the abscissæ of the resulting principal and focal points, which we denote by (E) (E^*) and (F) (F^*), we therefore get

$$(E) = E'' + \cdot470468 = N^3 + 1\cdot905158$$
$$(E^*) = I'' - 1\cdot32419 = N^3 + 3\cdot32143$$
$$(F) = (E) - 6\cdot82064 = N^3 - 4\cdot91548$$
$$(F^*) = (E^*) + 6\cdot82064 = N^8 + 5\cdot74437.$$

The focal length (f) of the objective therefore $= 6\cdot82064$, and the distance of the focal point (F) from its anterior surface $= 4\cdot91548$. If the image is to be at a distance of $p^* = 200$ mm., as is approximately the case in most of the modern instruments, the distance p of the object from the first principal plane can easily be determined from the relation

$$\frac{1}{p} + \frac{1}{p^*} = \frac{1}{f}.$$

We find $p = 7\cdot06146$ mm.; therefore $p - f = \cdot24$. As coefficient of magnifying power we get

$$m = \frac{f}{f - p} = -28\cdot3.$$

The negative sign indicates the inverted position of the image. The total magnifying power of the Microscope depends, of course, upon the action of the eye-piece, which will be more fully discussed subsequently.

If, in order to complete the objective, we now add the first and strongest double-lens, and here, also, assume the distance such that the posterior principal plane coincides with the anterior surface of the second lens, then, for this combination,

$$u^0 = -\frac{1}{\phi^0} = -\cdot 43333 \ldots \ldots ,$$

$$u' = -\frac{1}{(f)} = -\frac{1}{6\cdot 8206} = -\cdot 146637,$$

and

$$f' = 1\cdot 905158;$$

and the result of the calculation gives us the abscissæ-values of the principal and focal points, which we will denote by E and E^*, F and F^*,

$$\begin{aligned}
E &= E^0 + \cdot 609039 = N^0 + 1\cdot 6907697 \\
E^* &= (E^*) - 1\cdot 80008 = N^0 + 3\cdot 175196 \\
F &= E - 2\cdot 18042 = N^0 - \cdot 48965 \\
F^* &= E^* + 2\cdot 18042 = N^s - \cdot 69656 .
\end{aligned}$$

Consequently, the focal length amounts to only 2·18, and the distance of the focal point from the anterior surface ·49 unit of length. The distance of the object diminishes to ·5 with a tube-length of about 200 mm.

To make these numerical relations clearly apparent, the objective-system, which we have supposed with its principal and focal planes, is represented in Fig. 7, enlarged five times (the millimetre taken as unity). The cardinal planes (F) (E) (E^*) (F^*) of the two posterior lenses are indicated by dotted lines, and the principal planes of each double-lens by shorter lines.

The distance of the object is, obviously, a quantity dependent upon the focal distance; it increases and decreases, *cæteris paribus*, with the latter, and is nearly equal to it in the strongest objectives. It does not, however, stand in direct connection with the focal length of the objective; it is well known that objectives of different makers often vary considerably in respect to object-distance, though the powers are equal. The question, as to what circumstances influence this quantity, is, therefore, not without practical interest, and it may be easily answered from the preceding consideration, if the focal distance ($N^0 - F$) is determined from the formulæ

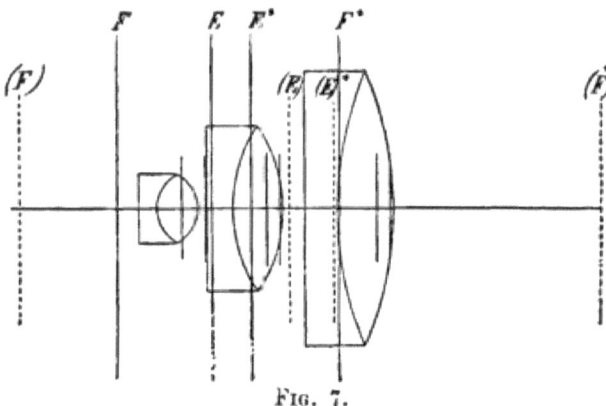

Fig. 7.

for the cardinal points of the system, and the values of the quantities l and k are substituted. If f, as before, is the focal length of the whole objective, ϕ^0 that of the first double-lens, and (f) that of the two other lenses of the system, we get

$$f = -\frac{1}{k} = -\frac{1}{u^0 u' t' + u^0 + u'} = \frac{\phi^0 (f)}{\phi^0 + (f) - t'};$$

and, similarly, we get, for the first principal point,

$$E = E^0 + \frac{l-1}{k} = E^0 + \frac{u' t'}{u^0 u' t' + u^0 + u'} = E^0 + \frac{t' \phi^0}{\phi^0 + (f) - t'}.$$

The subtraction of this latter value from the focal length gives as focal distance

$$N^0 - F = \frac{\phi^0 [(f) - t']}{\phi^0 + (f) - t'} - E^0 = \frac{\phi^0}{1 + \frac{\phi^0}{(f) - t'}} - E^0.$$

If the lenses and their principal points and focal lengths are given, then $N^0 - F$ evidently attains a value which is the larger the smaller t' is, since the denominator $1 + \frac{\phi^0}{(f) - t'}$ increases and decreases with t'. The focal distance is, therefore, greatest when t' is as small as possible, *i.e.*, when the first lens is brought as near as practicable to the second lens. This approximation of the lenses which naturally finds its limit in the contact of the two lenses, necessitates at the same time a diminution of the focal length, since the expression found for the latter varies with t'. The approximation of the lenses gives, therefore, a double advantage: a greater magnifying power, and, at the same time, a greater focal distance. This not only holds good for the first and second lenses, but also for the second and third; for, since the formulæ just obtained are applicable to any two systems of cardinal points, the combination of the third lens with the two anterior ones must lead to the same results with regard to the magnitude t' which here appears. This follows also from the formula for the focal length, applied to (f), which is the shorter the less the distance of the principal points (here I' and E'). The smallest value of (f) corresponds to the greatest object-distance (if the allowable supposition is made that $(f) > t'$), as a glance at the value of $N^o - F$ shows.

It is evident that the approximation of the principal points, above discussed, may be extended to every combination of refracting surfaces, and, therefore, also to single crown and flint lenses. Since with the latter the distance of the principal points from the surface of contact varies with the thickness, it follows that they must be made as thin as possible so as to increase the focal distance; by these means, at the same time, a higher magnifying power is obtained.

The focal distance and the object-distance, therefore, under similar circumstances, vary inversely as the distances of the refracting surfaces.

By this it is not meant that the sum of the distances can be taken as the absolute measure; it is self-evident that the stronger lenses, in this respect also, have a preponderating influence.

The refractive indices of flint and crown were taken above at 1·6 and 1·5. It is, *à priori*, clear, that an alteration of these figures will not be without influence upon the position of the optical cardinal points, and it is not difficult to understand the general nature of the effect of the increase or decrease. If the relation between the crown and flint remains the same, and still more if the crown only is regarded as variable, an increase of the indices of refraction will increase the optical action of the single lenses, and involves shallower curvatures with equal effect. It acts, therefore, in the same way as an approximation of the refracting surfaces, since it diminishes the ratio of the lens-thickness to the radii of curvature or to the optical effect. With equal power of the objective, the focal distance is, therefore, the larger, *cæteris paribus*, the higher the refractive index of the substance of the lens; with equal curvatures the magnifying power increases with the refractive index, while the object-distance varies according to circumstances.

To assist in the estimation of these influences we have collected in the following table the most important elements of our objective-system for the cases given in the first two columns. The form of the lenses and the refractive index of the flint (1·6) are here regarded as constant, but the distances e_1 and e_2 of the double-lenses (e_1 between the anterior and middle, e_2 between the middle and the last) as well as the refractive indices n_1, n_2, and n_3 of the crowns are regarded as variable. We will hereafter select from these combinations that which appears best adapted for the correction of the aberrations.

Given.	e_1	Focal length.	Focal distance.	Object-distance.	$E^0 - E$.	Amplification.
$e_2 = 0·1487$ $n_1 = n_2 = n_3$ $= 1·50$	0·1538 1·0000	2·1804 2·4682	0·4896 0·3908	0·51379 0·4220	1·48442 0·80137	90·725 80·0297
$e_2 = 0·50$ $n_1 = 1·50$ $n_2 = n_3 = 1·52$	0·1538 1·0000	2·2469 2·5528	0·4719 0·3692	0·50026 0·4022	0·99712 0·21753	88·0096 77·3452
$e_2 = 0·50$ $n_1 = n_2 = n_3$ $= 1·52$	0·1538 1·0000	2·1469 2·4522	0·4271 0·3346	0·453 0·368	0·94787 0·1179	92·1555 80·5594

2.—THE EYE-PIECE.

Eye-pieces are, in the first place, intended so to refract the diverging pencils of rays, which form the real objective-image, that they may all arrive at the pupil of the observer's eye. They have also to form a virtual image (as perfect as is possible under the given conditions) of the real image, which is presented to them as the object. It is, therefore, plain that a single lens will fulfil the latter condition in a satisfactory manner only for the centre of the field of view, and on account of chromatic aberration will produce a confusion at the peripheral points, since the pencils of rays emerging from these points pass through the margin of the lens. For this reason eye-pieces of modern Microscopes consist, as a rule, of two lenses so combined that their aberrations approximately neutralize each other. The combination of these lenses admits of certain modifications from an optical point of view, of which we may specially mention the following:—

a.—*The Campani (or Huyghenian) Eye-piece.*

The so-called Campani eye-piece consists of two plano-convex lenses, the lower one of which is called the field-lens, the upper the eye-lens. The former collects the converging rays coming from the objective before they have united into an image, and forms the image within the focus of the eye-lens, so that it can be viewed through the latter as through a magnifying lens.

For the mathematical consideration of the eye-piece it is perfectly immaterial in what manner it is combined with the objective. We may, for instance, first bring the field-lens into the calculation, and then combine the eye-lens with the resulting system; or, conversely, first determine the action of the eye-piece as a whole, and then its action in combination with the objective. We will choose the latter.

Let the focal lengths of the field-lens and eye-lens be respectively 40 mm. and 30 mm., their thickness 3 mm. and 2 mm. If, then, we denote the two pairs of principal points by E^0, I^0, and E', I', and the vertices of the refracting surfaces in their order from below upwards by N^0, N^1, N^2, N^3, and assume the index of refraction of the lenses to be 1·5, and their distance from each other as 43 mm., we get

$$E^0 = N^0, \quad I^0 = N^0 + 1,$$
$$E' = N^2, \quad I' = N^2 + \frac{2}{3};$$

consequently

$$t' = E' - I^0 = 45,$$

and, hence, for the resultant principal and focal points $E\ E^*$, $F\ F^*$,

$$E = E^0 + 72 = N^3 + 24, \quad E^* = I' - 54 = N^3 - 55\tfrac{1}{3},$$
$$F = N^3 - 24, \quad F^* = N^3 + 7\tfrac{1}{3},$$
$$E - F = F^* - E^* = 48.$$

With reference to this system of cardinal points, which represents the eye-piece as a whole, the image which is actually formed is, of course, not to be regarded as the object, but as the image which the objective would form alone (without the aid of the field-lens); it must be regarded as a virtual image, since it lies behind the first refracting surface of the system. Its position is determined by the final virtual image of the eye-lens, which for each observer is situated at the distance of distinct vision from the eye; it must, therefore, be different for different eyes, since this distance is variable. In order to have a definite idea on this point, let us suppose that the virtual image lies in a plane situated 200 mm. from the last surface of the eye-piece; the posterior focus p^* must accordingly be taken as equal to $144\tfrac{2}{3}$ mm.; whence the conjugate value $p = 36{\cdot}041$ mm.

The real image of the objective would consequently have to be formed at a distance of 12·041 mm. from the last surface of the eye-piece, in order to be seen by the eye placed at the given distance of 200 mm. The amplification m', which is produced by the eye-piece under these circumstances, is found by the formula

$$m' = \frac{f}{f - p} = 4{\cdot}01.$$

If, then, the microscope-tube is of such a length that the posterior focus of the objective, taken above at 200 mm., corresponds to the position of the objective-image relatively to the eye-piece as just described, then the total magnifying power of the Microscope is evidently m' times that of the objective as above found. For any other length of the microscope-tube, however, the posterior focus p^* of the objective must first be calculated, and the magnifying power then reduced in accordance therewith. If, for instance, the last surface of the eye-piece is at a distance of 200 mm. from the first surface of the objective, and the second principal point of the latter at a distance of 2·4011 from the anterior surface, we obtain for p^* the value $200 - (12{\cdot}041 + 2{\cdot}4011) = 185{\cdot}558$, and, from this, an objective amplification $m = 1 - \dfrac{p^*}{f} = 81{\cdot}584$, which latter number multiplied by 4·01 gives the total magnifying power of the Microscope.

The magnifying power of a Microscope is not a definite amount which can be fixed once for all. It is dependent upon the condition of the eye of the observer, and cannot be considered a constant quantity for the same observer, since the eye is, within certain limits, capable of accommodation. The magnifying power of the eye-piece, in particular, undergoes considerable fluctuations; for instance, the number 4·01, which corresponds to a vision of 200 mm., decreases to 1·93 for short-sighted persons whose limit of distinct vision is 100 mm. Neither does the magnifying power of the objective remain constant, as the position of the real image also varies with the range of vision, and consequently the ratio of the anterior to the posterior focus. Short-sighted persons bring the objective-image nearer to the eye-lens and thus move the final virtual image closer also; this lengthens the posterior focus of the objective, and shortens the anterior. The magnifying power of the objective is consequently increased, while that of the eye-piece is decreased. Since, however, the latter takes place in a far greater degree, the final virtual image is so much the smaller, and the combined amplification is consequently so much the less, the shorter the distance of distinct vision. For comparison the magnifying powers are given below for different ranges of sight. The tube-length from the first to the last refracting surface is taken at 200 mm., and the objective amplification, when p^* is equal to 200 mm., at 88·0096 (*vide* table on p. 34).

Distance of distinct vision.	p^*	Objective amplification.	Eye-piece amplification.	Combined amplification.
200	185·558	81·584	4·01	327·152
150	194·551	85·586	2·97	254·19
100	198·462	86·921	1·93	167·75

From this comparison we see, in the first place, that the eye-piece amplification is approximately proportional to the range of vision, while the objective amplification shows small fluctuations in the opposite direction; whence it follows that the combined amplification also varies with the range of vision. But the mathematical relation is not given by it, and a glance at the last column is sufficient to show that the ratio of the magnifying powers differs by several units from that of the distances of distinct vision. It would be a matter of calculation to find out whether, and to what degree, these differences continue, when, instead of the last surface of the eye-piece, an adjacent point in the axis, which nearly corresponds to the position of the eye during observation, is selected as the point of emergence for the determination of the distances of vision. Arithmetical examples would show to what extent the coefficients of amplification found by various observers differ, when they are reduced to the same conventional distance of vision.

We consider it, however, advisable to put the question to the test of a more complete mathematical examination by referring the action of the objective and of the eye-piece to a single system of cardinal points. As the combination is effected exactly as before, we simply collect the results of the calculation. We will denote the new principal points by E^0 and E^*, the focal points by F^0 and F^*, and the first and last refracting surfaces of the Microscope, whose distance is equal to 200 mm., respectively by N^0 and N^*. Then

$$E^0 = N^0 - 1{\cdot}1283, \qquad E^* = N^* + 6{\cdot}727,$$
$$F^0 = N^0 - {\cdot}5174, \qquad F^* = N^* + 6{\cdot}116,$$
$$\text{Focal length} = {\cdot}611.$$

The resulting system is, therefore, as might have been expected, a diverging one, *i.e.*, it forms only virtual images of real objects. The focal points are consequently (in contrast to all former combinations) nearer to each other than the principal points.

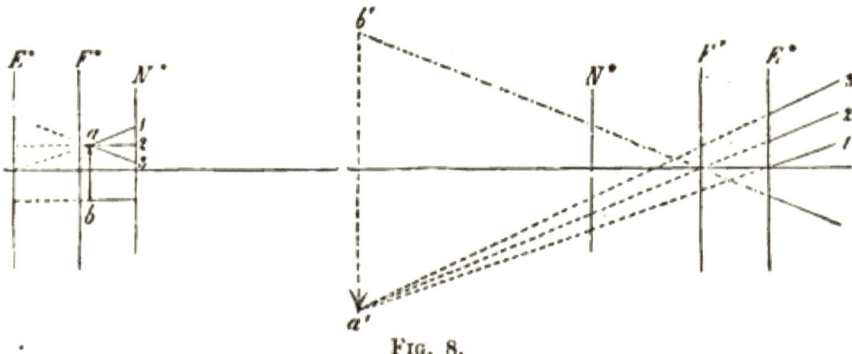

FIG. 8.

Fig. 8 shows how the last paths of the rays are to be constructed in such a system; $a\,b$ is the object and $a'\,b'$ the virtual image. The lines indicated by 1, 2, 3 correspond to each other; they cut the principal planes E^0 and E^* at equal distances from the optic axis, and their direction is determined by the familiar laws of refraction.

If, then, p^* is the distance of the image from the second principal plane and f the focal length, the coefficient of amplification is given by the formula, applicable for all cases, $m = 1 - \dfrac{p^*}{f}$, in which the negative sign indicates the inversion of the image. If m is taken as positive, this expression becomes

$$m = \frac{p^*}{f} - 1 = \frac{p^* - f}{f},$$

from which latter formula the following relation may be directly derived:—If the eye is situated in the second focal point of the Microscope, so that the distance of distinct vision is expressed by $p^* - f$, the magnifying power is exactly proportional to this distance. In other words, the virtual images which the Microscope forms with different adjustments for short- and long-sighted eyes, are seen at the same angle when viewed from the second focal point, and must, therefore, exactly coincide.

The combined action of the Microscope has been fully explained above. We have still to trace the actual course of the rays of light within the eye-piece and subsequent to their emergence from the eye-lens. Since the objective-image, which serves as a virtual object for the eye-piece, must be taken (with a vision of 200 mm.) as 12·041 mm. from the plane surface of the eye-lens, the position of the image formed by the field-lens can easily be calculated. If its focal length is 40 mm., as above assumed, we obtain from the given distance of the virtual object ($= 49 - 12\cdot12041 = 35\cdot959$) a conjugate focus of 18·936 mm. This is the distance, reckoned from the second principal point of the field-lens, at which the true image actually appears, and where, therefore, the diaphragm and the eye-piece micrometer must be applied in order to be seen as distinctly as the image.

The pencils of light which leave the objective have their base upon its second principal plane; their inclination to the axis is determined by the rays of direction, which emanate from the corresponding points of the object and may be regarded as the axes of the pencils. Since these axes all emanate from one point, that is, from the second principal point of the objective, they must also, after refraction in the eye-piece, intersect in one point, and this latter point is the image of the former. Its position is determined for the field-lens by the relation already given, $\dfrac{1}{p} + \dfrac{1}{p^*} = \dfrac{1}{f}$, in which, in accordance with former assumptions regarding the tube-length and the position of the principal points, $p = 200 - (48 + 2\cdot4011) = 149\cdot5989$, and $f = 40$ mm. We get $p^* = 54\cdot598$ mm., and since we regard the point determined by it as the virtual object of the eye-lens, the calculation gives as abscissa-value of the final real image-point, $N^* + 5\cdot938$, where N^* is the last refracting surface of the eye-piece.

The axes of the pencils, which correspond to the different points of the object, all intersect therefore in one point, which lies about 6 mm. above the eye-lens. This point may be termed the *eye-point*. A plane drawn through it perpendicular to the axis will, in general, be cut by each separate pencil of light in an ellipse; the whole cone of pencils, however, cuts the axes in a circle, the diameter of which, with given optical constants, depends upon the angle of aperture of the Microscope. If the latter is, say, 60°, the former will be somewhat greater than $\frac{1}{2}$ mm.; both increase and decrease simultaneously. As the diameter of the pupil of the eye is considerably larger, it is evident that it need not necessarily be brought to the eye-point, but merely in proximity to it, in order that all the rays of the emergent pencils of light may reach the retina.

Since the eye-point coincides within a small fraction of a millimetre with the second focal point, its position is obviously indicated by the real image of the source of light, which is formed above the eye-piece. If, for instance, we employ as source of light a window close at hand, or its image in the mirror of the Microscope, we see in the plane of the eye-point, with a suitable adjustment of the mirror, the clearly defined diminutive image of the window-frame or other objects which lie in the region of the effective rays. The diameter of this image, which may easily be determined by the method of construction applied in Fig. 8, corresponds therefore to the smallest transverse section of the emergent pencil of rays in the plane of the eye-point, or, otherwise expressed, the transverse section in question is given by the aperture-image above the eye-piece. We are now in a position to work out the construction of this image for a particular case with the constants above found, and, at the same time, to illustrate by a diagram the connection between the angle of aperture and the size of the image-surface.

In Fig. 9 let $a\,b$ be the source of light, the extent of which is limited by $m\,m'$. Let the angle at which the rays issuing from the points a and b intersect in the focus of the Microscope be 60°, and the angular aperture of the objective at least as large. If we draw a line $i\,r$ parallel to the axis from the point i, where the ray $a\,g$, which passes through the focus, cuts the anterior principal plane, it represents the path of the incident ray. A second ray, drawn from a to the anterior principal point, passes at a similar angle through the posterior principal point, and cuts the other ray in the point a, which is therefore to be regarded as the image-point. Similarly, we obtain β; $a\,\beta$ is therefore the image of the source of light. It will, of course, lie the nearer to the focal plane F^* the greater the distance of $a\,b$; its diameter, however, will remain unaltered as

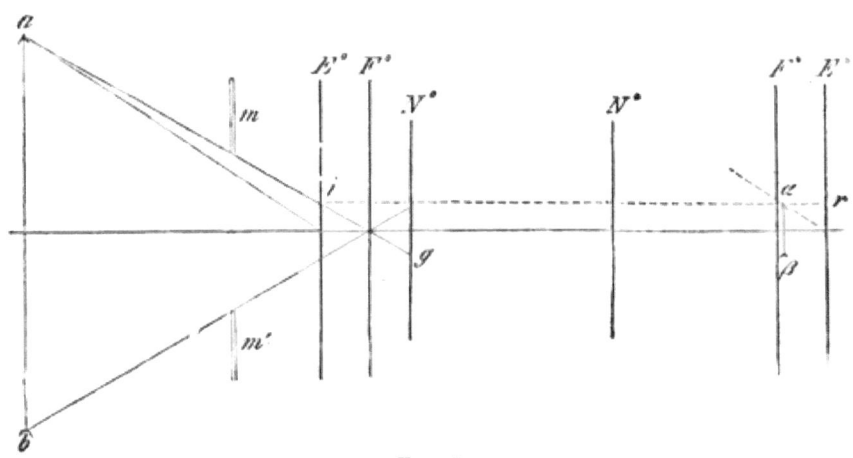

Fig. 9.

long as the angle of the incident marginal rays = 60°, as we have assumed. It is, therefore, only necessary to determine this constant diameter for the simplest case, which is when the distance of the source of light is infinite, and when, consequently, its image lies in the focal plane. Then half the diameter of the image is evidently given by the tangent of half the angle of aperture multiplied by the distance $E^*\,F^*$, that is, by the focal length of the Microscope. We get, therefore, for the whole diameter,

$$a\beta = 2 \times \frac{\tan 30°}{r} \times \text{focal length} = 2 \times \cdot 577 \times \cdot 611 = \cdot 7 \text{ mm.};$$

or, if the sine is substituted for the tangent, as is more correct (*vide* p. 24),

$$a\beta = 2 \times \cdot 5 \times \cdot 611 = \cdot 611 \text{ mm.}$$

The position of the eye-point is to a certain extent fixed beforehand in every Microscope; it must always be so chosen that, if the eye is applied at the usual distance from the eye-lens, all the emergent rays will contribute to the formation of the final virtual image, so that the whole field of view appears equally illuminated. The focal length of the field-lens must in every case be sufficient to admit of the formation of the image-point, corresponding to the second principal point of the objective, 10 to 20 mm. at least above the eye-lens; for, since the latter increases the convergence of the rays, the distance of the image-point in question is always somewhat greater than that of the eye-point.

A diaphragm is applied within the eye-piece in the plane of the image formed by the objective with the aid of the field-lens, and in conjunction with the eye-lens determines the size of the field of view. The diameter of the field is equal to that of the diaphragm multiplied by the magnifying power of the eye-lens, which has a different value according to the position of the eye and the distance of distinct vision.

b.—*The Ramsden Eye-piece.*

Modern Microscopes are usually furnished with the Campani eye-piece above described, as it is generally considered to be the most favourable for aplanatism. In special circumstances, however, other constructions of eye-piece may be applied with advantage, among which the best known is that of Ramsden.

The Ramsden eye-piece consists of two plano-convex lenses having their curved surfaces facing each other. The distance between them is so short that they act together like a doublet, whose magnifying power is equal to that of a single lens of greater curvature. The object must, consequently, be so adjusted that the objective-image is formed slightly in front of the lowest lens—between it and the objective.

The path of the rays through such a doublet may be traced with given distances and curvatures of the refracting surfaces as in the

Campani eye-piece. If, for instance, the two lenses, which we have described as the field-lens and the eye-lens, are combined after the method of Ramsden, and if the distance of the principal points turned towards each other equals 20 mm., we obtain for the focal length of the system 24 mm., and for the two principal points $E = N^0 + 18$ and $E^* = N^0 + 11$, in which N^0 indicates the first refracting surface. The second principal point is situated, therefore, 7 mm. in front of the first. It is, of course, obvious that the optician must bring the eye-point to a convenient position for observation by an appropriate choice of focal lengths.

The combination of two lenses to form a magnifying-glass, as in the arrangement of Ramsden, is advantageous, inasmuch as the objective-image is not first diminished and then magnified, but is magnified by each of them. The final virtual image (encircled by the diaphragm as by a frame) appears, therefore, if the lenses are equal in curvature and aperture to those of the Campani eye-piece, considerably larger than with the latter; in other words, the Ramsden eye-piece gives a larger field of view under circumstances in other respects similar, and as it allows of an almost entire elimination of image-distortion, it is especially serviceable for measurements with the eye-piece micrometer.

These advantages are accompanied by a slight drawback. The fact that the objective-image is very near to the surface of the lower lens causes the slightest defects in the polish, scratches, or particles of dust, &c., on this lens to appear in the field of view, and affect the distinctness of the image. This is not a serious inconvenience when the eye-piece is carefully cleaned. On the whole we prefer the Ramsden eye-piece to that of Campani.

c.—*The Aplanatic and the Orthoscopic Eye-piece.*

In the catalogues of some opticians we find, besides the ordinary Campani eye-pieces, aplanatic and orthoscopic ones, which are distinguished by a greater flatness of the field of view. Our experience has been limited chiefly to the use of the aplanatic eye-piece of Plœssl, which differs from the ordinary Ramsden eye-piece in being composed of two aplanatic double-lenses; the total length of the optical system is about 23 mm., and the magnifying power about $4\frac{1}{2}$. The image which is formed by this eye-piece is sharply defined throughout, at all points equally magnified (not distorted), and only very slightly curved.

The aplanatic eye-pieces of Schieck, Schrœder, and others are constructed on similar principles. In Schrœder's catalogue it is expressly stated that the aplanatic eye-pieces consist of two achromatic lenses, whilst the so-called orthoscopic eye-pieces of the same maker have only *one* achromatic lens. The orthoscopic eye-pieces of Kellner, on the other hand, are similar in optical action to the Campani eye-piece, but they consist of an achromatic bi-convex field-lens and an ordinary eye-lens. They give a large field of view free from distortion, without, however, appreciably increasing the performance otherwise. This is true also of the "periscopic" eye-pieces of Seibert and Krafft. The advantages obtained by any improved construction of the eye-piece—as, for instance, greater flatness of the field of view, and so on—are of minor importance. A real increase in optical power is, in spite of all assertions to the contrary, altogether out of the question.

The expressions "aplanatic" and "orthoscopic" appear to us to mean much the same thing. In order to eliminate the distortion of the image the incident pencils must be so refracted that their lines of direction cut the optic axis in the same point (*vide* Chap. VII. On the Flatness of the Field of View): the eye-piece must, therefore, be aplanatic. As in orthoscopic eye-pieces the freedom from distortion is especially commended, their principal value seems to consist in their being aplanatic.[1] Supposing they neutralize the curvature of the image-surface in a more perfect degree than the aplanatic eye-pieces of Plœssl and others, which we will assume to be the fact, the expression "orthoscopic" is not very distinctive, since it is used by some opticians as meaning "inverting" or "erecting"—that is, in quite a different sense. The so-called achromatic and periscopic eye-pieces are also avowedly aplanatic as a leading feature. By appropriate combination of the lenses, however, achromatism and greater extent of the field of view may be obtained at the same time.

The special advantage of the "holosteric," or "solid glass eye-pieces," is unknown to us. Most of them magnify very highly;

---

[1] In English works on Optics, "Flatness of Field" is distinguished from "Aplanatism": by the former is meant that the system forms a plane image of objects in a plane perpendicular to the axis; by the latter, that the system is free from spherical aberration.—[ED.]

that of Hartnack about 10 diameters, according to Frey; that of Tolles, according to Hagen,[1] 7 to 50 diameters and upwards—a quite superfluous magnifying power.

d.—*The Erecting Eye-piece.*

As the inverted position which the compound Microscope gives to images increases the difficulty of preparing objects on the stage, even if it does not retard research, modern opticians have endeavoured to effect a reinversion of the image katoptrically or dioptrically. The simple Microscope, which is generally used for dissections, ought to be abandoned for this purpose and be replaced by a better instrument, which would give the very important advantage of a greater focal distance. This idea has been realized as follows :—

To the ordinary eye-piece two other lenses were added, which were inserted in the microscope-tube with the convex sides upwards. The whole eye-piece is thus identical in arrangement with that of the terrestrial telescope, and therefore causes a reinversion of the objective-image. Instruments with these eye-pieces are to be obtained from Plœssl, of Vienna, of two different models, of which the smaller gives a linear magnification of 70—150, the other up to 300. Increase of the magnifying power is here obtained simply by drawing out the eye-piece tube.

Another method, which Oberhaeuser was the first to make use of, consists in employing a complete compound Microscope of low magnifying power as an eye-piece, bringing it so near to the objective that the real image formed by the latter is viewed by it. An object-distance is thus obtained, which admits of the freest use of dissecting instruments, &c., with moderately high magnifying powers. On the other hand, the distinctness of the images leaves much to be desired in comparison with that obtained with a simple Microscope. Such dissecting Microscopes are supplied by Hartnack, the successor of Oberhaeuser; they are provided with a rotating stage and give a magnifying power of 10—100 diameters.

Lastly, the inversion of the image may also be attained by reflecting prisms, which are applied either above the eye-piece

---

[1] "Archiv für mikr. Anat." Bd. vi. p. 215.

or within the microscope-tube. The action of a right-angled prism employed for this purpose is exhibited in Fig. 10. It produces a lateral inversion, but none in a direction at right angles to the plane of the paper; to complete the inversion, a second prism at right angles to the first must therefore be added. It is evident that such a combination can be replaced by a single prism, so constructed that two equivalent total reflexions take place within it. Both methods were variously employed by Chevalier, Nachet, Amici, and others. The most practical contrivance is the erecting prism (*prisme redresseur*) now supplied by Nachet, enclosed in a box, which may be attached or removed at will. This prism is shown in Fig. 11. Its under face is at right angles to the axis of the Microscope, and forms an angle of 58° with the upper face. The inclination of these faces is calculated to bend the optic axis to an angle of 30° from the horizon, which is convenient for working. The faces $a\,b\,c\,d$ and $b\,c\,e\,f$, which meet in the edge $c\,b$, are the reflecting ones; they form an angle of $81\frac{1}{2}°$, so that the planes of reflexion of rays incident from beneath cut one another, as calculation proves, at right angles. The other faces of the prism are without influence upon its optical action.

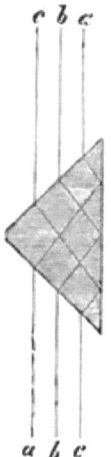

Fig. 10.

If we were willing to forego the convenience of removing the prism at will, the under face might with advantage be made convex, to replace the eye-lens—a construction which Nachet adopted in his older erecting Microscopes with two prisms.[1] The inversion of the image would be accomplished in this case without loss of light.

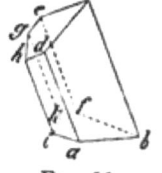

Fig. 11.

The improved erecting prism (*prisme redresseur perfectionné*, Fig. 12), which Nachet has introduced in his latest catalogue, is combined with a special form of eye-piece for the enlargement of the field of view. The new erecting eye-piece of Hartnack is similarly constructed; according to Frey it consists of a complex prism above the eye-lens, and gives a very bright though somewhat small field of view.

Fig. 12.

[1] Harting: "Mikr." 2nd ed. iii. p. 228.

c.—*The Spectral Eye-piece.*

Microspectral apparatuses have been recently made, which not only serve for the investigation of microscopic objects, but have also proved useful for other spectro-analytical researches. They consist of arrangements to be applied to the eye-piece of the Microscope, and may therefore be termed *spectral eye-pieces*.

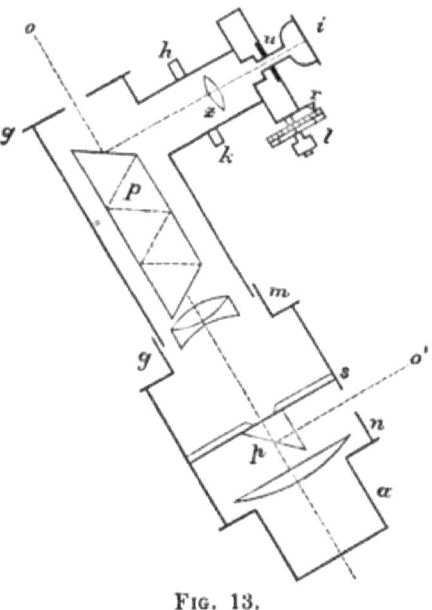

Fig. 13.

The best apparatus of this kind is that constructed by John Browning, of London, after his own design and that of Mr. Sorby. It consists of an ordinary eye-piece, of which the tube a (Fig. 13) fits into the body-tube. Between the field-lens n and the eye-lens m is a prism p, which lies in contact with the diaphragm s. The diaphragm is in the form of a slit, which may be diminished in length or breadth by a special contrivance. At the side is an opening o', through which the light from the objects serving for comparison-spectra falls upon the prism and is totally reflected through the series of prisms above to the eye of the observer at o. This apparatus may be used for observing the spectrum of the Algæ chlorophyll-green placed on the microscope-stage, and simultaneously the comparison-spectrum of a solution of chlorophyll in alcohol in a small test-tube placed in front of the aperture at o' on a specially devised stage.

The analyzing prism is applied above the eye-lens m, and is removable; it is composed of two flint- and three crown-glass prisms, and acts in such a manner that the light reaches the eye in a direction parallel to the axis. If we look through this prism down the Microscope in ordinary daylight, and adjust the opening of the diaphragm properly, a spectrum of unusual clearness and brilliancy is seen, in which the lines from B to G appear with extraordinary sharpness. With sunlight the spectrum extends from A to the two lines H, and appears as though covered with innumerable fine lines.

The best part of the spectral eye-piece is the ingenious apparatus for measurement contrived by Browning, which is applied to the side tube at h. It consists of (1) a mirror i, which throws the light into the tube; (2) a blackened plate u, on which is drawn a bright cross formed of two lines at right angles; (3) a lens z, which may be so adjusted that the image which it forms of the cross is reflected by the upper surface of the analyzing prism to the eye of the observer; (4) a screw l, by means of which the blackened plate u can be so moved that the image of the cross traverses the whole spectrum, and can therefore be adjusted to any particular point of it. As this screw is connected with a micrometer milled-head, it is possible to determine what adjustment, for instance, corresponds to the Fraunhofer line B, or C, &c. In order to determine conversely the position of a given line in the spectrum (*e.g.*, of an absorption-line of a solution of chlorophyll) it is only necessary to set the cross to it and read off the position of the screw on the milled-head.

The "spectral apparatus for Microscopes" of S. Merz, of Munich, is of a simpler construction, but it is at the same time less effective. It is made after the model of the English apparatus above described, though with the omission of the arrangement for measuring, and of the special stage and mirror for the comparison of the objects. The latter must therefore be adjusted in front of the aperture o' upon a separate stand, and the comparison of their absorption lines with those of the object under observation is the only method of determining the data required. Allowing for this defect, which is appreciable only in particular circumstances, the eye-piece of Merz will be found to answer every purpose.

3.—Means of Dividing the Pencils of Rays.

For some time past English and French opticians have made multocular Microscopes, by which several persons are enabled to observe one and the same object simultaneously. Although the importance of such instruments for science is more than doubtful, we must not pass over in silence the principles on which their construction is based.

If we disregard the earlier attempts of this kind, the multiplication of the image is effected in an essentially identical manner, namely, by the splitting up of the pencils of rays—usually immediately after their passage through the objective—into two or more equal parts, each of which is directed through a separate tube to the corresponding eye-piece. This splitting up is effected either dioptrically or katoptrically; dioptrically through an achromatic prism, as shown in Fig. 14, katoptrically by combinations of prisms, which divert the pencils in different directions by total reflexion (Figs. 15 and 16). The optical action of these

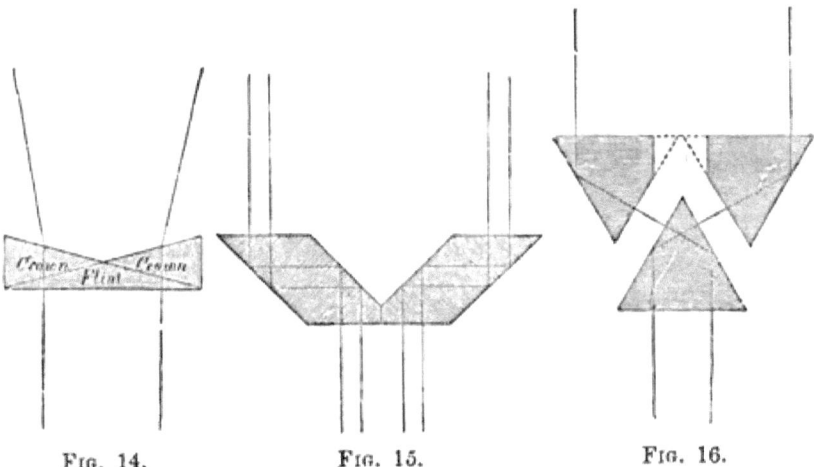

Fig. 14.   Fig. 15.   Fig. 16.

prisms in dividing the pencils into two parts is evident from the construction; it is also obvious that if in Fig. 16 the upper prisms (or the faces at which the second reflexion takes place) are turned round 90° a complete reinversion of the objective-image must be caused. The optician has it therefore in his power to give an erect or an inverted position to the final virtual image.

If a division of the cones of light into three or four pencils is to be effected, the refracting or reflecting surfaces, which produce the division, must be inclined to the median line in the same number of directions. The edges in Figs. 14 and 15 will therefore become solid angles with three or four limiting faces; the middle prism in Fig. 16 will become a three- or four-sided pyramid. The same effect would be attained by repeating the splitting-up of one or both of the divided pencils with other prisms of similar construction.

The multiplication of the objective-image is, of course, accompanied by a corresponding diminution of light; and a further loss of light occurs by absorption in the media employed. On this account only low-power objectives are applicable to multocular Microscopes, and even these, in the opinion of competent judges, leave much to be desired.

[Two short paragraphs in the original are here omitted, relating to Stereoscopic Binocular Microscopes, in which brief mention is made of three forms—Nachet's, Wenham's, and Hartnack's (illustrated by Fig. 17).—ED.]

V.

CHROMATIC AND SPHERICAL ABERRATION.

1.—CHROMATIC ABERRATION.

It is well known that the different coloured rays of which white light is composed are unequally refracted in their passage through a refracting medium. In a given system of lenses there is, therefore, for each separate colour a different position of the cardinal points, and, consequently, also of the final image. The confusion which results from this circumstance is known as *chromatic aberration*. Its elimination is rendered possible, as we assume to be known, by the property of refracting substances of so acting upon the different colours as to disperse them in a very unequal degree, though the difference in the refractive power is very slight. With crown-glass of mean refractive index 1·5342, and with flint-glass of mean refractive index 1·649, we get, for instance, as the index of refraction for the extreme rays,

	Crown-glass.	Flint-glass.
Red	1·5258	1·6277
Violet	1·5466	1·6711.

The difference between these extremes is with crown-glass ·0208, and with flint-glass ·0434. These two kinds of glass differ, therefore, in dispersive power, in the proportion of more than two to one.

Remembering that concave and convex lenses act with opposite effects, it is evident that the dispersion of colour produced by a convex lens of crown may be completely neutralized by the addition of a concave one of flint, without thereby cancelling the deviation of the rays. For if the deviation were equal but opposite in both lenses, the focal lengths therefore being equal, the flint having dispersive power twice as strong, would not only neutralize the colour-dispersion of the crown, but would produce an opposite effect in about the same degree. A ratio between the focal lengths may therefore be found, which would render the two lenses *achromatic* for the extreme rays, whilst still possessing the properties of a convex lens.

If we denote the focal lengths of the flint-lens for red and violet rays by F'_r and F'_v, and the corresponding focal lengths of the crown by F''_r and F''_v, the condition of achromatism is

$$\frac{1}{F'_r} + \frac{1}{F''_r} = \frac{1}{F'_v} + \frac{1}{F''_v}.$$

If the flint is plano-concave and the crown bi-convex, and if the radii of the three spherical surfaces are each $= R$, as we have assumed, on substituting for the focal lengths their values, the above equation becomes

$$- (n'_r - 1) \frac{1}{R} + (n''_r - 1) \frac{2}{R} = - (n'_v - 1) \frac{1}{R} + (n''_v - 1) \frac{2}{R}$$

in which n'_r, n'_v, n''_r, n''_v are the refractive indices corresponding to the focal lengths similarly denoted. By multiplication of all the terms by R we obtain

$$2(n''_v - n''_r) = n'_v - n'_r,$$

that is, the dispersive power of the flint must be double that of the crown if a double-lens constructed on the above assumption is to be achromatic. As soon as

$$2(n''_v - n''_r) > n'_v - n'_r,$$

the influence of the crown, and in the converse case that of the flint, preponderates.

With a double-lens a perfect union of rays of different refrangibility can be obtained for two definite colours only of the spectrum, for instance, red and violet. For since the dispersion in the crown and flint varies for different pairs of colours, it is not possible, even if the red and violet images exactly coincide, for the images produced by the intermediate rays to coincide with them. The differently-coloured images never appear exactly of the same size; some exceed the others more or less, and produce the coloured fringe which may always be observed with white light. Hence an approximate elimination of this coloured fringe, sufficient, however, for practical purposes, can only be attained for pencils of rays of a definite inclination. As soon as this inclination is altered, as occurs, for instance, by oblique illumination, chromatic aberration immediately appears. In the full sense of the word, therefore, a double-lens is never achromatic; and, even supposing that it were to give an absolutely colourless image with axial illumination, it would nevertheless be more or less over-corrected for oblique light.

The same holds good, of course, for systems of lenses; it can only be a question of approximating as nearly as possible to achromatism by a skilful combination of flint and crown lenses. In addition to the proper ratio of the focal lengths, the selection of the kinds of glass is important, as those are especially suitable in which the partial dispersions, that is, the relative dispersions of the same pairs of colours (Fraunhofer's lines), differ in the flint and crown as little as possible.

Moreover, it is obvious that in Microscopes in which the eye-pieces consist of single lenses, and which are consequently not achromatic, the flint lenses of the objective must have a relatively greater dispersive power, as they have to counterbalance not only the crown lenses united to them, but also the eye-piece. The objective must therefore be, what is usually termed, *over-corrected*. In most cases the influence of the eye-piece is very slight. It must be borne in mind that if the cones of light incident from an object-point are so refracted that the differently-coloured objective-images are displaced *laterally*, a compensation between the objective and the eye-piece becomes impossible. But this lateral displacement always takes place when the incident cones of light are inclined to the axis on one side, *e.g.*, when the mirror is moved out of the axis. We will discuss this point fully later on (*vide* On the Testing of the Microscope), in connection with others, which will facilitate its comprehension. Here we need only remark that the blue fringe encircling the field of view, or on the outline of a rather large object, does not prove that the objective is really over-corrected, and still less that it possesses a *chromatic* preponderance in proportion to the under-corrected lenses of the eye-piece.

2.—Spherical Aberration.

Spherical aberration, or aberration caused by the spherical form, is due to the fact that the focal length of the marginal rays is always less than that of the central rays, and diminishes with the distance from the axis. If the surface of the lens be divided into concentric zones, each zone represents a different focal length, and therefore also a different position of the image. The combination of flint and crown serves also to correct these defects. It is possible to establish such a ratio of the refractions in an achromatic double-lens that the marginal rays, which are deflected from the axis by the concave flint to a greater extent than the central rays, shall be more strongly refracted by the crown in an opposite direction, their focus then coinciding with that of the central rays. A double-lens in which all the rays are brought to a focus in one plane is called aplanatic.[1]

Every achromatic pair of lenses can be made aplanatic also. For since achromatism depends, as is apparent from the equation given at p. 51, upon the focal lengths and dispersive ratios, and aplanatism, on the other hand, upon the curvatures, the two conditions are not antagonistic. If the anterior surface of the flint-lens is given, the radii of the other surfaces can be computed so that all aberration will disappear. If, for instance, a double-lens with a plane anterior surface and equal curvatures is achromatic but not aplanatic, that is, if the influence of the flint upon the marginal rays is too weak, all that is necessary is a slight increase in the curvature of the surfaces of contact of the two lenses, and a corresponding alteration in that of the posterior surface of the crown, the ratio between the focal lengths

[1] Many authors give a somewhat more extended meaning to this expression. They call a system of lenses aplanatic, when not only spherical but also chromatic aberration is eliminated as far as possible. We follow Radicke ("Handbuch der Optik") and other mathematicians on this point.

remaining therefore unaffected. In the opposite case, the radius of the surfaces of contact must be increased, and then the posterior surface of the crown must be determined from the equation for achromatism. For the practical optician it is most advantageous that the different kinds of glass should be so constituted as to allow the radii of the crown to be almost equal, the shortest focal lengths being obtained by such combinations with the strongest practicable curvatures.

In the strictest sense of the word, aplanatism is just as little attainable as achromatism. Marginal rays are acted upon by flint-lenses in proportion to their distance from the axis. If, therefore, the focus of the extreme marginal rays coincides with that of the central, the rays passing nearer to the axis will always show traces of aberration.

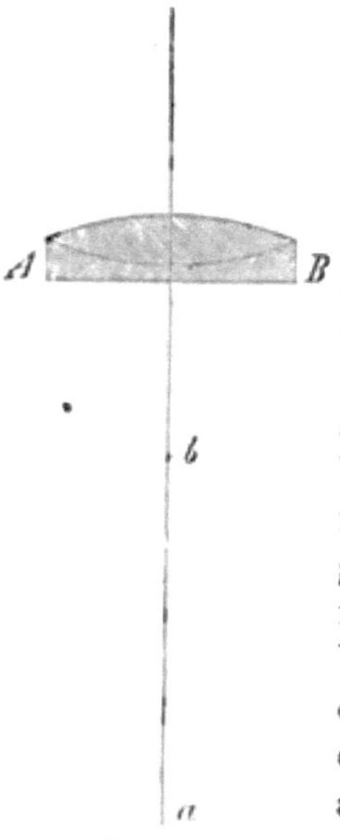

Moreover, calculation as well as experiment shows that a double-lens, which is sufficiently aplanatic for rays incident in a parallel direction, ceases to be so when the rays converge or diverge; and that, in general, aplanatism can only be obtained for definite distances of the (real or virtual) object-points. If, for instance, the double-lens AB (Fig. 18) is aplanatic for the point a lying somewhat beyond the focus, then, if the object be brought nearer, an over-correction is observed, which gradually increases and then decreases until it almost vanishes when the point b is reached. With still greater proximity of the object, or on its withdrawal beyond a, aberration in an opposite direction, that is, under-correction, appears.

FIG. 18.

The two points a and b, the rays from which pass through without aberration, are called the *aplanatic foci* of the double-lens; at the same time it must be observed that this term is not in all respects applicable, as the aplanatism in question extends only to a zone of limited dimensions, although of importance for the formation of the image. This is especially the case with lenses of deep curvature. The importance of these aplanatic foci has, therefore, been much over-estimated in recent times, the residual aberration which is present having been regarded as the effect of a prevailing under- or over-correction (*i.e.*, extending over the whole extent of the lens), and this idea has been transferred without limitation to the case of high-power objectives with large angles of aperture. Calculation, however, shows that in this case the ratio of the aberrations is a much more complicated one. "Spherical aberration may, on a strict investigation of its conditions, be divided in general into a series of independent factors, which, as they increase with the greater inclination of the rays towards the axis, pursue a very unequal course. Complete elimination is theoretically possible only for the two first factors. As soon as the angle of aperture exceeds a very small amount, the correction of spherical aberration can only be attained by compensating the ineffaceable higher terms by intentionally introduced residua of the lower ones. The increase of the unavoidable deficit, which this compensation necessarily leaves unremedied on account of the dissimilar course of the separate parts, determines the limit which must be set to the angle of aperture if this deficit is to remain without injurious effect in the microscopic image."[1]

In the manufacture of objectives with large angles of aperture (*e.g.*, 60° to 80°) the optician is therefore referred chiefly to his own experience; he must discover by repeated experiment not only the most suitable kinds of glass, but also the correct ratios of the radii of curvature and of the distances of the lenses, and must endeavour, in this purely empirical way, to eliminate, as far as possible, both the aberrations. It is clear that a preliminary theoretical testing of different combinations will always afford many valuable hints; for instance, it is an indispensable condition in high-power modern objectives with angular apertures of 100°, and upwards, that a single, nearly hemispherical, front-lens should be combined with a system of strongly over-corrected lenses. The discovery of this type of construction must be regarded as the true foundation of all recent improvements in objectives.

Whatever may be the path of the rays for the peripheral parts of the objective, it can always be so far improved by the alteration of a lens-distance (as, for instance, by moving the correction-collar) that at least the extreme marginal zone and the central part of the objective will work together. But the intermediate

[1] Abbe: "Archiv für mikr. Anat." Bd. ix. p. 425.

zone remains over-corrected in defective constructions, and all further approximation or separation of the lenses will only move the residuum of aberration between the centre and the margin, so that either this or that zone is temporarily more or less free from aberration, the one always at the expense of the other. In such cases more cannot be attained even by special "correction" contrivances or similar means; the *difference* of correction still remains, as it is based on the refractive powers of the lower lenses of the objective, and cannot therefore be removed by any less strongly refracting lenses placed above them. The real capacity of the Microscope can never be increased by such means. The advantages they furnish may be just as well secured by proper construction of the objective itself, and the errors which cannot be thus overcome remain notwithstanding the application of correction-adjustments.

The mode in which the optician has to combine all the factors which influence the path of the rays, in order to produce as favourable a result as possible, cannot be fixed once for all, however much the general laws of the course of the rays be studied. To come as near as possible to perfect aplanatism is, as a rule, left to experimental trials by alteration of the lens-distances, and by happy selection of the single lenses. Opticians of the older school endeavoured to attain this end to some extent by making each pair of lenses aplanatic; modern opticians, however, direct their attention to the total effect, and correct the aberrations of the anterior lenses by the opposite aberrations of the posterior ones. This last method leads, undoubtedly, to a higher degree of perfection, especially with high powers; the former, however, affords certain practical advantages for low powers. When each pair of lenses is aplanatic, the last pair may be used alone or in combination with the next, as the object-point then exactly replaces the virtual image formed by the front-lens, and there is consequently no derangement of the aplanatism. The anterior pairs, after the removal of the others, can obviously give only an indistinct image.

It is clear, from what has been adduced, that similarly-designated objectives as to power, even from the same manufactory, can only agree approximately. In consequence of the impossibility of reproducing the same curvatures, the optician must actually test every combination of lenses, and when he has eliminated the aberrations as far as possible, the magnifying power may vary considerably from what was intended. Experience teaches how to keep these unavoidable discrepancies within tolerably narrow limits.

The combination of lenses, which we have previously assumed for the determination of the cardinal points, has still to be examined with respect to both the aberrations. This examination can, of course, be only a purely theoretical one; it is not intended to discuss the different assumptions upon which this determination is based, but rather to furnish an example of an approximate calculation for similar combinations of lenses.

The data for the estimation of chromatic aberration are, however, wanting, as the dispersive ratios of flint- and crown-glass for the assumed indices of refraction are not known. We will therefore confine ourselves to spherical aberration, and will introduce the further simplification, that the marginal rays, whose course is to be traced, proceed from one point of the optic axis.

Let $A B$ (Fig. 19) be the first objective-lens, a an axial point of the object whose distance from the lens is δ, $a\,p$ a marginal

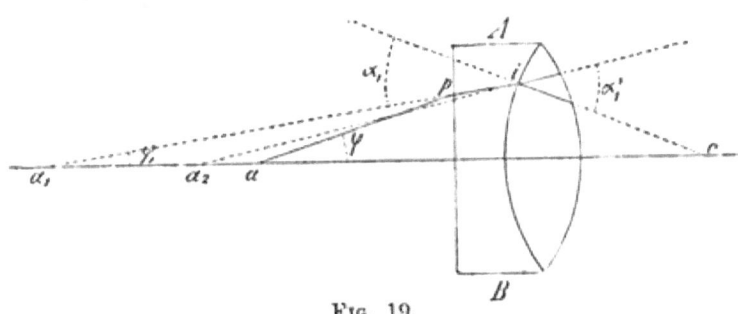

Fig. 19.

ray, which forms the angle ϕ with the axis, and therefore also with the perpendicular to the first refracting surface, $a_1\,p$ its direction after the first refraction; then the angle ϕ_1, which the refracted ray makes with the axis, is determined by the refractive index of the flint-glass. Consequently, the distance of the point a_1 from the refracting surface, or the focal length corresponding to the object-distance δ, which we will indicate by f_1, may be regarded as given. We have therefore

$$f_1 = \delta \frac{\tan \phi}{\tan \phi_1}.$$

The refracted ray is then to be regarded as incident on the second refracting surface—the surface of contact of the two lenses; the angle a_1, which it makes with a straight line drawn from the centre of curvature through the point of incidence, is the angle of incidence. Its magnitude is given by the trigonometrical ratio

$$a_1 c : ci = \sin a_1 : \sin \phi_1,$$

or, denoting by r the radius of curvature, and by d the thickness of the flint-lens,

$$(f_1 + d + r) : r = \sin a_1 : \sin \phi_1 ;$$

hence

$$\sin a_1 = \frac{f_1 + d + r}{r} \sin \phi_1.$$

The resulting angle of refraction, which may be called a_1', is obtained from the known refractive ratios of the flint- and crown-glass. In this case the direction of the ray after the second refraction, together with the angle ϕ_2 which it makes with the axis, are to be regarded as known. In the triangle $a_1 \, i \, a_2$ the sum of the angles a_1 and i is equal to the exterior angle ϕ_2, or, since the angle $i = a_1' - a_1$,

$$\phi_2 = \phi_1 + a_1' - a_1.$$

Hence, for the distance f_2 of the point a_2 from the second refracting surface, we get

$$f_2 = a_2 c - r, \text{ or, since } a_2 c : ci = \sin a_1' : \sin \phi_2 ,$$

$$f_2 = r \cdot \frac{\sin a_1'}{\sin \phi_2} - r.$$

Similarly may be calculated the angle of incidence a_2 for the last refracting surface of the double-lens, and from it the angle of

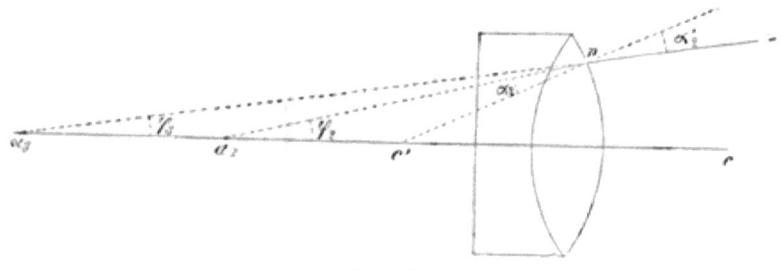

FIG. 20.

refraction a_2', and the angle which the refracted ray makes with the axis (ϕ_3). The triangle $a_2 \, c' \, n$ (Fig. 20) gives

$$\sin a_2 = \frac{a_2 c'}{c' n} \sin \phi_2 = \frac{f_2 + d - r}{r} \sin \phi_2,$$

and similarly, in the triangle $a_3 \, a_2 \, n$ the sum of the two acute angles equals ϕ_2, therefore

$$\phi_3 = \phi_2 - (a_2' - a_2).$$

Finally, we obtain from the triangle $a_3 \, c' \, n$ the length $a_3 \, c'$, and hence, by the addition of r, the distance of the point a_3 from the last refracting surface, that is, f_3, we get

$$f_3 = r \cdot \frac{\sin a_2'}{\sin \phi_3} + r.$$

By this equation the path of the marginal ray through the first double-lens is determined. These formulæ are also applicable for the other double-lenses, for which the calculation must be repeated exactly. The point a_3 is to be taken as the object-point for the anterior surface of the second lens; its distance $= f_3 + e_1$, where the latter quantity denotes the distance of the two lenses. The angle ϕ_3 is the angle of incidence for the plane anterior surface.

By the last refraction the angle ϕ receives a negative value, i.e., the ray of light is turned again to the axis, and therefore exhibits the opposite inclination to it. Consequently the sign $+$ becomes $-$ in the expression for the corresponding f_3, as shown by the construction, and we get

$$f_3 = r \cdot \frac{\sin a_2'}{\sin \phi_3} - r.$$

When the influence of the flint-lenses is excessive, the distance of the marginal rays from the axis, after their passage through the objective, is increased, and either their radii must be greater, or their indices of refraction lower.

If we now suppose that the extreme marginal rays make an angle of 30° with the axis (e.g., with an angular aperture of 60°), and if we denote their focal length, calculated from the posterior surface of the lens, by f_3, and that of the central portion of the lens by (f_3), the calculation for the first double-lens of the objective gives the following ratios:—

FIRST OBJECTIVE-LENS.

Object-distance. δ	Focal Lengths. f_3	(f_3)	Difference.
·4	4·00417	3·7328	·27137
·4724	4·63716	4·59605	·04111
·50026	4·9034	4·87685	·02655
·5075	4·96517	4·9504	·01477
·51634	5·0475	5·0431	·0044

From this table it is evident, in the first place, that the first double-lens is approximately aplanatic for an object-distance of ·51634 mm., but appreciably under-corrected for shorter distances, since the marginal rays, produced backwards, cut the axis at a greater distance than the central rays. It is, however, assumed that the rays proceeding from the object-point reach the anterior surface of the lens without deviation, which, it is well known, is not usually the case. The ray passes through the fluid in which the object is placed to the cover-glass and thence into air, and is thus twice refracted before meeting the objective. The divergence of the marginal rays is thus increased; they appear finally to emerge from a point which lies above the real object-point, and which is the more remote from it, the thicker the cover-glass and the greater the angle which the rays make with the axis. We will hereafter investigate more exactly the effect thus produced; we need only premise that, in consequence thereof, a somewhat shorter object-distance must be assumed for the marginal rays, if the calculation is to hold good for the cases which ordinarily occur.

If we select ·5 mm. as the object-distance (*vide* p. 31) of the given combination of lenses, a diminution of about ·005 mm. for the marginal rays would produce a perfect equality of the focal lengths f_3 and (f_3). Of course this slight reduction of the object-distance is not equivalent to the influence of ordinary cover-glasses; but, as no other combination approaches so nearly to aplanatism of the objective, we may take it as the basis of further calculations, and assume for the second and third double-lenses the distances and refractive indices (*vide* table on p. 34) which correspond to the object-distance ·5. The focal lengths of the marginal and central rays will then be expressed as follows:—

SECOND DOUBLE-LENS.			THIRD DOUBLE-LENS.		
δ	f_3	(f_3)	δ	f_3	(f_3)
5·03069	23·802	24·403	24·302	177·203	177·2

The spherical aberration appears to be completely eliminated; but this holds good only for marginal rays of 30°, and does not prove the combination of lenses to be aplanatic to the greatest extent for the other zones of rays, nor, therefore, for the whole effective cone. The inner portion of the cone of rays might possibly give a more favourable image if the distances of the lenses were varied. The working out of the calculation for all the inner rays to every 5° or 10° would determine this point.

Voluminous as these calculations are for each different case, until at length data are found which give a minimum of aberration, they should unquestionably be undertaken if the problem could thus be solved once for all, of producing objectives of the greatest possible perfection in any desired quantity. But it is one of the most difficult problems of the optician's art to satisfy even approximately the demands of the mathematician in the construction of lenses. In most manufactories[1] the unavoidable inaccuracies of workmanship are so great that the calculations, in fact, retain a practical value only for general guidance, and all beyond is left to the skill and experience of the workman.

As already pointed out, in the construction of the double-lenses, when the optician wishes to produce either a slight degree of under- or over-correction, or perfect aplanatism, he must select those pairs of lenses whose combined action approximates to that of his standard lenses. The proper combination of the double-lenses to form the objective demands care and judgment. Many modern objectives are built up each complete in itself, so that for different amplifications (with the same tube and eye-

[1] In the "Archiv für mikr. Anat." Bd. ix. p. 414, Abbe states that for some time past C. Zeiss, of Jena, has been manufacturing both high and low powers, equal to any produced hitherto, strictly according to calculated formulæ.

piece) different objectives must be employed.[1] In the higher powers the front lens is generally mounted, so that its distance from the others can be regulated to allow for the influence of the cover-glass or the obliquity of the incident rays.

The correction of the aberrations is therefore effected in practice, as in theory, by the method of approximation. In practice, the testing begins with the last refractions in the system, and proceeds downwards to the primary one; in theory, the ray of light is traced from the anterior surface of the objective up to its last surface.

Special attention is due to the effects of *spherical aberration in the eye-piece*. Since the cones of light which reach the points of the objective-image are much diminished, and meet only an

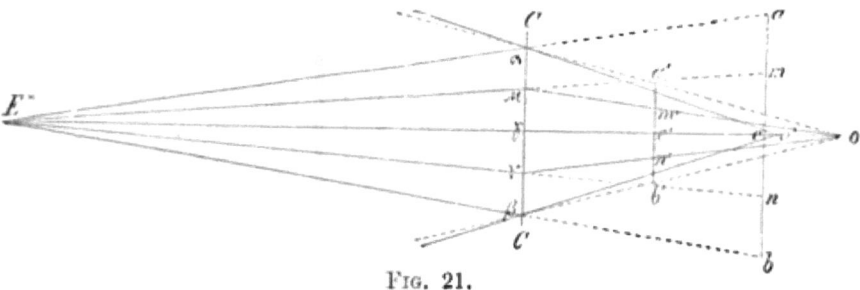

Fig. 21.

exceedingly small portion of the field-lens, it is not single rays which experience the stronger refraction of the margin of the lens, as in the objective, but the whole image-forming pencils which pass through this margin. Spherical aberration is not exhibited through the obliteration of single image-points, for the aberrations within a pencil of rays are hardly perceptible; it influences, however, the direction of the axes of the pencils after refraction, and hence the relative position of the image-points. As a demonstration of this action, let E^* in Fig. 21 be the second principal point of the objective, in which the axes of the image-

---

[1] The high-powers of Hartnack, Bénèche, Kellner, Plœssl, &c., are constructed on this principle. The two posterior lenses alone give a very indistinct image, which is over-corrected both spherically and chromatically. The blue mist surrounding the outlines is generally so considerable, that it might be *à priori* doubted whether so much aberration could be compensated in a satisfactory manner by the addition of the anterior lens. It cannot, however, be disputed that in practical Optics the best results have hitherto been obtained with such combinations of lenses.

producing cones of light intersect, CC the infinitely-thin field-lens, ab the (unformed) objective-image, and $a'b'$ the real image of the field-lens. If the field-lens were aplanatic, the cones of light directed towards a and b (represented in the figure by simple lines) would cut the optic axis in the same point in which the more central cones, for instance, $E^* m$ and $E^* n$, intersect. Any points m, n .... in the plane ab would therefore have corresponding positions in the real image $a'b'$, for the proportions are

$$am : mc = a\mu : \mu\gamma = a'm' : m'c',$$

and, similarly,

$$bn : nc = \beta\nu : \nu\gamma - b'n' : n'b'.$$

The objective-image would therefore undergo an entirely uniform diminution.

This uniformity is, however, destroyed through the stronger refraction of the peripheral pencils. While the pencils directed

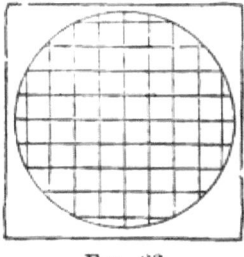
Fig. 22.

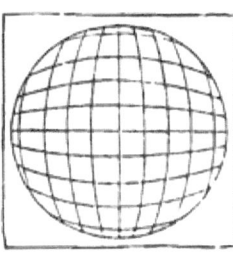
Fig. 23.

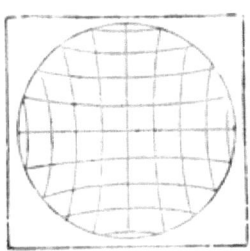
Fig. 24.

to the points m and n are refracted to o, those proceeding to a and b intersect in o'. In consequence of this, the points a' and m' on the one side, and b' and n' on the other, as shown by the figure, are brought nearer to each other than would be the case if the diminution were uniform. The same reasoning applies, of course, to any other points which lie near to each other in a radial direction. We therefore arrive at the general conclusion, that the surface-elements of the objective-image are the more diminished, in consequence of the spherical aberration of the field-lens, the greater their distance from the optic axis. Accordingly, the objective-image of a net-work of squares, for instance, Fig. 22, would appear in the real image of the field-lens as in Fig. 23.

The effect is exactly the opposite if, under the same conditions, a lens is placed so that it forms virtual images instead of real ones. This will be apparent if, in Fig. 21, the lines drawn towards o and o' are produced backwards; then, in consequence of the spherical aberration, the peripheral image-points will move beyond the position they would occupy if the lens were aplanatic. The margins of the image will therefore appear more magnified than the central parts, and the network will consequently present the appearance of Fig. 24. This is the distortion which is observed in a greater or less degree with every non-achromatic magnifying-lens, and which was formerly erroneously called the *curvature of the field of view*. Since it is obvious that the eye-lens causes a similar distortion, acting, of course, in an opposite direction to the field-lens, a ratio may always be found by which the final virtual image of the Microscope will exhibit a nearly uniform enlargement, that is, will appear tolerably flat. The so-called *flatness of the field of view* is dependent on the practical attainment of this ratio. We shall return to this point in a special chapter.

Spherical aberration of the field-lens and eye-lens always implies, therefore, under the conditions of microscopic vision, a distortion of the real image (in the sense that the margins are slightly less enlarged than the centre), and a distortion of the final virtual image in an opposite direction. It is quite immaterial in this case whether the image-points lie in a plane or in a curved surface; the (actual) *curvature* of the image-surface is a phenomenon of quite another kind, which was erroneously regarded by the early microscopists as the cause of the distortion. We shall discuss this point more fully in the proper place; the only question here is, whether, and in what way, spherical aberration influences the image-surface itself? or, in other words, what changes will occur in the image formed by an aplanatic lens, if a non-aplanatic one of equal focal length (for central rays) is substituted for it? The question is easily answered. A non-aplanatic lens acts on the peripheral pencils of light, since they meet only a small part of the lens, exactly in the same way as an aplanatic one of shorter focal length, and therefore, *cæteris paribus*, brings the real image-points somewhat nearer, and forms the virtual ones at a greater distance. Spherical aberration of the field-lens acts, moreover, so to say, attractively upon the margins of the real image, while spherical aberration of the eye-lens has a repellent influence upon those of the final virtual image. Attraction and repulsion decrease, of course, from without inwards, and in the centre become *nil*. Curvature of the image-surface is therefore due to spherical aberration, and is more or less apparent by turning the convex side of the aplanatic system upwards or downwards.

VI.

INFLUENCE OF THE COVER-GLASS.

A GLANCE at Fig. 25 will show that two rays which proceed from a point a, after their passage through a medium bounded by parallel plane surfaces ($m\ n$ and $p\ q$), will appear to have come from a point a', more or less distant from a. If both rays are equally inclined to the refracting surfaces, the two points a and a'

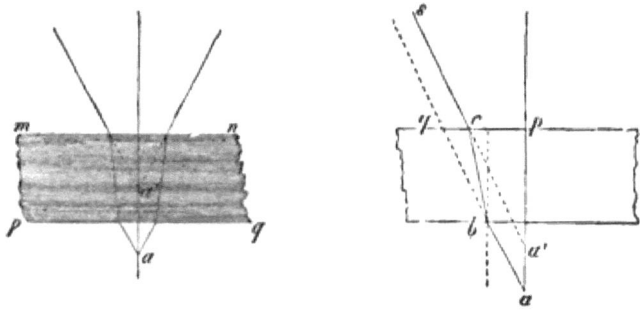

Fig. 25.   Fig. 26.

will lie in a straight line perpendicular to the surfaces; if, on the other hand, they are unequally inclined, the line joining them will form an oblique angle with the perpendicular.

A pencil of light, whose rays diverge from the object-point a, will consequently be no longer homocentric on emerging from the cover-glass, but will consist of an infinite number of superposed cones, whose apices vary with the obliquity of the incident cone in relation to the cover-glass. The real object-point is therefore represented virtually as a line, whose length evidently varies with the thickness of the cover-glass and the angle of aperture of the objective.

If ab (Fig. 26) is an incident ray which passes through the cover-glass in the direction bc and then passes into the original medium, cs is parallel to ab. The position of the point a', in which the emergent ray, produced backwards, cuts the perpendicular ap, can therefore be trigonometrically determined. Then, if a is the angle of incidence, a' the angle of refraction, and D the thickness of the cover-glass, we get in the first place $bc = \dfrac{D}{\cos a'}$. The triangle qcb gives, therefore,

$$qc = bc \cdot \frac{\sin(a - a')}{\cos a}.$$

Since $a'c$ is parallel to aq, we obtain from the triangle aqp

$$aa' : qc = ap : pq = \cos a : \sin a;$$

consequently

$$aa' = qc \cdot \frac{\cos a}{\sin a}.$$

If we substitute for qc and bc their values, we get

$$aa' = D \frac{\sin(a - a')}{\sin a \cos a'}$$

and, by an evident reduction,

$$aa' = D \left(1 - \frac{\tan a'}{\tan a}\right).$$

If a is taken as 40°, and the refractive index of the cover-glass 1·5, then, if the surrounding medium is water, we get

$$aa' = \cdot 168106 \times D,$$

and if the surrounding medium is air,

$$aa' = \cdot 565037 \times D.$$

In the most usual case (*i.e.*, water below, air above) the latter expression is increased to a small extent, varying with the distance of the object-point from the lower surface of the cover-glass, and when the distance = 0, this is also zero.[1]

---

[1] The lateral displacement which takes place when the light is incident obliquely has too little practical interest to require further discussion here. The obliquity of the illumination is entirely without importance, for the contour-image and the delineation of details are dependent upon factors unconnected with the path of the rays. Cf. on this point the chapter on the Theory of Microscopic Observation.

It is evident from what has been stated (pp. 60, 61), as to the path of the marginal rays, that a diminution of the object-distance, which in the case of rays of 40° inclination amounts to about $\frac{1}{5}$ mm. more, even with a cover-glass of only $\frac{1}{3}$ mm., must exercise a noticeable influence on the position of the corresponding objective-image, and, at the same time, on the correction of the aberrations. An objective which produces sharp images of uncovered objects will, on the application of a cover-glass, define the more indistinctly the greater the alteration of the object-distance thereby caused. Similarly, an objective intended for immersion in water will prove less aplanatic for observations in air, unless the error is rectified by alteration of the distances of the lenses. From this it is evident that objective-systems with angular apertures of 120° to 160°, as now supplied by opticians, are far more sensitive to such influences than those in which the angles of aperture amount to only 60° or 80°. On this account the high-power objectives of most recent construction are furnished with means to alter the distance of the front lens from the two others, to the extent necessary for proper compensation. As a rule, the lenses must be the nearer together, the greater the apparent approximation of the object caused by the cover-glass. Contrivances for effecting this compensation are applied to the high-power objectives of various German, French, and English makers. The mechanism is shown in Fig. 27. The metal stud s indicates the different positions of the lenses. The medium position a, and the two extreme positions b and c, are shown separately. The approximation of the lenses should be effected preferably in such a way that the front lens remains fixed, instead of moveable, as formerly, which is attended by the practical advantage that the object does not disappear from the field of view during the adjustment.

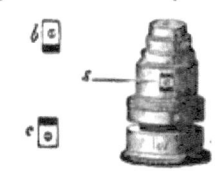

Fig. 27.

Mohl has written in detail in his "Mikrographie" upon the influence of cover-glasses on objectives of medium power, and we refer to his statements with regard to the older systems of Plœssl and Amici. With regard to the recent instruments which we have examined, we have not found that the differences of thickness usually met with in cover-glasses (say $\frac{1}{3}$ to $\frac{1}{6}$ of a millimetre) influence the microscopic image to any considerable degree.

Hence, probably, the opticians now construct their medium objectives for cover-glasses of about ¼ mm. in thickness, whilst the objectives of Amici were expressly intended for use with cover-glasses of various, and in some cases very considerable, thickness (up to 1·5 mm.).

VII.

THE FLATNESS OF THE FIELD OF VIEW.

On observing a straight line through a lens, or system of lenses, its image appears to be straight only when it passes through the centre of the field of view; in every other position it will appear curved, with its convex side towards the centre, and the curvature varies with the distance from the centre. A mesh-work of squares (Fig. 28) will accordingly appear as represented in Fig. 29, and

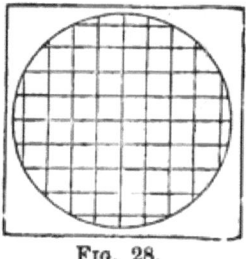

Fig. 28.

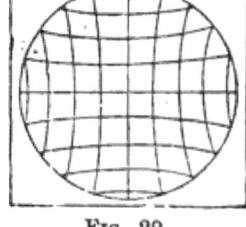

Fig. 29.

every other object in the virtual image will appear more or less distorted in a similar manner.

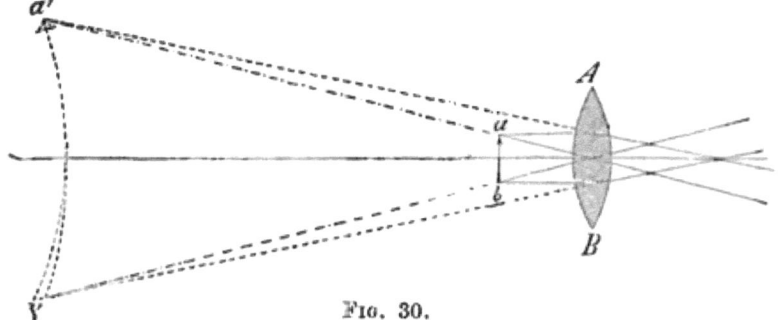

Fig. 30.

It was formerly customary to explain these distortions as due to the fact that the points of the virtual image $a'\ b'$ (Fig. 30), which the lens $A\ B$ forms of the plane $a\ b$ (assumed to be at right angles to the optic axis), lie in a curved surface convex to the object, and that the peripheral parts of this surface, in consequence of their greater distance, are somewhat more magnified. Similarly, it was considered that the opposite distortion of real images (Fig. 31), in which the magnification decreases from the centre outwards, was occasioned by a corresponding curvature of the image-surface (Fig. 32) [cf. Harting, "Das Mikroskop," first edition, pp. 134 and 278]. Harting and others have since changed their views (cf. loc. cit., second edition, vol. i. pp. 100 and 141); but as the traditional idea crops up here and there, and is, moreover, contained in all the older text-books, we will repeat without alteration in this second (German) edition our investigations on this point.

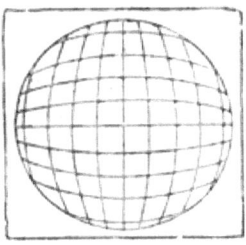

FIG. 31.

That the old explanation is entirely erroneous is plain from what has been said above (pp. 62–65). It was there shown that such

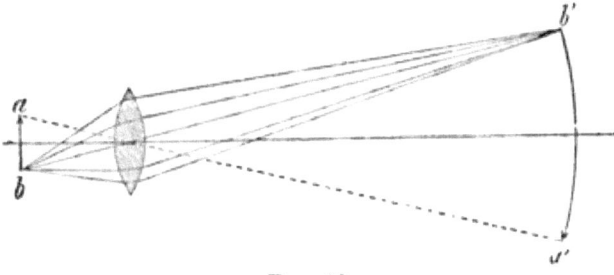

FIG. 32.

distortions are not to be explained by the curvature of the image-surface, but by the stronger refraction of the peripheral pencils, and are therefore due to the spherical aberration of the image-producing lenses; it was at the same time pointed out that a real curvature of the image might take place quite independently of this, and that it varies with the spherical aberration of the lenses, according as their convex sides are turned upwards or downwards. Nevertheless, it may not be superfluous if we return once more to this subject, in order to explain, in as simple a case as possible, not only the distortion by the eye-lens, but also the curvature of the virtual image.

Let $a\ b$ (Fig. 33) be the field-lens image, in which the diminished cones of rays (which are represented in the figure by simple lines) converge to the point o; let $A\ B$ be the eye-lens, whose centre of curvature is in o; and let the eye of the observer be adjusted for infinite distance. Then, since the pencils converging to o are incident in the direction of the radius, they undergo no deviation at the lower surface, and the refraction is confined to removing the point of convergence to an infinite distance in the direction of the axis of the pencils. The virtual image $a'\ b'$ resulting from this (which is to be considered at an infinite distance)[1] agrees, of course, completely with the object, *i.e.*, there is a perspective coincidence of the single object-points with their images. Rectangular net-work should therefore appear rectangular also in the image. Nevertheless, this virtual image is considerably curved, because the marginal points of the object are further from the refracting surface than the centre.

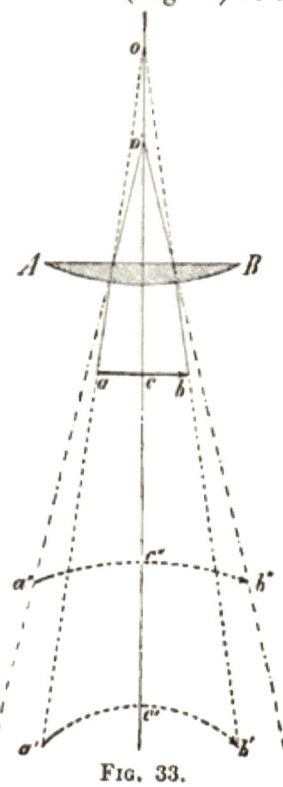

Fig. 33.

By the second refraction at the upper surface of the eye-lens the parallelism is obviously not destroyed—the image remains at an infinite distance. But the pencils are bent aside from their course the more in proportion to the angle which they form with the axis of the Microscope. The amplification which the virtual image thus undergoes is expressed approximately by the refractive index. It is not, however, uniform throughout. If we suppose the object $a\ b$ to be divided into any number of equal parts, the pencils proceeding from the points of division meet the eye-lens and (if produced backwards) the plane of projection of the virtual

---

[1] Our figure is accurate for the assumed finite distances. The virtual image is in this case brought nearer in consequence of the second refraction by the $\left(\dfrac{n-1}{n}\right)$-th part of its former distance, in which n denotes the refractive index. For the others, the course of the rays is exactly the same.

image $a'\ b'$ at equal distances. A uniform magnification would therefore clearly imply that the deflected pencils of light cut the axis of the Microscope in the same point. This supposition is not, however, correct, since the point of intersection of the peripheral rays always lies nearer to the eye-lens than that of the central rays. The final virtual image is thus necessarily more or less distorted. For if c'' (Fig. 34) is the image of the object-point c lying in the axis, and p'' that of any point p of one of the divisions, a point q at twice the distance would, in the case of uniform magnification, have to be formed in the image also at twice the distance, therefore in q''. In reality, however, the image-forming pencil of light is revolved somewhat more round its point of incidence r in the upper surface of the eye-lens, since it cuts the axis at a slightly less distance. The point q'' lies therefore somewhat further outwards, as the dotted line indicates; consequently, in the virtual image, $c''\ q'' < 2 \times c''\ p''$, and $p''\ q'' < c''\ p''$. This means that the magnification increases with the distance from the axis. The increase is limited, however, as is readily seen, to the radial direction; the tangential direction is affected only because the displacement of the image-points in a radial direction causes a proportional change in their distances. Straight lines in the object, which (if necessary produced)

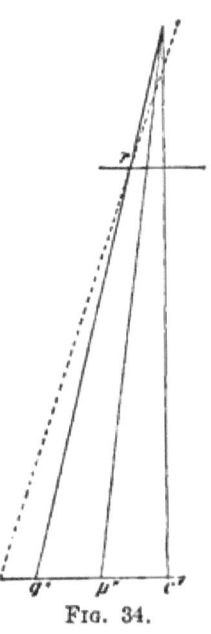

Fig. 34.

intersect in the centre of the field of view, must therefore appear in the image as straight lines; in every other direction, however, an irregular magnification involves their curvature, which is the stronger the greater their distance from the centre.

We may, consequently, regard it as established, that curvature and distortion are two entirely different phenomena, which are not to be confounded. Distortion depends, we repeat, upon the spherical aberration of the refracting surfaces; curvature, on the other hand, upon the unequal distances of the respective object-points (in which the image of the first surface is to be regarded as the object for the next succeeding surface). Even if spherical aberration is eliminated, it still does not follow that the microscopic image is rendered plane, and, conversely, a perfectly plane image may appear more or less distorted. Both defects must be corrected, if the flatness of the field of view, in the traditional sense of the expression, is to be complete.

Distortion as well as curvature can be eliminated in two different ways: first, by making the eye-piece aplanatic and orthoscopic by the addition of plano-concave flint-lenses; secondly, by skilful selection and combination of simple plano-convex lenses for the field-lens and the eye-lens, the opposing aberrations being so regulated that they cancel each other.

As the principles upon which aplanatism is based have been explained in a previous chapter, we need not return to the subject. It is evidently a much easier task to construct aplanatic eye-pieces than aplanatic objectives, because it is possible to arrange the curvatures of the lenses in accordance with the requirements of calculation. The points we have to discuss are, how aberration can be corrected in an ordinary eye-piece, and how, in general, the curvature of the refracting surfaces influences the curvature of the image?

The following will afford the necessary data with regard to the first point:—Let $b\,q$ (Fig. 35) be the path of a peripheral pencil

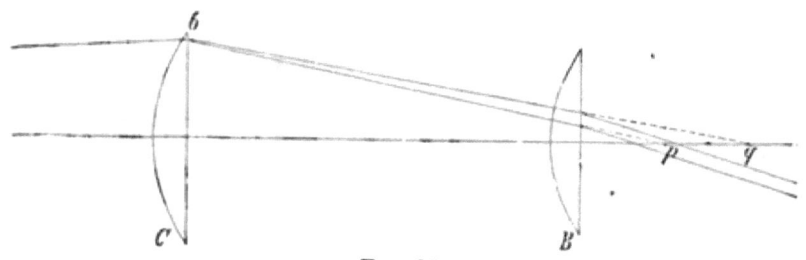

FIG. 35.

after refraction through an aplanatic field-lens, or, otherwise expressed, the direction which a refracted pencil would take in order to give a perfectly correct real image, i.e., corresponding precisely with the object; further, let $b\,p$ be the path of the pencil after its passage through a single positive lens of equal power but non-aplanatic. It is then evident that the aberration of this last pencil is only completely cancelled if, after its passage through the eye-piece, it appears to come from a point which corresponds to the direction of the pencil $b\,q$ after a refraction free from aberration in the aplanatic eye-lens. If the eye is adjusted for infinite distance, the two pencils must therefore proceed parallel after their emergence from the eye-piece, which assumes that parallelism is produced by the refraction at the lower lens-surface. We will extend our remarks on this simple case.

Let us suppose the eye-lens adjusted so that its centre of curvature lies centrally between the points p and q; then both pencils will be refracted towards each other almost without aberration, and with equal deviation, since they are incident at equal and very small angles (about $1°$—$1·5°$). For parallelism, obviously, the point b, from which the rays proceed, should lie in the focal point of the refracting surface. Since the corresponding point of the field-lens image must be in exactly the same position to be seen in the virtual image at infinite distance, it follows that aplanatism can only be secured on the hypothesis that the eye-lens is at the distance of its focus from the field-lens, and when, accordingly, the real image appears in the plane of the latter. An increase in this distance must, with similar positions of the centre of curvature (that is, with a shallower field-lens or a deeper eye-lens), necessarily involve convergence, while a decrease will necessitate divergence of the emergent pencils. In the first case the aberrant pencil would be refracted towards one assumed to be free from aberration, and its prolongation backwards would correspond to a point lying too near the axis,—the virtual image would appear to be less magnified towards the margin. In the second case, the contrary effect is produced; the pencil, produced backwards, meets a point too far distant from the centre,—and the image shows an opposite distortion.

It is otherwise, if the centre of curvature of the eye-lens lies nearer to it than the points p and q. The angle of incidence of the aberrant pencil is, in this case, noticeably greater than in the preceding, consequently, aberration is also traceable in the eye-lens; and the pencil is refracted more strongly towards the axis than the one assumed to be free from aberration. If the point of emergence b of the two pencils were at the distance of the focus from the spherical surface, they would diverge after their passage through the eye-piece, whilst in the previous case they were parallel. In order to produce parallelism, therefore, the distance must be chosen somewhat greater than the focus, in proportion to the angle of incidence of the aberrant pencil. In the Campani eye-piece the real image is not, therefore, formed in the plane of the field-lens, but above it.

By similar reasoning it may be shown, that the displacement of the centre of curvature to the right (which a shallow eye-lens presupposes) requires, as the condition of aplanatism, that the distance of the two lenses shall be less than the focus of the upper lens. In this case an arrangement is applied to the eye-piece, which agrees in principle with that of Ramsden, in that the real image is formed *in front of* the lower lens.

It is therefore possible to combine single lenses in different ways to form an aplanatic system under the given conditions. A strict fulfilment of the conditions is, in general, possible only for particular pencils, just as the elimination of spherical and chromatic aberration in the objective is limited theoretically to particular inclinations and colours. An eye-piece which is aplanatic for the violet rays, cannot, as a rule, be so for the red, since both the position of the centre of curvature and the distance of the lenses depend upon the refrangibility of the rays. The point p in the figure is evidently removed the further to the left the greater the refrangibility of the rays, while q, as we may suppose, does not change its position; moreover, the focus of the eye-lens is different for each colour. The confusion thus produced in the case of the refraction at the lower surface of the eye-lens, depends, of course, upon the dispersive power of the glass, but it is further dependent upon the actual position of the point p, in which the prolongations of the differently-coloured pencils cut the axis. If p lies for all colours to the left of the centre of curvature, an eye-piece which is aplanatic for violet or central rays will move the marginal points of the red image a little too far outwards, because the point of divergence b for this colour is still situated within the focus. The field of view would therefore appear with a red fringe. This chromatic aberration is, however, equalized by the refraction at the upper lens-surface, for here the violet rays are inclined more strongly towards the axis than the red rays. It is evident, also, that this counter-action will preponderate if all the rays are refracted to the axis by the combined action of the lens. The red margin of the field of view is consequently converted into a blue one in the final image.

The same result is obtained, if the point p, with regard to all the colours, is situated to the right of the centre of curvature; but, since in this case the spherical aberration of the red pencils, incident at a greater angle, increases with the distance of the two lenses, there is a limit where it equalizes the chromatic aberration. Hence the blue fringe of the field of view is changed to a red one. As this limit is, with given curvatures, dependent solely upon the dispersive power of the glass, a glass may be selected which will admit of the production of aplanatism for red and violet light at the same time. And if, for instance, crown-glass possesses approximately this property for a particular combination of lenses, it must be possible to reduce to a minimum the remaining aberrations by means of slight alterations in the distances and curvatures. We therefore arrive at the conclusion that the elimination

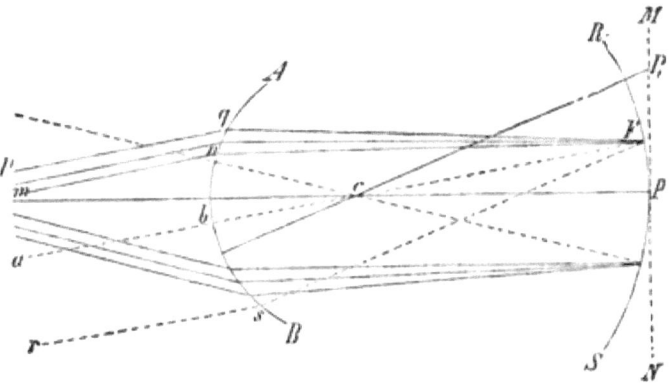

FIG. 36.

of dispersion is compatible with the simultaneous elimination of spherical aberration.

As regards the *curvature of the image-surface*, the usual explanation as to the field-lens image having its convex side downwards is erroneous. Exactly the opposite takes place. We may convince ourselves of this by placing a flat plate of glass covered with fine dust, or a micrometer plate, upon the diaphragm of the eyepiece, and by comparing its margins in the microscopic image with those of the field-lens image. We shall find that the margin of the latter image requires a somewhat lower adjustment of the eye-lens than the similarly-situated points of the glass plate. The field-lens image is therefore curved upwards. On theoretical principles it cannot be otherwise. If we suppose a refracting surface AB (Fig. 36), which brings to a focus in F al. rays between pq and rs which are parallel to the axis ac, the rays of the peripheral pencil $pqmn$ will also, of course, be refracted towards this point. This applies equally for every other cone of rays and the pencil parallel to it. Moreover, the distance of the different focal points of the refracting surface, measured upon the axis of the cone passing through the centre of curvature, and consequently also the distance of the centre of curvature itself, is a variable quantity. The focal points, therefore, lie in a spherical surface whose centre coincides with that of the refracting surface.

If we now suppose the cylindrical pencils converted into slightly converging ones, so that the points of convergence of the differently-inclined pencils are equidistant from the centre of curvature of the refracting surface, their real image-points are brought somewhat nearer to the refracting surface; they form, however, afterwards as before, a curved surface with the same centre of curvature. Applied to the field-lens, this means that if the objective-image, forming the virtual object, has a curvature which is convex above, its centre coinciding with that of the field-lens, then the image produced by the refraction at the spherical surface has a curvature parallel with it.

It can be further shown that a curvature is produced in the same sense, even if the objective-image is in a plane. Let us assume, for the sake of simplicity, that it lies in the tangential plane MN (Fig. 36) of the focal surface, and let us denote the distance from the refracting surface of the object-point P by p, that of the corresponding image-point by p^*, the distance of the object-point by P_1 and its image-point by p_1 and p_1^*, the focal length by f, and the radius of the field-lens by r.

Then $\dfrac{1}{p^*} = \dfrac{1}{f} + \dfrac{1}{p}$, and therefore in the given case

$$p^* = \dfrac{pf}{p+f} = \dfrac{1}{2} f = \dfrac{1}{2} p,$$

and if the refractive index is taken as 1·5, $p^* = \dfrac{3}{2} r$. The quantity p_1 is determined by the proportion $(p_1 - r) : (p - r) = 1 : \cos \phi$, that is, if ϕ denotes the angle $P_1 c P$, and the radius of the focal surface is taken as unity. From this results

$$p_1 = r + \frac{p - r}{\cos \phi} = r + \frac{2r}{\cos \phi};$$

consequently

$$p_1^* = \frac{p_1 f}{p_1 + f} = \frac{\left(r + \dfrac{2r}{\cos \phi}\right) \cdot 3r}{r + \dfrac{2r}{\cos \phi} + 3r} = \frac{3r}{2} \cdot \frac{2 + \cos \phi}{1 + 2 \cos \phi}.$$

The distances p^* and p_1^* are therefore, respectively, $\dfrac{3r}{2}$ and $\dfrac{3r}{2} \cdot \dfrac{2 + \cos \phi}{1 + 2 \cos \phi}$; consequently the distances of the image-points from the centre of curvature, which we will call d^* and d_1^*, are $\dfrac{1}{2}r$ and $\dfrac{3}{2}r \cdot \dfrac{2 + \cos \phi}{1 + 2 \cos \phi} \cdot r$. By an evident reduction we obtain

$$d_1^* = \frac{1}{2}r \cdot \frac{4 - \cos \phi}{1 + 2 \cos \phi};$$

and hence

$$d^* : d_1^* = 1 : \frac{4 - \cos \phi}{1 + 2 \cos \phi}.$$

If the two image-points were situated in a plane at right angles to the axis of the Microscope, the ratio of their distances from the point c would obviously be given by $1 : \dfrac{1}{\cos \phi}$. We will now consider whether the second term is greater in the first or in the second expression. If we take $\beta = 1 - \cos \phi$, there results

$$\frac{4 - \cos \phi}{1 + 2 \cos \phi} : \frac{1}{\cos \phi} = \frac{3 + \beta}{3 - 2\beta} : \frac{3}{3 - 3\beta}.$$

In this latter form the two values represent two fractions, in which the numerators are greater than the denominators. The numerator and denominator of the first fraction are also greater than those of the second by β, consequently $\dfrac{3 + \beta}{3 - 2\beta} < \dfrac{3}{3 - 3\beta}$. The image-surface is therefore curved in this case also, and its convex side is directed upwards. Under the given conditions this curvature is little more than zero; for if, for instance, we take $\phi = 4°$, and

$r = 20$ mm., the quantity $p_1{}^*$ will be about ·02 mm. less than in the plane image. But this is, in reality, further increased by the fact that the refracting surface is not, as we have supposed, aplanatic, but is accompanied by the aberration due to spherical form. It acts, as we have explained, upon the peripheral pencils of rays, just as an aplanatic surface of shorter focal length, and, in consequence, brings the corresponding image-points somewhat nearer. The curvature must therefore be perceptible, even if the spherical aberration is taken at only ·1 of the focal length.

The action of the eye-lens has already been explained for a particular case. We found that the image of a flat surface formed by it must be curved, because its peripheral points are further distant from the refracting surface-elements than the central points. Such a curvature must, of course, always take place when the spherical aberration acts in the same manner as in the case of single lenses. The final virtual image of an ordinary eye-piece cannot possibly be flat, unless the objective-image itself is curved and its convex side turned downwards. The most favourable position of the eye-lens will always be that in which the pencils of rays proceeding from the field-lens are incident upon the spherical surface at the smallest possible angle.

With the Ramsden arrangement of eye-piece, the conditions which influence the curvature are more complicated. Since the lower lens is here turned with its plane surface to the real objective-image, the virtual image formed by it would be curved downwards, if spherical aberration (which, as has been shown, acts in an opposite direction) did not preponderate. The combined effect of the eye-piece is therefore, in general, very nearly the same as in that of Campani.

From a practical point of view, moreover, a slight curvature of the field of view is unimportant. No eye-pieces are known to us (not excluding the aplanatic and orthoscopic ones of Plœssl, &c.) in which it is completely eliminated. The chief and proper aim of the optician is to eliminate as far as possible the distortion of the differently-coloured images, and to regulate the remaining deviations so that the red and violet image-points coincide perfectly, at least in the central portion of the field of view. If this is attained for a definite length of tube, the eye-piece will reproduce, as distinctly as possible, any objective-image with an equal length of tube.

VIII.

THE CENTERING OF THE SYSTEMS OF LENSES.

In our previous consideration of the path of the rays of light in the Microscope, we have throughout assumed that the refracting surfaces form an exactly concentric system, i.e., that their centres of curvature all lie in one straight line. The optician will always endeavour to realize this assumption; but the best workmanship results in an approximation only, and in the high-power objectives there are always unavoidable errors. The question therefore arises, how defective centering influences the clearness of the image, and what errors are of especial importance?

According to the usual view, exact centering is one of the conditions of aplanatism. Harting[1] states that even a slight inaccuracy will necessarily be very detrimental; and Mohl[2] affirms that a distortion of the image on one side is caused by it. This view seems at first sight plausible and comprehensible; it is, nevertheless, opposed to theory as well as to observation, as we will show.

Let us assume a perfectly concentric objective-system composed of three double-lenses, which is as far as possible aplanatic for all points of a field of view of given diameter, and inquire what influence a slight lateral displacement of the lenses would exert upon the combined effect of the system. Let E_1, E_2, and E_3 (Fig. 37) be the pairs of principal planes of the three double-lenses, F_2 and F_3 the corresponding focal planes of the two posterior lenses, and o_1, o_2, and o_3, the optic axes, the first of which we suppose to coincide with the axis PQ of the

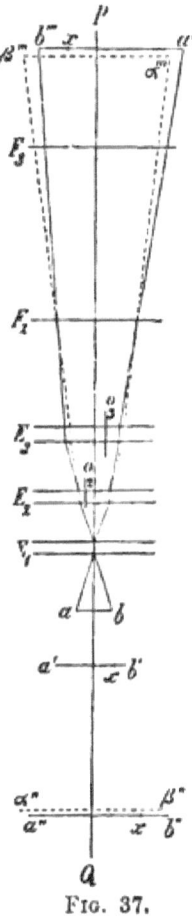

Fig. 37.

---

[1] Harting: "Das Mikroskop," p. 275, and 2nd ed. i. p. 304.
[2] Mohl: "Mikrographie," p. 176.

Microscope, while the two others are displaced to the right and left, in a parallel direction; and let $a\,b$ be the object, whose separate points give equally perfect images with accurate centering. If, with the help of the rays of direction, we construct the path of the pencils which proceed from the marginal points a and b to the objective, and compare it with the path as it would be in the accurately concentric system, the errors due to the displacement can readily be comprehended. In the Fig. the ordinary lines refer to the excentric and the dotted lines to the concentric system. The virtual or real images, which the separate double-lenses form, are denoted by $a'\,b'$, $a''\,b''$, and $a'''\,b'''$ in the first, and with similarly-accented Greek letters in the second. The position of these images relatively to the axis is, of course, previously determined by the direction of the rays, if necessary produced backwards; their distances are regulated by the focal lengths, and are, of course, in both cases the same; for greater clearness they are, however, represented in the Fig. as somewhat different. It is evident that the virtual image $a'\,b'$ of the first lens suffers no alteration by the displacement. Considered as an object to the second lens, its right-hand margin, b', comes into a somewhat unfavourable position, since it is in this case further from the optic axis o_2 by the amount of displacement. The lens therefore forms a less perfect image of this portion of the margin, as may reasonably be supposed. The entire remaining portion (by far the greater) fulfils, on the other hand, afterwards, as before, the condition that none of its surface-elements, for which the second lens is as far as possible aplanatic, fall outside of the field of view. If we make $b'\,x$ equal to the amount of displacement, then in the image $a''\,b''$ a corresponding portion $b''\,x$ will be unsatisfactory, while $a''\,x$ retains its original distinctness. Now, however, the whole image, as the Fig. shows, lies somewhat further to the right than before the displacement, and at a distance which is in the same ratio to $b''\,x$ as $a''\,b''$ to $a''\,b'' - a'\,b'$. To verify this point, it is only necessary to draw through b' and the upper principal point of the second lens, in its two different positions, two lines, produce them downwards to β''' and b'', and compare the triangles meeting in b'.

The virtual image $a''\,b''$ assumes, therefore, in general, a different position to the axis o_3. The relative position, of course, remains the same only when it moves exactly as far as the axis itself.

The real objective-image $a'''\ b'''$ attains the greatest possible distinctness in this special case, but since the part $b''\ x$ of the object is already indistinct, it is indistinct also in the image. If, on the other hand, the displacement of the virtual image is greater or less than that of the axis, a corresponding marginal portion of the red image appears under less favourable conditions in the first case on the left, and in the second on the right side, and is, in consequence, less sharp. The part $b'''\ x$ is therefore somewhat indistinct again at the edge.

With every displacement, therefore, a corresponding portion of the margin of the objective-image disappears, so that, in reality, if the erroneous centering of the flint- and crown-lenses is also taken into account, the whole periphery of the field of view must suffer more or less, only a central portion is preserved in its original purity, and this portion may, nevertheless, have any excentric position.

In order to obtain a few data as to the extent to which the displacements represented in our figure influence the microscopic image, we will assume that the first lens magnifies three linear, and the second and third four and six linear respectively—a proportion which approximately coincides with actual observation. The displacement of the axis o_2 would amount to ·25, and that of the axis o_3 to ·5 mm. Therefore $b'\ x = ·25$, $b''\ x = 1$, $b'''\ x = 6$ mm.; further $\beta''\ b''$ (in a direction at right angles to the axis) $= (4 - 1) \times ·25 = ·75$ mm., and consequently a portion of $6 \times (·75 - ·5) = 1·5$ mm. in breadth, measured from b''', is rendered indistinct in the objective-image. This portion, however, if the diaphragm in the eye-piece is supposed to be immovable, falls upon its margin; for since $a''\ b''$ appears displaced by ·25 mm. further to the right than o_3, the objective-image lies, with reference to o_3, at a distance of $6 \times ·25 = 1·5$ mm., and therefore, with reference to PQ or to $a'''\ \beta'''$, the field of view of the eye-piece in the concentric system, at a distance of $1·5 - ·5 = 1$ mm. further to the left. (In the Fig. it is the reverse.) If the diameter of this field of view, *i.e.*, of the diaphragm in the eye-piece, is 18 mm., there still remain 5 mm. visible of the indistinct portion $b'''\ x$, while on the other side, between a''' and a''', 1 mm. of the aperture is unoccupied, and can only be filled if the object $a\ b$ is increased $\frac{1}{12}$ mm. at a. This increase becomes somewhat less distinct in the image $a'\,b'$, but suffers no further loss through the succeeding refractions.

Defective centering of the objective-system may further arise from the fact that the optic axes form small angles with the axis of the Microscope—and this is a second case which we have to discuss. Let E_1 and E_2 (Fig. 38) be again the principal planes of the first and second double-lenses of a perfectly concentric system, $P\,Q$ the common optic axis, $a\,b$ the object and $a'\,b'$ the virtual image of the first lens, and therefore the object to the second. If the latter is now turned round one of its principal points, without otherwise altering its position, while the optic axis $o_2\,o_2$ changes its position with regard to $P\,Q$, the portion $a'x$ of the virtual image comes into a more unfavourable position, just as in the preceding case. The action is, therefore, similar to the case of a lateral displacement of the optic axis; but the error due to the difference of angle increases with the distance of the object. It should be added, that in consequence of the inclination of the lens, the image formed by it appears to be inclined also, from which circumstance the final objective-image, even if the inclination were compensated for by opposite aberrations, must necessarily be deteriorated. In general, then, the deviation of the optic axes acts more disadvantageously than their mere lateral displacement; at the same time it should be observed, for example, that an angle of about 2° is required in order that with an object-distance of 7 to 8 mm., as was supposed for the second lens, the above assumed displacement of ·25 mm. may be equalized. It must, consequently, be decided by experiment which kind of deviation is easier to correct.

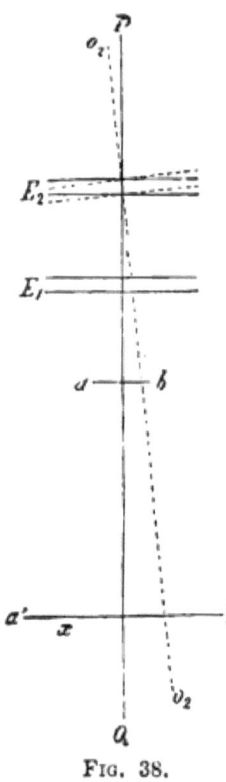

Fig. 38.

If both kinds of error are present in a system of lenses, the effect is exactly the same as if first a displacement of the axes parallel to their own direction had taken place, and then a deflexion, or *vice versâ*. The arithmetical expression of the errors is, of course, simply the sum of both the effects, which we have just treated of separately.

In accordance with what we have already stated, it is therefore quite correct to say that defective centering exercises an injurious influence upon the microscopic image. If, however, we disregard the small inclination of the images to the axis, this influence is confined, in the first place, merely to the margin of the field of view, and reaches its centre only in the case of stronger inclinations (which, of course, ought to be avoided). The familiar phenomenon, due chiefly to the action of the eye-piece, by which the distinctness of the microscopic image diminishes towards the edge of the field of view, may, in consequence of these aberrations, be still further intensified.

Finally, we will examine one further point, to which we shall refer in discussing the Testing of the Microscope. We have to show what changes in position the image of any object-point undergoes through imperfect centering, if the image-forming objective is turned round a given axis (PQ, Figs. 37 and 38). For this purpose we start from the results obtained above, according to which a displacement of the lenses generally causes a displacement of the objective-image also. In the example given, where the optic axis was moved ·25 and ·5 mm. to the right and left, this displacement amounted to 1 mm. to the left. If we suppose the objective to be furnished on the left side with a sign, it is evident that whilst it turns round the axis PQ the displacement will always take place in the direction which is determined by this sign. The centre of the circular objective-image therefore assumes successively the positions denoted by 1, 2, 3, 4, with regard to the centre o (Fig. 39) of the field of view (bounded by the eye-piece diaphragm), in which we may imagine crossed wires, $a\,b\,c\,d$, to be extended. It therefore describes a circle, of which the point o is the centre, and the distance of the displacement is the radius. Similarly, every other point in the image (since its position relative to the centre must remain the same) describes a circle of equal

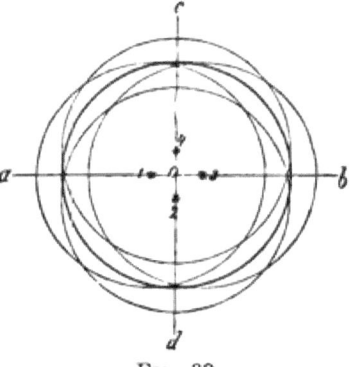

Fig. 39.

diameter, whose centre coincides with the corresponding image-point of the concentric system.

Similar phenomena are also observable, if, instead of the whole objective, single defectively centred lenses are turned round their axis. The displacements may, indeed, prove still more considerable, because aberrations, which in other cases had a counteracting effect, have, in consequence of the rotation, to be added to the others. Moreover, no such evident relation, as usually supposed, exists between the inclination of the individual axis and the resulting displacement of the objective-image. When, for instance, Harting states that every irregularity is magnified exactly as much as the object itself, and that therefore a difference of 10 mic. in an image magnified 500 linear becomes 5 mm., it is an entirely erroneous assertion, which finds a sufficient refutation in the example itself. On the contrary, it is always conceivable that the defects of centering may cancel each other, so that the resulting image does not suffer any displacement, although it loses in distinctness.

Moreover, in practice the ratios between the different amounts of displacement are without significance, since the centering of the objectives is always accomplished lens for lens, for which purpose are usually employed not the dioptric images, but the reflected images which the lens-surfaces form of a luminous object placed on one side. [Cf. chapter on Testing the Centering.]

IX.

BRIGHTNESS OF THE FIELD OF THE MICROSCOPE.

THE brightness of the field of a Microscope may be defined as that amount of light which is brought to the unit of surface of the retina, when a uniformly luminous surface of known intensity serves as the object. The ratio of this quantity of light to that which the same illuminating surface brings to a unit of surface of the retina with the naked eye represents its arithmetical expression. It is not a quantity dependent upon the distance from the eye, for in the retinal image a change in the distance of any object becomes larger or smaller in the proportion in which the aperture of the cones of light, while proceeding from the single points, varies. The greater amount of light contained in the larger pencils of rays is therefore always distributed over a surface of the retina which is larger in the same proportion.

It might be supposed that the same ratio would exist for the eye in microscopic vision, and that hence the power of the Microscope would vary in the same ratio with the angular aperture of the incident rays. This brightness of the field would then be equal to unity, i.e., we should see the objects equally bright as with the naked eye. And if any combination of lenses were able to receive and direct to the eye a cone of rays with an aperture still greater than the above-mentioned ratio requires, the field of view would naturally appear to be more brightly illuminated, in proportion to the surplus, than when seen with the naked eye. Such a case can, however, never occur, as will be shown, even with the lowest magnifications; for the higher magnifications it is *primâ facie* an impossibility. The diameter of the microscopic image (and therefore also the retinal image) increases with higher magnifying power in a much more rapid proportion than the angles of aperture of the incident cones of rays, which reach the objective from the object-points, and thence the eye. In vision with the naked eye this angle amounts to 28—34 minutes, that is, about half a degree, if we fix the diameter of the pupil at 2 to $2\frac{1}{2}$ mm. and the distance of distinct vision at about 250 mm. In microscopic vision, on the other hand, it amounts to 100° and upwards in proportion to the power and peculiarities of the objective. From every point in the plane of adjustment, an amount of light, in this case, reaches the eye which is greater than that received by the naked eye in the approximate proportion of $100^2 : (\frac{1}{2})^2 = 200^2 : 1$. But this amount of light is distributed over a surface of the retina m^2-times greater, where m is the coefficient of linear amplification. The brightness of the field of view resulting from this, which we denote by v, is therefore expressed by $v = \dfrac{200^2}{m^2}$; or, in general, if we take the angle of aperture of the objective as equal to ω, and that of the naked eye equal to $\frac{1}{2}$, $v = 4\left(\dfrac{\omega}{m}\right)^2$. The luminous power is consequently equal to 1, if $m = 2\omega$; it is less than 1, if $m > 2\omega$, which is obviously the usual case. It is, of course, here assumed that the incident cone of rays is large enough to fill the whole aperture of the objective. This condition is, however, very seldom fulfilled in microscopic observations, since the diaphragm limits the inclination of the incident rays. In place of ω, therefore, the angle $a\,p\,b$ (Fig. 40) enters, at which the diaphragm $a\,b$ is seen from the point p in the plane of adjustment. If the angle $= \delta$, the above expression becomes $v = \left(\dfrac{2\delta}{m}\right)^2$. With the same illumination, the brightness of the field of view is, accordingly, in inverse proportion to the squares of the coefficients of linear amplification. If, for instance, we select a diaphragm, so that $\delta = 30°$, the coefficients of amplification will be 240, 300, 360, 420, &c., for the more powerful objectives, and the relative brightness $\tfrac{1}{16}$, $\tfrac{1}{25}$, $\tfrac{1}{36}$, $\tfrac{1}{49}$, &c. But if, on the other hand, we increase the value of δ to double or treble the assumed value (by using a larger diaphragm), the brightness attains—provided always that ω is at least of the same magnitude —four times and nine times, respectively, the above fractions; these fractions will have, therefore, equal denominators, and the numerators 4 and 9.

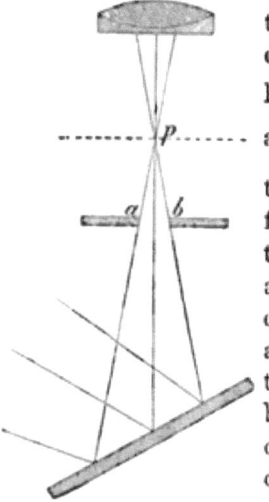

Fig. 40.

For larger angles of aperture the expression $v = \left(\dfrac{2\omega}{m}\right)^2$ is not quite correct. The quantity of light in the different cones of rays is, strictly speaking, not in proportion to the squares of the angular magnitude indicated, but to the area cut off by the cone from a sphere having its centre at the apex of the cone, that is, to the area of the active calotte. For it is evident that a self-luminous radiant emits rays in all directions, and therefore illuminates equally at all points a spherical surface, whose centre lies in the source of light itself. It depends, therefore, upon the extent of this spherical surface which is effective in a given case; an n-times as great an area of this calotte always corresponds to an n-fold amount of light. If we work out the calculation for different apertures of the incident cones of rays, and take as unity the amount of light for $\omega = \tfrac{1}{2}°$, we obtain, for example, the values given in the following table. The fourth column contains the actual areas for the case **when the hemispherical surface is taken as equal to 1,000,000; the fifth contains the corresponding tangential area.**

Assumed objective-amplification.	Angular aperture.	Corresponding calotte.	Area of calotte when the hemisphere = 1,000,000.	Corresponding tangential area.
—	0·5°	1	9·5	—
—	1°	4	38	—
—	2°	16	152	—
—	5°	100	952	953
—	10°	400	3806	3827
10	20°	1599	15192	15545
17	30°	3586	34074	35000
30	40°	6348	60307	66237
25	50°	9862	93692	106721
33	60°	14102	133975	166666
70	100°	37601	357215	710140
150	120°	52631	500000	1500000

On comparing **these calottes** with the nearly corresponding combined amplifications (i.e., such as are practically obtainable), the result we arrive at, as stated above, is a gradual diminution of the amount of light for the higher powers. But, with low powers, the possibility, however, seems to **obtain** that an object observed through the Microscope appears **to be** more brightly illuminated than when seen with the naked **eye.** If **we assume** that the angle of aperture 30°, contained **in the above table,** corresponds to an objective-amplification **of seventeen times,** which **agrees** with experiment, we obtain, **with an eye-piece amplification of** about three times, an image-area **which is 50^2 2,500** times larger than **the** object, **whilst** the amount of **light** utilized **is** about 3,500 **times** greater than in ordinary **vision.** The brightness **would** accordingly be increased in the ratio **of** 35 to 25. But the question arises whether a cone of light with an aperture of this size emerges sufficiently narrow from the objective to be received by the **pupil** of the eye; for it is evident the above conclusion becomes **illusory if** that **is not** the case.

For the examination of this question we will construct the path of rays in the Microscope according to the diagram above explained (p. 41). Although, owing to want of space, the distances there given do not coincide with the actual dimensions, the conclusions to be deduced are generally accurate. Let P (Fig. 41) be a radiant point situated on the axis, which emits a cone of rays, as shown in the figure, to the anterior limiting surface N^0 of the refracting system. The corresponding cone of rays proceeding from the posterior limiting surface N^* has now to be constructed. For this purpose we trace only the marginal ray Pg, which limits the angle of aperture. Its prolongation (in this case backwards) meets the anterior principal plane E^0 in the point i; consequently the line drawn parallel to the axis through i cuts the second principal plane E^* in the point a, which belongs also to the emergent ray. Further, if we draw through the anterior focus a line parallel to $i\,P$ to meet the anterior principal plane, and,

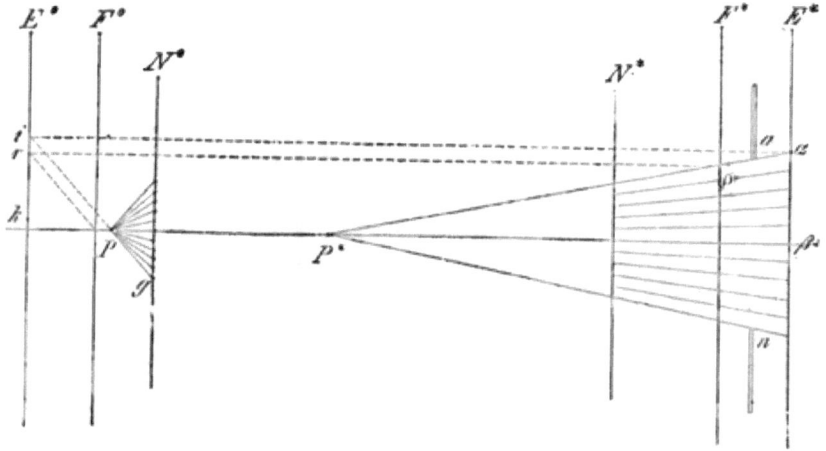

FIG. 41.

from the point of intersection, r, a second line parallel to the axis, the latter meets the posterior focal plane F^* in a point ρ, which likewise belongs to the emergent ray, and consequently determines its direction. The prolongation backwards of the line $a\,\rho$ gives at the same time the position of the virtual image-point in the axis. If we add also the marginal ray which proceeds symmetrically on the other side, the emergent cone of rays is evidently completely determined. The question now arises,—What is the ratio of the two angles of aperture? As a glance at the figure shows, the incident cone of rays has an aperture $\omega = 2 \times$ angle $i\,P\,k$, and the emergent cone an aperture $\omega' = 2 \times$ angle $a\,P^*\,\beta$. Since $i\,k = a\,\beta$, and since $k\,P$ is equal to the anterior, and $\beta\,P^*$ to the posterior focal distance, we obtain for the trigonometrical tangents of the two angles

$$\tan i\,P\,k = \frac{a\beta}{p}\,;\ \tan a\,P^*\,\beta = \frac{a\beta}{p^*},$$

in which p and p^*, as before, denote the two conjugate foci. But the squares of these tangents are proportional to the areas of transverse sections of the two cones at the distance 1 from the apex. For these surfaces we therefore obtain the proportion

$$\frac{a\,\beta^2}{p^2} : \frac{a\,\beta^2}{(p^*)^2} = (p^*)^2 : p^2\,;$$

but the same proportion indicates also the areal magnification, because $\frac{p^*}{p}$ is equal to the coefficient of linear amplification. Hence we arrive at the conclusion, *that the areas of transverse sections of corresponding cones of light at the distance 1 from the apex are proportional to the squares of the coefficients of linear amplification.* If, now, the emergent cone completely fills the aperture of the pupil $a\,n$, as assumed in the figure, the aperture of the incident cone is fixed thereby; for as soon as the aperture is further enlarged, as the above relation involves, the transverse section of the emergent cone becomes larger than the aperture of the pupil, and the surplus of light received by the objective therefore remains functionless. Since the surface of the transverse section of the incident cone is always somewhat larger (though when treating of a small aperture only infinitesimally so) than the corresponding spherical calotte at an equal distance from the vertex, we are led to the conclusion, *that a luminous surface observed through the Microscope cannot, under any circumstances, appear brighter than when seen with the naked eye.*

In all these calculations of the brightness, it has, of course, been tacitly assumed that the cones of light proceeding from the object reach the eye without loss. The losses caused by reflexion and absorption are therefore not taken into account. Their amount in a given case cannot be readily determined with accuracy, but so much is certain, that they form only a small fraction of the calculated brightness with low and medium magnifications, and consequently make no essential difference in the accuracy of the mathematical expression.[1]

X.

THE OPTICAL POWER OF THE MICROSCOPE.

The delineating power of the Microscope does not, as is known, bear a strict proportion to its magnifying power. While some Microscopes exhibit the form and details of a given object with an amplification of, say, 100 linear, others require 150—200 to render the same details apparent. On these differences is based the idea which is generally termed the *optical power* of the Microscope; it embraces all the properties which influence the distinctness and brilliancy of the microscopic image. Of the nature of these properties, however, conceptions are prevalent which are in many ways vague, and which require to be carefully sifted, and in part formulated on an entirely new basis. We will therefore separate the different considerations which must be estimated.

---

[1] According to the explanation above given (p. 40), the magnitude of the aperture-image above the eye-piece is with given optical constants dependent only upon the magnitude of the effective objective-aperture; its diameter, d, is given by the formula

$$d = \frac{2 \sin \delta}{r} \times \text{focal length},$$

in which δ denotes the angle at which half the diameter of the source of light is seen from the focus of the Microscope. If, consequently, in ordinary vision the aperture of the pupil is also equal to d, the image-forming cones of rays have the same aperture as those reaching the eye from the Microscope. We may therefore say that the field of view in microscopic vision appears as bright as it would to the naked eye if a diaphragm were held before it, having a clear opening just admitting the image of the source of light which appears above the eye-piece. With this proposition the theorem propounded by Abbe ("Archiv für mikr. Anatomie," Bd. ix. p. 438) agrees in all essential points.

1.—DEFINING AND PENETRATING POWER ACCORDING TO THE EARLIER AUTHORS.

With reference to the optical power of the Microscope, microscopists are accustomed to distinguish two different properties which are manifest in the testing of the instrument, viz., the so-called *defining* power, and the *penetrating* or *differentiating* power. By the defining power of a Microscope is understood the capacity of exhibiting clearly and distinctly in the image the form and outlines of the object; and by penetrating power, the capacity of displaying fine structural details, such as layers of cell-membrane, the markings of diatoms, &c. This distinction was first established by Sir W. Herschel for telescopes, and was applied, later on, by Goring to the Microscope. Herschel stated that telescopes with rather large apertures, even if otherwise defectively constructed, are particularly adapted to render visible obscure nebulæ and constellations, which cannot be distinguished with smaller instruments of the best kind; on the other hand, the latter will show closely adjacent bright points separated from each other, which would appear in the former as if forming a single point.

The penetrating power, or piercing power, is accordingly dependent upon the size of the aperture; the defining power, or outline power, upon the accuracy of construction of the instrument, *i.e.*, upon the correction of the aberrations. It is also evident that every optical instrument, as well as the naked eye, gives sharper images the nearer the foci of the different-coloured and differently-inclined rays on the retina; and that, on the other hand, slight differences in the luminous power of dim object-points are more apparent the greater the quantity of light reaching the retina from such points. Hence animals with large pupils see at night more distinctly than man, and obscure objects are rendered more distinct to the human eye in proportion to the enlargement of the pupil.

The distinction between defining and penetrating power, which Herschel established for the telescope, is therefore fully confirmed for every optical apparatus, the naked eye not excepted; but it must not be forgotten that the penetrating power varies proportionally to the aperture of the pencil of rays issuing from the object and reaching the eye, not with that of the refracting apparatus. It follows that if the incident pencils only partially fill the aperture of the objective, the unoccupied portion is entirely ineffective.

Let us now investigate whether, and in what sense, the idea of penetrating power, as above explained, can be applied to the Microscope, and whether the properties usually indicated by that expression really correspond to that idea. In the case of telescopes, the pencils of rays which emanate from the points in the (generally self-luminous) object fill the entire aperture of the objective; an enlargement of the aperture acts consequently like the enlargement of the pupil of the eye. This obtains when the light coming from the object is very faint; on the other hand, when the light is too strong, errors will be perceived, since all optical appliances with large apertures are accompanied by stronger aberrations. Now, as an increase of light is possible to any extent in microscopic observation (for direct sunlight can be employed if desired), the case of the larger aperture of the objective does not come into consideration from this point of view. The diaphragms of the illuminating apparatus are so small, that the rays converging towards the points of the stage generally form cones of light of 20°—30° aperture. After undergoing a deviation in the object, these rays group themselves again into pencils, which appear to come from the separate object-points, and depict their image in the Microscope. The aperture of these pencils is somewhat larger or smaller than that of the incident rays, these differences are, however, so small that they may be disregarded in our investigations. The pencils which reach the objective only partially fill its aperture, if this be, say, 60°—80°, and under the given conditions are not dependent upon the magnitude of this aperture. The analogy between the telescope and the Microscope drawn by Goring and others is therefore not applicable. What has been termed penetrating power in the latter (that is to say, the power of rendering perceptible fine structures, lines, or points) is in reality quite distinct from the space-piercing power of the telescope, as rightly pointed out by Harting. If it were desired to retain the parallel, the differentiating power of the Microscope should rather be compared with the property of the telescope of resolving double stars.

2.—Importance of the Angle of Aperture.

Since the incident pencils are not dependent upon the angle of aperture of the objective, we have still to inquire how the magnitude of this angle influences the microscopic image. It might be supposed that the action of the larger aperture is dependent upon a favourable distribution of light and shadow, because globules of oil, starch-grains, &c., exhibit a broad marginal shadow with a moderate amplification, which disappears if a more powerful objective with larger aperture is used; or that through an extension of the luminous parts an equalization of shadow and light might, to a certain extent, be established, in the sense that with striated objects, for instance, the lines of shadow and light would be equalized in breadth, and hence more easily distinguishable. Such an explanation would be untenable, for, in the first place, the luminous parts of many test-objects preponderate so much in the image that a diminution would be advantageous; but yet their resolution is only possible by means of objectives of large angular aperture,—therefore the distribution of light and shadow depends not only upon the objective, but also upon the size of the diaphragm, as will be more fully shown later on. The angle of deflexion of the rays, which corresponds to the limiting points of the inner shadow and the penumbra, is always determined by formulæ in which the two angles of aperture ω and δ enter only as the sum or difference; and though the limiting points themselves are not thereby mathematically given, their position is nearly the same as if the refraction had taken place in an equivalent infinitely-thin lens in the plane of the object. It is, at any rate, permissible in our discussion to substitute such a lens for the refracting surface, such as a diatom-frustule, without influencing the accuracy of our conclusions. It is, moreover, quite immaterial in theory whether an objective of 60° aperture be combined with a diaphragm of 30°, or, conversely, an objective of 30° with a diaphragm of 60°,—the difference in the image will be hardly appreciable. We will demonstrate this by a diagram.

Let $a\,b$ (Fig. 42) be the equivalent lens, c its centre, and $ff, f'f'$ its cardinal focal planes on either side, which consequently intersect the perpendicular passing through c in the cardinal focal points. Let $r\,c\,v = 60°$, the aperture of the objective, and $p\,c\,q = 30°$,

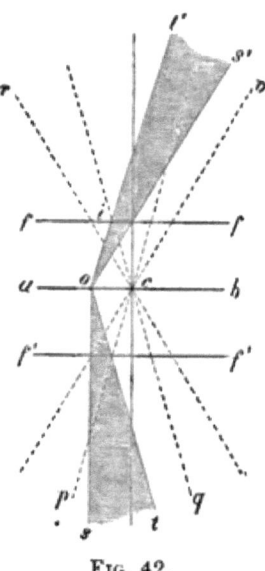

Fig. 42.

that of the diaphragm; then the path of the rays may be easily determined in accordance with the principles laid down in the Introduction. The ray drawn through the focal point parallel to $v\,c$, after refraction, proceeds parallel to the axis; the ray incident from below, in a direction parallel to $q\,c$, is so refracted that it cuts the upper focal plane in the same point as the prolongation of $q\,c$. By this ray the marginal rays of the pencils (both incident and emergent), which contribute to the illumination of the point o, are determined. All rays which are deflected to the right or left to a greater angle than those shown in the figure, meet either on the margin of the diaphragm or on that of the object. We can just as easily construct the effective pencils for any other points. For instance, the limiting point of the central shadow is determined by a line drawn parallel to $v\,c$ through i.

If we suppose the figure turned round 180°, it will evidently illustrate the course of the rays for the combination of the diaphragm of 60° with the objective of 30°. The limiting lines between light and shadow remain as before; and if the pencils proceeding from the object remain homocentric, after the refraction in the objective, as assumed for both cases, there is nothing that could explain any appreciable alteration in the image, either in distinctness or amount of light. It is, therefore, perfectly correct to say that the magnitude of the angle of aperture in objectives has, in respect to their dioptric functions, neither more nor less signification than the magnitude of the aperture of the diaphragm. It may be further inferred that the advantages which the former possesses are either of a purely practical nature, or are due to circumstances which have no connection with the dioptric formation of the images.

In order to determine all cases practically possible, we must refer once more to the construction of the path of the rays in Fig. 42. If $t'\ o\ s'$ is the pencil proceeding from the object, then its marginal rays, and consequently all the others, meet the peripheral portion of the objective, and the centre of the objective receives nothing. If we turn the figure round, and regard $s\ o\ t$ as the pencil coming from the object, all inclinations from $0°$ to $15°$ are represented; it will therefore occupy half the aperture of the objective. If, then, an objective with a large angle of aperture is less aplanatic for central rays up to $15°$, and upwards, than for outer rays of $45°$—$60°$, which is actually the case, then the inclination of the incident pencil essentially influences the clearness of the image; and if, in practice, the optician can eliminate the aberrations of pencils incident obliquely, which meet only the margin of the objective more easily than the direct ones, then a large angular aperture is of decided advantage.

There is, however, as has been shown by Abbe, a second consideration of far greater weight and more important signification, in respect to which the magnitude of the angle of aperture exercises an influence which may be mathematically proved and experimentally fixed. The point in question has not only reference to a corrective against accidental incompleteness of construction, but also to a *specific function* of the angular aperture in relation to the rays deflected in the object-plane, and interfering in the upper focal plane, or in its neighbourhood, which may be proved by the fact that they alone show in the microscopic image the fine structure of the object, such as the striæ of diatoms, &c.

The theory of this image-formation by interference will be subsequently explained when treating of the optical action of objects. We will here limit ourselves to the elucidation of a few facts which prove beyond doubt the signification of the angle of aperture for the delineation of fine structure. The facts are the results of careful comparisons instituted by Abbe with a series of objectives, as perfect as possible, of different focal lengths and angles of aperture, testing their capacity by various kinds of test objects, among which were coarse and fine powders, and also fine systems of lines ruled in extremely thin layers of silver. He thereby proved:—

(1.) As long as the free aperture of the objective remains so large that its diffraction-effect does not appreciably prejudice the sharpness of the image, the magnitude of the angle of aperture is without influence upon the representation of the contours of microscopic objects, that is, the boundaries between the unequally transparent parts in the field of view, when these parts are not less than about 10 mic.

(2.) On the other hand, as soon as the objects become smaller in their details than the stated limit of minuteness, a noticeable difference constantly appears in favour of the larger angle of aperture. At the same time it is immaterial whether the details in question are produced by unevenness of the surface, or by differences in the transparency of the substance.

(3.) The smaller the linear dimensions of the details in question, the larger must be the angle of aperture of the objective, if they are to be perceptible with any particular kind of illumination (for instance, either entirely central, or as oblique as possible), independently of the greater or less relief and sharpness of the details of the object, and also of the focal length and magnifying power of the objective.

(4.) When the details of the object take the form of striæ, systems of lines, &c., the same angular aperture invariably exhibits with oblique illumination perceptibly finer details than with central illumination, whether the constitution of the object admits of the possibility of shadow effects, or completely excludes them.

(5.) A structure of this kind, which a particular objective does not resolve with direct (axial) illumination, is still invisible if the plane of the object is inclined at any angle whatever to the axis of the Microscope, though it may be completely resolved by oblique light when lying at right angles to the axis. When the plane of the object is inclined to the axis of the Microscope, resolution is, however, at once obtained if the incident pencil is directed perpendicular to this plane. Consequently, the increased effect of oblique illumination depends exclusively on the obliquity to the axis of the Microscope, and not on the obliquity to the plane of the object.

The angle of aperture therefore represents a *specific power* which is not dependent on the other properties of the objective, nor on its magnifying power,—the power of rendering perceptible in the microscopic image fine structural details, which, at a lesser inclination of the incident rays, remain invisible. This may be called the power of differentiation or resolution, in accordance with the literal meaning of those terms. On the other hand, however, the facts above mentioned prove just as clearly that the dioptric reunion in one image-surface of the pencils issuing from the object does not explain the delineation of very fine structures, but that considerations external to the Microscope must necessarily be taken into account.

As already pointed out, it is the diffraction (or an equivalent deviation due to refraction or reflexion) of light in the object which produces the image-forming rays of the fine structural details. These diffracted rays produce well-defined diffraction or interference images in the upper focal plane of the objective where they interfere, and remain in the microscopic image, and therefore take part also in the final virtual image.

If we shut out these interference images by diaphragms, or if the angular aperture of the objective is too small to admit at least the first pencil of rays produced by diffraction, as well as the undiffracted light, the corresponding details disappear in the microscopic image,—a valve of *Pleurosigma* shows neither squares nor lines, and fine rulings upon glass appear as a homogeneous surface.

The signification of the angle of aperture consists, therefore, in the fact that it renders possible the production of the **interference image**, in which alone the finer structures are included.[1]

[1] Our former discussions on this point were not exactly based, as Harting asserts ("Mikr." 2nd ed. i. p. 278), upon the theory of illumination, differing somewhat from his own, but entirely upon the supposition, generally recognized up to that time, that the Microscope image is a dioptric one. So far as this is the case, the proposition which was then put forward is still correct, viz., *that, disregarding the amount of light and amplification, the Microscope developed no other power than that which progressed at an equal pace with the elimination of the two aberrations:* and the same may be said in general also of the conclusions derived therefrom. By the demonstration, however, that the phenomena of diffraction play so important a part in the formation of the image, the case has changed essentially in this as in so many allied questions, and it is useless here to repeat the views which, according to the new explanation of the matter, could have only a very secondary practical importance.

On the other hand, it may not be superfluous here to lay stress upon the fact that hitherto none of the treatises on the Microscope known to us have scientifically established, even as a fact, the effect produced by the angle of aperture, much less explained it.

3.—Fusion of the Interference and Dioptric Images.

If the optical power of the Microscope is distributed between two images, produced in essentially different ways, of which one represents the defining, the other the resolving power, then the combined effect obviously does not depend merely upon the perfection of each picture, regarded separately, but also upon the exact superposition of the one upon the other. Since the diffracted rays, on account of their inclination to the axis of the Microscope, meet the marginal parts of the objective with direct (axial) illumination, while the others, according to the deflexion in the object, occupy, some the middle, others the peripheral parts, it follows that an accurate superposition of both images is only possible *if the objective is uniformly free from aberration in the whole extent of its aperture.* And similarly with oblique illumination, even to a greater degree, since here also the deflected rays may meet the central part of the objective. If this condition cannot be satisfied, or if the objective has considerable errors of construction, a perfect union of the deflected and non-deflected rays, which issue from the same point, is impossible in the final image. It may then happen that the diffraction image is formed in a different plane from the dioptric one, or may appear displaced laterally with regard to it, and systems of striæ, which lie in the same plane in the object, require a different focal adjustment to render them perceptible.

In addition to these propositions of general importance, a few special observations on particular cases may be of service. It is customary to value the performance of an objective according to its resolving power with oblique illumination; consequently, the optician endeavours to increase the angle of aperture as much as possible, and to adapt the whole construction to particular test-objects. The defect of such a proceeding is at once evident. For, obviously, the limit of differentiating power for a given objective lies in such details as allow of at least the first diffracted pencil being admitted under the most favourable circumstances, as well as the direct rays. The two effective pencils touch the margin of the objective in two diametrically opposite points: the direct pencil on the side opposite to the mirror; the diffraction pencil on the other side,—the whole middle portion of the aperture is therefore inoperative. The objective may therefore be free from aberration in the narrow effective circumferential zone, a condition which is easily attainable by mere alteration of the distances of the lenses, even with constructions in other respects defective, and the two images coincide; the action of the objective will therefore, in this special case, be satisfactory, or possibly very favourable. It only implies, however, that the angular aperture is large enough to admit at least two of the pencils contributing to the formation of the image, one diffracted, and one undiffracted, and that it is possible to correct a narrow peripheral zone with sufficient accuracy. The observer thus determines merely the *limit*, not the *power*, of differentiation for cases which ordinarily occur.

The formation of the image with axial illumination takes place similarly. In some cases the direct rays pass through the middle and also very differently situated parts of the objective; but the diffracted pencils, which reproduce the details of the object at the limit of differentiating power, touch the margin of the free aperture, and produce, therefore, a well-defined diffraction image, so far as the marginal zone is accurately corrected. If, then, the dioptric image is *very indistinct*, and the coincidence of the two images very inexact, the delineation of fine structures remains, of course, unaffected, and the superficial student, who attaches undue importance to this point, would estimate the quality of the objective too highly.

From these special cases we again arrive at the result, that an accurate fusion of the interference with the dioptric images, combined with a sufficient sharpness of delineation, is only possible if the objective is uniformly aplanatic throughout its aperture.

4.—Ratio of Aperture and Focal Length.

The consideration of the fact that the magnifying power of a Microscope is dependent on its focal length, whereas the diameter of the details just perceptible in the microscopic image is solely dependent on the angle of aperture, leads necessarily to the conclusion that both these factors, equally indispensable for the delineation of small objects, must bear a fixed arithmetical ratio if their combined action is to be harmonious. For instance, if the angular aperture of the Microscope is such that the delineation of fine lines, whose interspaces are 1 mic., is just possible, it would evidently be a fruitless labour to increase the amplification much beyond the amount which experience has shown to be sufficient for their observation. Similarly, the selection of too low a magnification must be regarded as a mistake, since then the surplus of resolving power, which is given by the angle of aperture, is completely lost to the eye. An improperly chosen ratio between the focal length and the angular aperture involves, therefore, in the one case, *a useless magnifying power*, not represented by anything in the image; in the other, a *useless differentiating power*, which resolves more details than can be seen.

The practically best proportion between the two factors cannot be determined, once for all, with mathematical precision, on account of the differences of construction. A table of corresponding magnitudes cannot, therefore, be given here, and we must confine ourselves to the explanation of an example of the points which, under certain conditions, give rules for construction. For this purpose we assume, as Abbe[1] states, that the angle of aperture of a dry system should not exceed 110°. The limit of differentiation is thus arithmetically determined, and the distance d of the smallest details, whose delineation is just possible with extremely oblique illumination, amounts[2] to ·3 mic. In order to render such details visible, it is necessary to magnify them to about 120 mic., and in order to observe them conveniently, 180 or even 210 mic. would not be too high.[3] Consequently, we get for the case of simple

---

[1] Of the absolute accuracy of this angle we do not feel convinced. It is based upon the supposition that the curvature of the lens-surfaces is exactly spherical. We are, however, informed by practical opticians that it is possible, in the polishing of a zone of the lens between the centre and the periphery, to give a somewhat less decided curvature than the exact spherical form, by which an enlargement of the angle of aperture would be feasible.

[2] The angle of the illuminating rays was here taken at 80° (generally too high), and half the angle of aperture equals 55°. We therefore get the equation

$$d = \frac{\text{wave-length}}{\sin 80° + \sin 55°}.$$

[3] These numbers vary according to the sharpness of outline of the image. With the naked eye, the divisions of an eye-piece micrometer, even if the interspace is ·1 mm., may be perceived with favourable illumination; but a

recognition of the details, a *faultless* magnification of 400, and for convenient observation, one of 600—700 diameters. The errors of workmanship of the instrument raise these figures somewhat higher, probably to 500 and 800. The simple perception of the image takes place, then, with an eye-piece amplification of about 4. The amplification by the objective is still $\frac{500}{4} = 125$; for its focal length, we get from the formula

$$f = \frac{p^*}{m+1}, \text{ if } p^* = 190 \text{ mm.}, \frac{190}{125} = 1\cdot 5 \text{ mm.}$$

Should a somewhat higher eye-piece amplification be admissible, say five linear, then the magnifying power of the objective = 100, and its focal length 1·9 mm. In both cases the further surface-extension of the image, for convenient observation, may be effected by means of deep eye-pieces.

Similarly, for an immersion objective, whose angle of aperture = 180° in air, we get half a wave-length, or ·28 mic., for the absolute limit of differentiation. The magnification required for *accurate* delineation of such details is in round numbers 430, and, in consequence of the unavoidable errors of construction in the present state of workmanship, this may be increased to 600 or 700. With an eye-piece amplification of four linear, we get for the objective a magnifying power $m = 150$ to 175, from which f can be estimated, according to the suppositions above given, at the still considerable magnitude of 1·26 to 1·09 mm. If, then, the combined amplification were 800, which is certainly sufficient, the focal length would still not be perceptibly less than 1 mm. We conclude that objectives of $\frac{1}{30}''$, &c., can possess no advantage beyond the very doubtful one of useless amplification. Consequently, we shall not be wrong in estimating all amplifications exceeding 1,500 to 2,000 as valueless for scientific purposes.

dimension perceptibly greater (often as much as 150 mic.) is usually necessary with dioptric or interference images. In many cases an amplification at least twice as great is required for convenient observation; this would therefore amount to 1,000.

XI.

DIFFRACTIONAL ACTION OF THE APERTURE OF THE LENSES.

If we view a luminous point, say the reflexion of the sun on a thermometer bulb, through an aperture of about ·5 mm., *e.g.*, a needle-hole in a card, it will appear as a bright circular disc, surrounded by alternately bright and dark rings. If we substitute for the luminous point a bright image-surface of fine markings, the diffraction figures of adjacent luminous points will encroach on each other similarly to the dispersion circles in the dioptric image of non-aplanatic systems of lenses. The action is therefore identical; the fine details disappear in both cases, and the image appears faint.

A similar diffractional action takes place in high-power objective-systems, in which the clear aperture decreases more and more with the focal length. This action is entirely independent of other deficiencies of construction; it takes place in objectives that are as free as possible from aberration, just as necessarily as in those of inferior construction, and attains such a degree in the highest powers, that any further increase of magnification must be entirely useless.

The absolute measure of these disturbances may be theoretically determined.[1] We can prove that its influence is universally proportional to the magnitude of the last aperture-image above the eye-piece, and that its action is exactly equivalent to viewing the microscopic picture, regarded as free from the diffractive effect, through a small aperture of the size of the aperture-image. From what we have already shown, the diameter d, of the aperture-image, is found by the formula

$$d = 2\,\frac{\sin \delta}{r}.f,$$

---

[1] See Helmholtz: "Pogg. Ann., Jubelband," p. 557, and specially p. 579. A generally applicable mathematical explanation of these phenomena has been promised by E. Abbe ("Archiv für mikr. Anat." Bd. ix. p. 432).

where δ represents half the angle of aperture, and f the posterior focal length of the Microscope. If $\delta = 180°$, then $d = 2f$, or if, instead of f, we substitute the magnifying power m for a range of vision of 250 mm.,[1] $d = 2 \cdot \dfrac{250}{m}$. With a combined amplification of $m = 1,000$, d will equal ·5 mm.; and if $m = 5,000$, $d = $ ·1 mm. The diffractional action in this case may be exemplified by making an aperture of about ·1 mm. with a needle in a sheet of tinfoil, and viewing through this diaphragm a wire-gauze held near a candle-flame; the contour of the object will then appear indistinct through the influence of the diffraction phenomena.

This one fact suffices to show that every attempt at an unlimited increase of the combined amplification, which many opticians have aimed at in modern times, must be fruitless. The diffraction effect is so great, with a magnification of even 3,000 linear, that all hopes of attaining higher figures without a corresponding sacrifice of distinctness must be abandoned. In reality, the amplification—in the strict sense of the word—is reduced to a much smaller figure, on account of the conditions of the formation of the structure-image, which will be discussed later on.

XII.

ILLUMINATION.

It is not our intention to describe all the different illuminating devices issued by the opticians; this information will be found *in extenso* in the well-known micrographical works of Mohl, Harting, and others. It appears to us more important to elucidate the generally very vague ideas prevalent regarding the influence of

---

[1] We know that $m = \dfrac{p^{\bullet} - f}{f}$; therefore $f = \dfrac{p^{\bullet} - f}{m}$, or, since f is very small compared with $p^{\bullet}$, $f = \dfrac{p^{\bullet}}{m} = \dfrac{250}{m}$.

illuminating apparatuses,—properties being attributed to them which they cannot possibly possess.[1]

Our special task is, therefore, to elucidate the theoretical principles necessary for the comprehension and application of the different kinds of illumination.

1.—Illumination by Transmitted Light.

The most important and only indispensable part of every illuminating apparatus for transmitted light is the mirror. In order to determine its action theoretically, let us imagine it exposed to an unlimited source of light, equally luminous at all points. Every

---

[1] It may not be superfluous to specify a few of the views which have been published regarding the influence of the illumination on the observation of microscopic details. They will illustrate how vague have been the ideas which opticians and microscopists have held on this point. Schleiden says, in his "Elements of Scientific Botany" ("Grundzüge d. wiss. Botanik"), p. 103: "It (the illuminating mirror) is made either plane or concave—concave so that the pencils of light emerging from it may exactly fill the aperture of the stage. . . . . Where possible, illumination with the plane mirror is preferable; it is true, the quantity of light is not so great, but the parallelism of the rays is decidedly more conducive to the certainty of the observation, for it appears that displacements in the image may be caused by the convergence of the rays from the concave mirror. I have often noticed this phenomenon, but I confess that I can give no explanation of it, since opticians leave us here entirely in the dark." Earlier authors, however, viz., Wollaston, Brewster, and Dujardin, considered converging light the most favourable, and the two last-named even thought that the object should lie exactly in the focus of the converging rays. According to Pritchard, difficult test-objects are well shown only with diverging light. Harting, who mentions these opinions, explains them by the different influence of objects upon the path of the rays of light, and offers on his own part the eclectic view, that illumination should sometimes be effected by parallel, sometimes by converging or diverging light, according to the requirements of the special cases and the nature of the object. Further on, when describing the means for controlling the direction of the rays, he mentions that the shape of the plane mirror is immaterial, though the concave mirror must of course be circular. Goring differed from him on this latter point, urging that the illuminating mirror should be elliptic and not circular, it would then appear circular when seen from above. He proposed a truly colossal size (nearly 5 inches in length by 4 inches in breadth), with a view to increase the brightness of the image.

We think the unprejudiced reader will be convinced, from the following explanation, that the incident rays which contribute to the illumination of the surface-elements of the field of view always converge, and that the precise shape of either mirror is quite immaterial.

element of the reflecting surface will therefore emit light corresponding with the directions in which it is incident. If we disregard the very small differences of loss which the rays reflected at unequal angles undergo, the mirror will act with regard to the object-plane of the Microscope precisely as a self-luminous surface, each surface-element acting as a self-luminous point. In this case it is quite immaterial whether the mirror is plane, concave, or convex, since the luminous power of the separate surface-elements is not dependent on their inclination to the optic axis, consequently the brightness of the whole mirror surface is the same, whether this inclination varies from element to element or remains constant. If $a\,b$ represents the diaphragm, any surface-element whatever of the object-plane p (Fig. 43) is, therefore, illuminated by rays which proceed from the points of the mirror surface between m and n. Of the whole cone of light diverging upwards, which each of these points emits (in the Fig. indicated for the point o), only an infinitely narrow part co-operates, whose base is p. The intensity of the illumination is, consequently, so far as the mirror surface is sufficiently extended, limited by the diaphragm $a\,b$, which also limits the aperture of the incident cone of light $m\,p\,n$ on it. The distance of the mirror surface from the object is immaterial; for although the brightness of a surface-element varies with the square of the distance, the quantity of light varies in the same proportion. The total quantity of light is the same in either case.

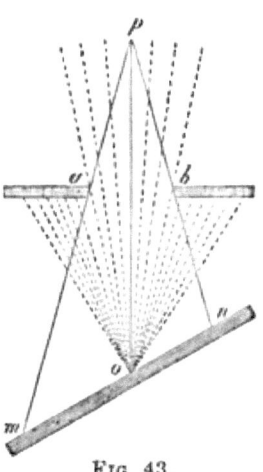

FIG. 43.

It follows, that a concave mirror will illuminate the object-plane equally well, whether its focus lies in the plane itself or not, on the supposition that its surface exceeds the limit determined by the diaphragm. For since the curvature of the mirror, as above shown, is entirely without influence, the position of its focal point has no optical importance.

Let us now consider how the condition of an unlimited and equally luminous source of light is practically fulfilled (by unlimited we mean exceeding the limits determined by the diaphragm). In order to satisfy the above condition at least approximately, when the mirror is directed to the sky the rays which illuminate the field, if traced backwards to the mirror and thence to the source of light, must reach this latter without diminution. If, for instance, $p\,m$ and $p\,n$ (Fig. 44) are two marginal rays, which just touch the edge of the diaphragm, they are reflected by the plane mirror $A\,B$ to s and t, and by the concave mirror $C\,D$ to s' and t', while the central ray $p\,o$ passes in both cases to r. If the rays s and t meet the window in the wall $f\!f$, and nothing obstructs the path, all intermediate rays will proceed to the source of light, the sky, and in an opposite direction the field of view. Whether, according to the form of the mirror, the rays diverge, converge, or proceed parallel, is immaterial, provided a sufficiently large portion of sky affords uniform illumination. If, however, anything obstructs the uniformity of the illumination, the form and position of the mirror become important. If, for instance, a part of the sky is more luminous than the adjacent portions, a concave mirror, which focuses the light in p, will give a stronger illumination. Similarly, for every other source of light of relatively small extent (a lamp-flame, a white wall, &c.) the concave mirror will, so far as regards strength of illumination, be preferable, and it should always be so adjusted that the source of light is focused in the field of view.

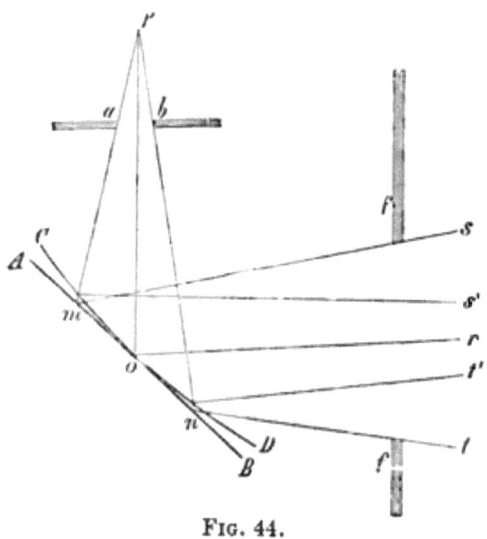

Fig. 44.

The size of the mirror has first to be considered in relation to the aperture of the diaphragm. Let us suppose the diaphragm removed; then the surface of the mirror forms the base of the incident cone of light, and its aperture varies with the distance of the mirror from the field of view.

All this is, of course, founded upon the simplest laws of optics. It is quite incomprehensible to us that recently-published micrographic works should still refer to parallel or divergent light, which is to be applied, according to circumstances, for the illumination of the field of view (*i.e.*, probably its surface-elements). The light which illuminates a given point of the field of view is always convergent, *i.e.*, the rays proceeding from the mirror intersect in that point, and diverge from it towards the objective. The converse supposition hardly merits confuting.

Secondly, as regards the different lenses or systems of lenses usually inserted in the path of the incident light, it may readily be shown that, like the form and position of the mirror, they exercise influence only in certain cases, while on the other hand they are inoperative in all cases where the mirror and the source of light may be regarded as unlimited. If p (Fig. 45) is a surface-element of the field of view, and $a\,b$ the diaphragm determining the aperture of the incident cone of light, then the first refracting or reflecting surface below the diaphragm acts as an unlimited source of light. For every point of it receives rays from all directions which come

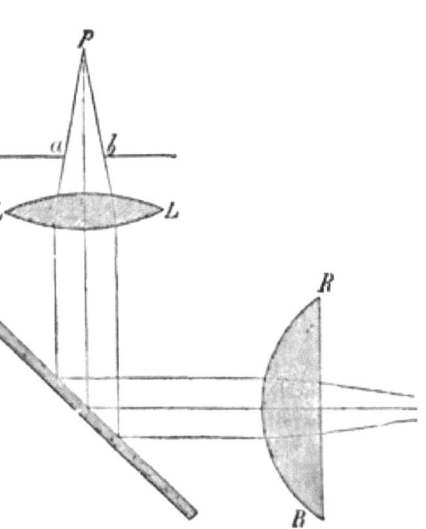

Fig. 45.

into account, and consequently emits rays in all corresponding directions. It is quite immaterial whether the pencils reach the surface directly or after various refractions or reflexions, for the difference of direction in the incident rays does not come into consideration, the real source of light being unlimited. We can, moreover, regard any plane below the diaphragm (the diaphragm itself included) as an illuminating surface, whether a deviation caused by refraction or reflexion takes place in it, or not. If, therefore, we place our illuminating surface temporarily in the plane of the diaphragm, every point of the latter acts as a self-luminous point, and the illuminating lenses $L\,L$ and $R\,R$ can clearly have no other action than that of causing the rays, which serve to illuminate the point p, to undergo several deviations on their path between the diaphragm and the original source of light, as is exhibited by the lines in the Fig.

The relations are altered when the aperture of the incident cone is limited by the mirror. The insertion of a convex lens produces in this case a stronger convergence of the rays, as shown in Fig. 46, and therefore, under certain circumstances, a greater angle of aperture. It is always possible to place the lens so that the effective cone of light, whose apex lies in p, has for its base the whole upper surface of the lens. The lens produces, therefore, the same result as a diaphragm situated in the same plane, which receives light from a relatively unlimited mirror surface, and, like the latter, furnishes more light in inverse ratio to its distance from the object-plane. The increase, however, in both cases reaches its limit as soon as the marginal rays, produced backwards, meet the periphery of the mirror.

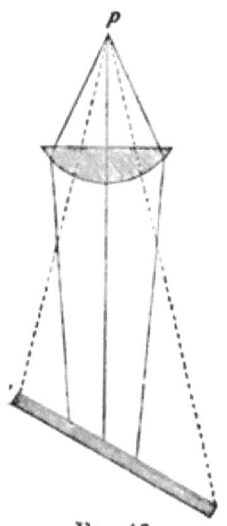

FIG. 46.

This consideration also applies to the case where the original source of light is limited, while the reflecting surface is unlimited. It is only necessary to substitute the former word for the latter, in order to make the preceding sentences literally correct for this supposition also. They may, *mutatis mutandis*, be extended to the very common case where the source of light as well as the mirror surface are limited, or where the illumination is not uniformly bright. The greatest possible brightness which can be attained by means of illuminating devices cannot under any circumstances exceed that which a sufficiently large mirror with a relatively unlimited source of light would alone afford.

It follows that if the illuminating apparatus is furnished with a diaphragm which neutralizes the inequality of the conditions for the differently-coloured rays (in the sense that all the rays, without exception, if produced backwards, would reach the source of light), no difference can possibly exist between achromatic and non-achromatic systems of lenses. If the source of light were of so small an extent that the blue marginal rays, for instance, could not meet it, together with the red rays, or conversely; if, moreover, it were not possible to regulate the illumination so that the differently-coloured rays would proceed from nearly equally bright points of the source of light (*e.g.*, a bright sky), and therefore, produced backwards, meet it,—under these circumstances an achromatic system would be preferable. Since, however, such restrictions seldom occur in practice when the instrument is provided with a concave mirror, and may certainly be avoided in most cases, the manufacture of achromatic illuminating devices appears to be unnecessary.

From what has been pointed out, the different illuminating devices are therefore effective in two directions only: they give to the cone of light, which illuminates the field of view, an equal intensity throughout, and, in the second place, extend its angular aperture. The statements with regard to any other influence they possess are purely imaginary; there is no foundation for such assertions as, for instance, that they cause the disappearance of the interference lines at the edge of the object, and resolve finer details better in proportion to their freedom from aberration, &c. With the same diaphragm and equal focal length, an ordinary condensing lens is just as effective as the most complex system of lenses, provided the diameters of the refracting surfaces and of the mirror correspond, for all colours, to the size of the diaphragm. On this assumption the angle of aperture of the incident cone will be the greater, the nearer the diaphragm to the object-plane. It is quite immaterial whether the focal point of the illuminating apparatus lies in the object-plane or not; its position is practically important only when, on close approximation to the objective, the marginal rays, produced backwards, no longer meet the mirror surface, and consequently the maximum of brightness is nearly reached when the image formed of distant objects by the illuminating apparatus appears in the field of view.

If the illuminating apparatus is to satisfy all requirements, it must also be provided with means, besides the diaphragm, applied above the last refracting surface, to shut out any desired portion of the cone. Many objects appear with the greatest distinctness if the central rays are cut off, whilst others appear more distinct if illuminated by oblique light incident on one side only. To facilitate the illumination, in addition to the usual graduated series of diaphragms, central stops are needed, which may be conveniently applied on a disc rotating over the diaphragm. It is advisable, on the grounds already discussed, to apply these stops, &c., *above* the illuminating apparatus.

The so-called *oblique illumination*, which is now much employed for the resolution of difficult details, consists of an incident cone of light, whose axis is more or less inclined to that of the Microscope. Such a cone may be obtained by placing the mirror or the diaphragm somewhat laterally to the optic axis, or by covering up a portion of the diaphragm, so that merely half, for instance, of the original cone is effective (Fig. 47). It is evident that if the angle of aperture of the diaphragm were equal to that of the objective, and a deviation did not take place in the field of view, every possible obliquity of illumination could be obtained by the latter method. Since, however, the first condition is rarely fulfilled (the diaphragm having almost always a relatively smaller aperture), and the last *cannot* be fulfilled, it is in many cases advantageous for the illuminating apparatus itself to admit of revolution round a horizontal axis, thus furnishing incident light of any desired inclination. Nevertheless, we do not estimate highly the value of such contrivances. A diaphragm with lateral motion, such as most modern Microscopes possess, combined with a suitable adjustment of the mirror, will be quite as serviceable in most cases.

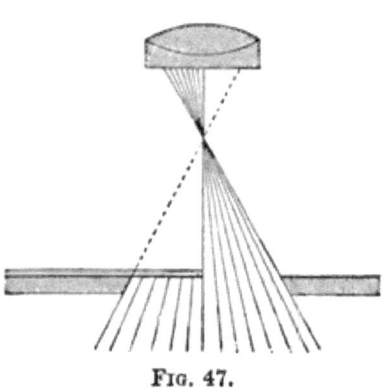

FIG. 47.

Among the really useful contrivances for illumination (especially for the testing of objectives, &c.), that recently described by Abbe ("Archiv für mikr. Anatomie," Bd. ix. p. 469) deserves to be recommended. It consists of two non-achromatic lenses, mounted in a short tube. The upper lens is plano-convex, and is greater than a hemisphere; when properly adjusted, its plane-surface lies slightly below the upper plane of the stage. The focus of the combination is only a few millimetres above the plane-surface of the upper lens, and is therefore very near to the object. The angle of aperture of the emergent rays is 120° in water, so that the marginal rays are inclined to the axis at an angle of nearly 60° in water.

It is now clear why for the production of the different kinds of illumination we need only a sliding diaphragm of suitable aperture, placed in the axis for central illumination, and more or less laterally for oblique illumination. The circular movement in Abbe's apparatus is effected by a rack and pinion; for general work, however, this mechanism is unnecessary, as the movement of the diaphragm can be effected sufficiently well by the hand. For further information we must refer our readers to Abbe's paper (quoted above), and will only add that the apparatus is manufactured by Carl Zeiss, of Jena.

2.—Illumination by Reflected Light.

The illumination of opaque objects has the advantage, that microscopic observation thus approximates to vision with the naked eye, for the final image on the retina is produced in both cases by rays which are reflected by the surface of the object. They are not, however, strictly identical; in microscopic vision, in consequence of the greater aperture of the effective cone of light from the object, the light and shadow are always distributed, *cæteris paribus*, somewhat differently. With this difference in the angle of aperture is also connected the difficulty of providing an incident cone corresponding with every cone that reaches the objective from the object-point, *i.e.*, of so regulating the illumination, that if the pencils are produced backwards, every ray meets the source of light. If, for instance, *A B* (Fig. 48) is the upper surface of a body with hemispherical prominences, and *g h* the effective portion of the objective, the cone *g h p* is

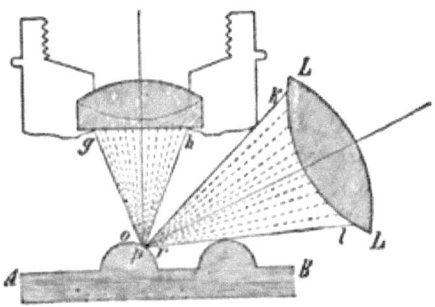

Fig. 48.

reflected at *p*, so that its marginal rays take the directions *p k* and *p l*. With the dimensions given in the Fig., these rays just touch on one side the surface of the object, and on the other the margin of the objective, and thus reach the unlimited source of light, the bright sky, through the window of the laboratory. Every obstruction to the incidence of the light may be overcome by the application of a condenser (*L L* in the Fig.), or a concave mirror, which will thus render exactly the same service as in the case of transmitted light. All intermediate rays reach the source of light, and consequently, in the opposite direction, the objective. The point *p* is thus illuminated by a complete cone of light, having an aperture equal to that of the objective.

Let us now suppose similar cones of light drawn to all points to the right and left of *p*; obviously a part only will reach the source of light, whilst some are reflected either towards the objective, or towards points of the surface of the metal-work. The cones are, therefore, only partially formed. The loss is greater the nearer we approach, on the one side, the vertex *o*, and on the other side, the edge of the hemisphere. In all cases, complete obscurity must be produced at *o*, since the incident cone of light coincides with the reflected one; similarly for the point *r*, situated on the right of *p*, illumination is, under the given conditions, quite impossible, for all rays reflected at *r* are incident beyond the effective portion of the objective. The hemisphere, therefore, appears bright only in the part surrounding *p* (Fig. 49, *a*), and even if the source of light were equally intense at all points equidistant from *o*—which is not practically the case, as the light is incident from the side—this bright spot would appear only as a narrow circular zone (Fig. 49, *b*).

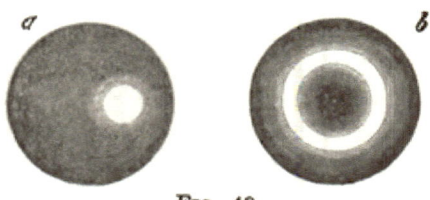

Fig. 49.

We will endeavour later on, when discussing the Theory of Microscopic Observation, to determine more exactly the distribution of light and shadow for certain object-forms; we shall here merely show that an illumination of this kind cannot, in general, be favourable for the determination of the form-relations of a given object, for even with a moderate aperture of the objective, a large portion of the surface is in the true shadow, or in the penumbra. As a rule, it is only applicable when the amplification is not more than 100—120 linear, so that the aperture of the incident cone (*k p l* in Fig. 48) is considerably larger than that of the objective. The application of a condenser enables us to produce such a cone, even if the Microscope is some distance from the window, provided the focal length of the lens is not too great in proportion to its diameter. And it follows, moreover, that an equivalent concave mirror, or a Selligue prism (a rectangular prism, in which the two surfaces of the right angle are convex) would yield the same results as a condensing lens; still, the latter would, in most cases, be more convenient.

The above-mentioned disadvantage, viz., that with this kind of illumination the greater part of the object is within the shadow, may, to a certain extent, be eliminated. For, since the distribution of light and shadow in the microscopic image depends entirely upon the position and aperture of the incident cone of light (assuming the objective-system to be given), it is possible, by skilful application of reflecting surfaces, to illuminate a larger portion of the object. If, for instance, the window of the laboratory, or the illuminating-lens, is the (secondary) source of light, an ordinary mirror, suitably adjusted on the other side of the Microscope, will duplicate the source of light. A spherical body exhibits a single excentric spot of light with one-sided illumination, but will now exhibit two; and similarly a second symmetrical line is added to the line of light, which is formed by a metal wire running parallel to the window. If the Microscope were placed in a semicircle of mirrors near the window, the effect of light would be approximately as if it were exposed to unrestricted side and top light. A similar effect, although not perfectly equivalent, would be produced by a small concave mirror of cylindric or parabolic section, applied at the end of the objective so as to focus the image of the window in the object-plane. To increase its utility, we might give it a suitable curvature in the direction of the axis, as exhibited in Fig. 50. The object-point p would thus be illuminated as if exposed to the open sky, except the portion at the apex determined by the angle $q\,p\,n$.

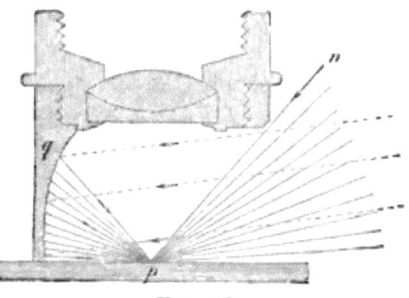

Fig. 50.

Such an illuminating apparatus[1] has not, to our knowledge, been

---

[1] The apparatus is known in England as the "Silver Side-reflector," or "Parabolic Side-reflector," and is mounted to slide on the front of the objective, like the Lieberkühn, but with a hinged joint that facilitates the

manufactured. The Wenham paraboloid,[1] and other devices, which have been popularized more especially in England, agree with it in providing side illumination of greater or less extent, though only within rather narrow limits of inclination. This applies also to the apparatus of Abbe, above mentioned, as employed for illumination with reflected light. In this case the peripheral rays of about 60°—48° inclination are totally reflected at the upper surface of the thin cover-glass upon the object; those of less inclination must be partially stopped off; when immersion objectives are employed, total reflexion no longer takes place. On the other hand, we have in the Lieberkühn mirror a counterpart of Fig. 50, since it produces exactly the radiation corresponding to the angle $q\,p\,n$, and therefore represents that part of the sky which in our case was shut out. This effect is obtained by means of a small concave mirror perforated in the centre, and sliding on the end of the objective; the front lens of the objective fills the opening in the mirror, and thus completes the reflecting surface.[2] The ordinary illuminating mirror reflects the light through a wide aperture in the stage to the Lieberkühn, the central rays being stopped off by a suitable opaque disc. The curvature of the Lieberkühn must be such, that the pencils from the object-points, if produced backwards, fall upon the illuminating mirror, which is only possible if the foci of the Lieberkühn and the objective lie near to each other, or coincide, though the latter condition is not strictly necessary.

adjustment. The apparatus was described and figured by the late Richard Beck in the "Transactions Micr. Soc." xiii. (1865) p. 117.—[ED.]

[1] Mr. Wenham's original device consisted of a parabolic mirror turned downwards, sliding beneath the stage in the manner of a cylindrical diaphragm. Mr. George Shadbolt suggested the internal reflexion from a glass paraboloid in preference to the ordinary reflexion from the parabolic mirror, and this was adopted by Mr. Wenham. The apex of the paraboloid is truncated suitably below the focus, and the truncated end is ground deeply concave to permit the adjustment of live-cells, or diaphragms, which are mounted on a movable stem passing through the axis of the paraboloid. Mr. Wenham subsequently discovered that by making the truncated surface plane, the paraboloid might be utilized as an animalcule cage, and to obtain various effects of dark ground illumination on suitably mounted objects when placed in immersion-contact with a slide of proper thickness. "Transactions Micr. Soc." iii. (1852) pp. 83 and 155, and iv. (1856) p. 59, and "Monthly Micr. Journ." ii. (1869) p. 28.—[Ed.]

[2] The reflexion at the front lens is so considerable, that the image of a globule of mercury formed by it again produces a diminished image in the latter, which is still clearly visible in the Microscope.

With this apparatus objects may be viewed on a dark or coloured ground; they may be placed in special holders (dark wells), &c.

By the Lieberkühn mirror the visible portions of the object are pretty equally illuminated,[1] in contrast to the usual effects of light and shadow produced by ordinary side-light.

---

[1] We meet here and there with vague ideas as to the action of the different illuminating devices. Many microscopists confine themselves exclusively to what they see, without attempting to explain the production of the effect. From the communication of Abbe, already several times quoted, we find that observers who are capable of passing judgment on this point, on account of their mathematical and philosophical attainments, recognize the accuracy of our exposition.

PART II.

THE MECHANICAL ARRANGEMENT OF THE MICROSCOPE.

We have thought it desirable to devote a special chapter to the mechanical arrangement of the Microscope, for the sake of those readers who have not the opportunity of referring to other micrographic works. We have limited ourselves, however, to a short explanation of the principles followed in the different manufactories of the better-known opticians.

I.

GENERAL RULES FOR THE CONSTRUCTION OF STANDS.

The Focusing.—The stand of the compound Microscope is, in the first place, intended to bring the objective to the proper distance from the object, and to retain it firmly in this position. For this purpose it must be so constructed as to admit of considerable alterations of distance between the objective and the stage (for instance, when changing the objective), and also be provided with means for delicate adjustment to a given level. The mechanism for these purposes allows of three modifications, each of which is used in a very different way. The focusing takes place (*a*) by movement of the body-tube towards the fixed object-stage, upon which the object under investigation is placed, (*b*) by movement of the stage towards the fixed body-tube, or (*c*) by a combination of both movements, where that of the body-tube serves for the coarse adjustment and that of the stage for the fine adjustment. Each of these arrangements has, of course, its advantages and disadvantages, and their relative estimation is, to a great extent, a matter of individual judgment. Mohl expresses himself decidedly against any movement of the stage, as tending to the development of general unsteadiness, and because the screw-micrometer, which is indispensable in his opinion, can only be applied efficiently to a fixed stage. The first reason is at any rate worthy of notice; the last is less evident on account of the slight use made of the screw-micrometer. On the other hand, with the fixed body-tube is combined the advantage that the head can always be kept at the same height when different objectives are employed, while in the opposite case its position depends upon the magnifying power. The differences of height which the focusing involves are inconvenient only with the low-power objectives, and may be overcome by suitable shortening of the body-tube. The more important reasons are therefore in favour of the perfect immovability of the stage in the direction of the optic axis.

The coarse adjustment is effected either by a pinion which may be applied in different ways, or by sliding the body-tube in a slightly elastic socket by the hand. The former method is now employed in most of the larger instruments, the latter in the majority of small Microscopes, both German and French. The socket surpasses the pinion in simplicity and smoothness of motion; but by long use the surfaces of contact become worn, and dust and oxidation injure them. The tube is therefore apt to slide either too easily or the reverse, unless its fitting is maintained in proper condition. The inconvenience of its sliding by its own weight can be avoided by slight pressure of the socket.

The principle of lever-movement, which is in most English instruments employed for different purposes, has been turned to account in the large Microscopes of Plœssl for the coarse adjustment. The arrangement in question is recommended as "a substantial and convenient construction."

The *object-stage* should be sufficiently large (at least $2\frac{1}{2}$ to 3 inches in diameter), plane, and as firm as possible. Projecting screw-heads, fixed spring-clips, &c., are undesirable; they interfere with the application and free movement of large glass plates in the plane of adjustment. The stage may be circular or square; its upper surface may be black or not; the objectionable glare of light, which is often seen in the Microscope when low powers are used, is generally due to the *object-slide*. Where the certainty of the observation may be impaired by this reflected light, it must be excluded either by the hand or other suitable means.

The aperture of the stage should be at least as large as the field of view of the lowest-power objective. As, however, for many researches a much smaller aperture is desirable, and for others a much larger one, of at least 1 inch in diameter, the stage itself should have a large aperture which may be reduced to the desired size by means of perforated discs, which drop into the central hole to a suitable depth. This arrangement is convenient for other purposes—for instance, the application of selenite plates in investigations with polarized light.

Slide-grooves attached beneath the stage are advantageous; the whole diaphragm-apparatus may be thus moved laterally at pleasure, or removed altogether, which is important in many physical researches.

Large stands should be constructed so that the stage and the whole upper part of the Microscope may be revolved together on the optic axis; the incident light then strikes the object at any desired angle in azimuth. This arrangement, for which we are indebted to Oberhæuser, is now generally adopted in large Microscopes. In the smaller stands, this system of rotation may be replaced by a revolving stage-plate, which should be provided with centering arrangements. For since the slightest excentricity of the axis of revolution produces marked alterations of position of the object during the turning—which readily increase till the object vanishes from the field of view, and on the other hand the optic axis of the different objectives never lie exactly in the same vertical—it is impossible to adjust the plate accurately once for all; it must therefore admit of being moved at right angles within certain limits. In micrometric and angular measurements the centering movements facilitate the exact adjustment of division-lines or crossed threads on a particular point, and are thus of great service in crystallographical investigations. This rotating stage-plate should be removable. If it is graduated on the circumference, as in the large instruments of Seibert and Krafft, it will be serviceable for the measurement of angles.

The Illuminating Apparatus.—In accordance with our discussion of the principles of illumination, the apparatus may differ in two ways: (*a*) with regard to the aperture of the incident cone of light, and (*b*) with regard to its inclination. A perfect illuminating apparatus must therefore include means for controlling both these conditions.

In most cases a plane or a concave mirror of ordinary size satisfies these requirements if it can be brought sufficiently near to the object, and admits, moreover, of lateral movement. If the light thus obtained is still too weak, the convergence of the rays may be increased by condensing lenses. The limitation of the incident cones of light is effected in both cases by diaphragms attached below the object-stage, or sunk in it.

Large Microscopes are generally furnished with plane and concave mirrors in one and the same mounting, which can be moved in all directions, or at least in the vertical plane. The smaller instruments usually have a concave mirror only, which revolves on its own centre, and is at the same time movable in the direction of the axis; this simplification in the mounting of the mirror necessitates the use of special means for oblique illumination. A still greater restriction of the movements of the mirror—as we find, for instance, in the small drum stands (stands with cylinder-base)—is, generally speaking, inconvenient.

With reference to the apparatus for the illumination of opaque objects, we must refer our readers to the statements made above.

Diaphragms.—It is obvious that a small diaphragm-aperture, say at the distance of a millimetre from the plane of adjustment, limits the incident cones of light like an aperture of double the diameter at double the distance. Theoretically it is therefore quite immaterial in what manner, or even in what plane between mirror and object, the diaphragms are applied, provided only the size of the aperture is in the proper ratio to its distance from the object. In practice, however, the diaphragm-apparatus should be capable of being easily and conveniently used; it should permit the most different gradations of the illumination without displacing the object; and, moreover, it should completely exclude all extraneous light reflected by the work-table or stand. For this purpose revolving discs with a series of apertures were originally applied below the stage, which were at the same time movable in a vertical direction. The apertures were so arranged that when the disc was revolved they coincided successively with the optic axis. An arrangement of this kind, accurately constructed, answers most purposes at the present day; it should not, however, be placed at too great a distance from the object-stage. One of the most convenient arrangements for the disc of diaphragms is to hollow out the stage from below, and construct the diaphragm-plate with a corresponding curvature. The larger stands of Zeiss are provided with calotte diaphragms on this plan.

Besides these revolving discs the *cylinder-diaphragms* introduced by Oberhæuser have recently come more and more into use. They are cylindrical caps with apertures of varying size, which drop into a larger cylinder sliding in a socket beneath the stage. They are convenient for adjustment at different distances beneath the object, and thus regulate the amount of light very perfectly.

We have already referred to various contrivances for shutting off the central rays.

The Base.—The chief purpose of the base is to give stability to the whole instrument; it must, therefore, be neither too small nor too light. As a model of firmness and convenience the *horse-shoe stand* of Oberhæuser may, perhaps, be named; it combines great solidity with the advantage that the illuminating mirror can be lowered to the level of the table, by which the necessary space is gained for other contrivances to be applied below the stage. Less convenient, because arranged only for axial illumination, and, therefore, applicable only for the smaller instruments, is the so-called *drum stand* of Oberhæuser; it rests upon a circular foot weighted with lead, which is connected with the stage by a cylindrical tube with an opening in front. Some opticians unite a similar foot by one or two pillars to the object-stage and to the whole upper part of the stand. Others, again, adopt a tripod base, of which the feet are sometimes fixed and sometimes arranged to fold up, &c. Small instruments, such as travelling and pocket Microscopes, are so constructed that the stand may be screwed on the lid of the case containing them. For daily use, however, we should not recommend such an arrangement.

Length and position of the body-tubes.—Most modern opticians on the Continent give the body-tube a length of 200—220 mm. from the anterior surface of the objective to the terminal surface of the eye-piece, so that the total height of the instrument amounts to 300—360 mm. Only the older large Microscopes of Plœssl and his imitators (Pistor, Schieck, &c.) differ essentially from this, reaching a height of about 450 mm., with a tube-length of about

300 mm., so that with a table of ordinary height they can only be used standing. The English and American instruments seem also to be remarkable for large size.

Large stands are sometimes arranged so that the body-tube turns with the object-stage on a horizontal axis, and may, therefore, be inclined at any desired angle; a more convenient position of the head is thus attained, and the body-tube may be used in a horizontal position. These advantages, however, seem to us to be more than counterbalanced by the inconvenience which the inclination of the stage involves. Where a horizontal position of the body-tube is required, which is seldom the case, the instrument may be laid down. The horse-shoe stands are excellently adapted for this purpose, the two ends of the horse-shoe forming, with the stage, a heavy tripod.

The desiderata of a good stand are, in our opinion, (1) a substantial and convenient shaped brass-work; (2) a firm and sufficiently large stage; (3) an illuminating mirror (plane and concave), rotating on its centre and movable laterally; and (4) a suitable arrangement of diaphragm. We should estimate as of secondary importance the movement of the mirror forwards and in a vertical direction (easily arranged, indeed, in most instruments), and especially the revolution of the stage round the optic axis. These are properties which might be supplied in a simple manner; the former by the application of a condenser allowing of oblique illumination on all sides, the latter by the much less costly revolving stage-plate. Where, however, the shape of the stand admits of free movement of the mirror, this is preferable and does not increase the expense. Finally, all other arrangements which have been mentioned above are of a minor character, and may be dispensed with in most researches.

II.

THE STANDS OF MODERN OPTICIANS.

We now proceed to describe a type-model stand that has met with much favour among the leading firms of Continental opticians.

Dr. E. Hartnack, of Potsdam, and A. Prazmowski, of Paris (formerly the firm of Hartnack & Prazmowski, successors to G. Oberhæuser):—The large older *horse-shoe stand* of Oberhæuser (Fig. 51), which these opticians have retained without essential alterations, and which many other opticians have adopted with more or less modifications, has already been referred to as an excellent type-model. It deserves this appellation, because it combines, as no other does, the advantages of the greatest efficiency with simplicity of construction. The coarse adjustment is effected by sliding the tube in the socket, and the fine adjustment by a micrometer-screw, which raises and lowers the socket. The stage revolves, and carries the optical body round with it; it is provided on its under surface with grooves in which the substage slides carrying the condenser or diaphragms; the mirror is movable laterally and vertically, one side being plane, the other concave.

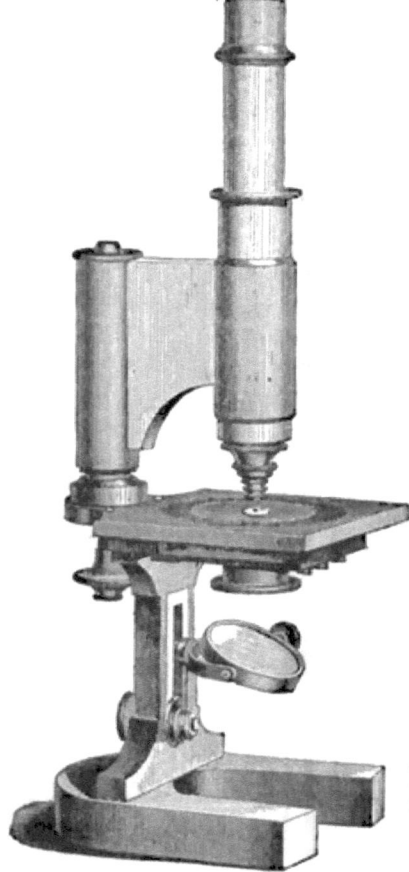

Fig. 51.

Recently these opticians have somewhat altered the construction of this stand, by supplying, amongst other things, rackwork for the coarse adjustment.

[We have thought it preferable to omit any translation of the other parts of this chapter, which are descriptive of stands, as the matter is dealt with somewhat incompletely, at least according to English views.[1]

[1] In addition to the description of the Oberhæuser Stand—which we have preserved as being the type-model of the authors—the following are described and figured by them :—

OPTICIANS.	STANDS.
Hartnack & Prazmowski	Medium horse-shoe.
,, ,,	New small (Fig. 52).
,, ,,	Students' (Fig. 53).
Nachet & Son (Paris)	Largest (Fig. 54).
,, ,,	"Grand Modèle" (Fig. 55).
,, ,,	"Grand Modèle, droit" (Figs. 56-7).
,, ,,	"Moyen Modèle, inclinant et droit" (Figs. 58-9).
,, ,,	"Nouveau Modèle, inclinant" (Fig. 60).
,, ,,	"Petit Modèle, inclinant et droit" (Fig. 61).
,, ,,	Simple and compound combined (Fig. 62).
,, ,,	Stereoscopic binocular (Fig. 63).
,, ,,	Double and treble-bodied (Fig. 64).
,, ,,	Binocular apparatus, two forms, stereoscopic, and for two observers (Figs. 65-6).
,, ,,	Inverted, two forms (Figs. 67-8.)
,, ,,	Laboratory dissecting (Fig. 69).
,, ,,	Pocket (Figs. 70-1).
,, ,,	Horizontal (Fig. 72).
,, ,,	Revolving nose-piece or objective-carrier (Fig. 73).
A. Chevalier (Paris)	A general reference to eleven stands.
Plœssl & Co. (Vienna)	Large and small Microscope, described with a reference to a large and small erecting dissecting Microscope, and a travelling Microscope.
F. W. Schieck (Berlin)	⎫
L. Bénèche (Berlin)	⎬ No detailed description.
C. Zeiss (Jena)	⎟
E. Leitz (Wetzlar)	⎭
Seibert & Krafft (Wetzlar)	Describes in detail their special form of fine adjustment without friction (Figs. 74-6). [See "Journ. R. Micr. Soc." iii. (1880), pp. 883 and 1047.]
G. & S. Merz (Munich)	No detailed description (Fig. 77, three forms).
Smith, Beck, & Beck (London)	Brief description (Fig. 78).

[The following are the authors' remarks on the *general* part of the subject:—]

It may not, however, be superfluous, as a supplement to what has been said, if we point out the importance of several points omitted in the introduction, which, according to our experience, deserve special consideration in a critical examination of stands. We place in the first rank of importance *accurate and durable metal-work*. It is by no means a matter of indifference whether the thread of the screws, the focusing arrangements, &c., are of durable construction. On this point we have been able to convince ourselves in the workshop of C. Zeiss, of Jena, how much more exactly the parts of the stand (as, for instance, the pillars and the socket, slides, &c.) may be made and fitted together if the metal is worked by the Fraising machine, instead of being turned or planed.

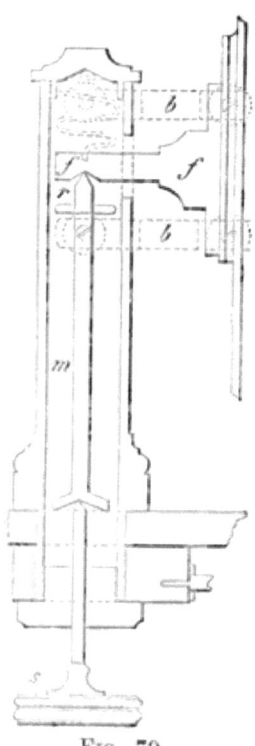

Fig. 79.

Of the special features which a good stand should possess, the chief are :—(1) The solidity of the mechanism of the focal adjustment, which always leaves much to be desired. The shifting of the image, the "back lash" of the screw, &c., are well-known disadvantages almost invariably present in the older stands. On this point also we have met with some important improvements in the workshop of Zeiss; but we must not expect the friction due to the existing modes of construction can be completely eliminated by slight alterations in the construction of the instrument. Whether the fine adjustment devised by Seibert and Krafft will be really durable, time must show. As far, however, as the principle of construction exhibited in Fig. 79 admits of a judgment on the point, friction is reduced nearly to a minimum. The micrometer-screw s acts upon the funnel-shaped head of the rod m, the upper end of which acts in a similar manner upon ff, the solid bar attached to the socket of the body-tube. The ring r, which serves as a guide-piece, lies loose in the hollow column, and as a rule does not touch the rod; its function is merely to prevent the point of the rod from slipping out of the notch in ff. If the cross-bars bb, which with the tube and column form the movable parallelogram, are of good construction and properly fitted, the focusing movement will presumably be very smooth and free from lateral play.[1] (2) An arrangement of the stage which will, among other things, admit of extremely oblique illumination, is very desirable. For this purpose the aperture of the stage should be bevelled out beneath, otherwise the obliquity of the rays will be limited by the margin. The substage, diaphragms, &c., must be removable, as arranged in many of the larger instruments. With regard to the other appliances which are mentioned in the catalogues, we fail to find, almost without exception, any contrivance which, in investigation with polarized light will allow not only the crossed Nicols, but also the differently-coloured selenite plates to be fixed, and the object alone rotated on a vertical axis.[2] The revolving stage-plate answers this purpose only when the necessary space for the reception of the selenite plates is reserved between it and the fixed stage.

With reference to the adjustment of the mirror, the older arrangements for this purpose are well known. In recent times Nachet, Zeiss, and others have applied a movement forwards, as well as from right to left—an improvement which in certain cases much facilitates oblique illumination.

The stands of German and French opticians have reached such a degree of perfection that they completely satisfy ordinary requirements. For unusual investigations, however—for instance, in physical researches—some additional appliances are sure to be necessary.

[1] The two cross-bars bb, on each side, are attached to the column by two screws (shown in dotted lines) permitting pivot motion in the vertical, without lateral play; the socket of the body-tube (joined to ff) is held at the front ends of the cross-bars between the *points* of four screws (shown in dotted lines): the friction is therefore confined to the motion of the cross-bars between the eight screws (*i.e.*, four on each side). When the focusing-screw s is turned, the solid bar ff is pushed upwards, or the spiral spring shown above it presses it downwards, and the cross-bars assume a diagonal position—similar to the motion of an ordinary parallel ruler, of which one side is stationary; ff therefore moves slightly backwards, from the normal position figured, carrying with it the optic axis. This motion of the optic axis presents difficulties of centering in combination with a rotating stage.—[ED.]

[2] This movement is applied to the best English and American stands.—[ED.]

PART III.

TESTING THE MICROSCOPE.

I.

TESTING THE OPTICAL POWER IN GENERAL.

As we have shown in a previous chapter that all properties which influence the optical power of a Microscope, and therefore also the sharpness and brightness of its images, are based firstly upon the absence of both kinds of aberration, and secondly upon the magnitude of the angle of aperture, we might therefore confine the testing of the instrument to these two points, and to the determination of the magnifying power. In estimating the quality of a Microscope a rigorous examination is never necessary, as the question simply is whether it compares favourably with other known instruments and satisfies in general the requirements we are justified in demanding. We will now apply ourselves to this practical question, and afterwards follow out the separate elements which enter into consideration on a more special testing.

1.—Absolute Power of Discrimination.

Let us inquire, in the first place, how the quality of a Microscope may be estimated, setting aside all peculiarities and faults of construction upon which it depends. This is usually done by means of the so-called test-objects, certain details of which are resolved by the better instruments with a given amplification, while those of inferior quality either do not bring them out at all or only indistinctly. We will not dispute that the data thus obtained are perfectly sufficient, on the whole, for a correct estimation of the capacity of a given instrument. We must not, however, overlook the fact that these test-objects themselves may easily lead to an entirely one-sided conception of the optical power, because the pencils of light, by whose interference the details are delineated, with the ordinary illumination pass through only a narrow peripheral zone of the objective, and consequently disclose nothing as to the whole central portion of the refracting surfaces. We learn therefore, as a rule, merely the *limits* of resolution, not the optical power in general. This also applies to the images of wire-gauze, produced by small air-bubbles, &c., if the meshes are diminished to the limit of discrimination. The microscopic image is here also an interference image which, as regards its details, is formed exclusively by the pencils of light which graze the margin of the aperture of the objective. Nevertheless, these images of wire-gauze (which Harting has recommended) offer certain advantages which the ordinary test-objects do not afford, and to which we are inclined to assign special value, especially for comparative observations, and hence we place them in the first rank in the following considerations.

Images of Wire-gauze as Test-objects.

The images of wire-gauze are recommended as test-objects preferably to all others because, in the first place, for every range of amplification they present one and the same object; for the virtual images of a spherical air-bubble, always formed with equal sharpness, can be diminished beyond the limit of discrimination, without alteration of the relations of light and shadow. The pencils of light proceeding from the wire-gauze—which, on account of the minuteness of the air-bubble, are to be regarded as composed of *parallel* rays—under all circumstances undergo (as shown in Fig. 80) such a refraction that, whilst completely filling a widely opened cone, they diverge towards the objective. (The Fig. is accurately constructed for refraction in water at the calculated angles.)

To this agreement with test-objects—which so far as regards the nature of the details is complete, the only difference being in magnitude—is added, in the second place, the important fact that

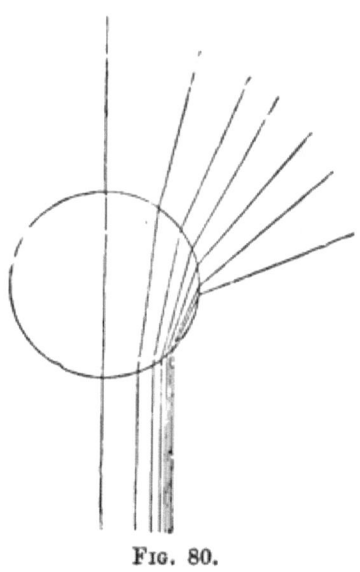

FIG. 80.

their production is very simple and inexpensive. Air-bubbles of every range of size are very easily obtained. By dissolving gum-arabic in water they are formed abundantly, because the air contained in the fragments of gum is not completely absorbed. They are also readily obtained by placing a drop of glycerine, or oil, upon a plane glass slide and beating it with a knife-blade until it begins to froth; it should then be protected by a cover-glass, care being taken to place at the edges some suitable substance (threads of cotton, slips of paper, &c.) to prevent undue compression. Minute bubbles of carbonic acid gas serve the same purpose, and are easily obtained by pouring a drop of dilute hydrochloric acid upon powdered chalk. With a large selection of air-bubbles of different size it is possible, with the same wire-gauze, by merely changing the object-distance slightly, to obtain images that reach the limit of discrimination for all powers.

In order to determine with certainty the size of the network at

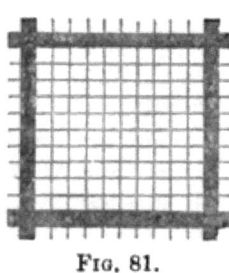

FIG. 81.

the limit of discrimination, we advise as follows:—A square piece of wire net, containing about 5×5 to 10×10 meshes, is enclosed by strips of black paper (Fig. 81), and is then placed between the mirror and the air-bubble in such a position that its image is just perceptible. If one side of the square is measured, it is only necessary to divide by the number of interstices contained in this side, to determine the distance of the interspaces—that is, the diameter of the meshes. For instance, if the entire square has a diameter of 7·5 mic., as represented in the Fig., the distance of the wires (disregarding on both sides half the thickness of a wire) is exactly one-tenth, that is ·75 mic.[1]

For fixing the wire-gauze a holder, consisting of a horizontal arm provided with a ring, which is movable on a vertical arm, is the most convenient; or it may be attached by means of wax to the lower end of a cylinder-diaphragm, and the distance regulated by sliding the carrier in the socket. This method of testing presents no difficulties; it is only necessary to carefully select air-bubbles that yield clear and sharp images.

We give on pp. 130-1 a table of the results of measurements made with different instruments. In the observations indicated by an asterisk, the limit of discrimination has been taken to be the point where the meshes of the network could be clearly recognized by an observer who was not familiar with the object. As the intensity of illumination is, of course, not altogether immaterial, most of the determinations were repeated several times with clear and with cloudy sky, and the mean values taken as the measure. The distance of the axes of the wires is taken as the diameter of the meshes, according to the method of calculation above referred to; it is, of course, equal to the sum of the diameter of one wire and an interspace.

Some of the data already given relating to the objectives of Oberhaeuser, Amici, Plœssl, Kellner, and Baader, are omitted from the table, as they do not admit of comparison, nor are they of any practical importance. To compensate for this, the enumeration of results is enriched by a greater number of examples from the recent publication of Otto Mueller;[2] as he also made his measurements as above described, his results can be compared with each other as well as with ours.

The objectives examined were, for the most part, of recent construction; some, however, were not so; on this point therefore we add a few special remarks. Both the objectives of Plœssl belong to an old, large Microscope; those of Merz, on the other hand, to

---

[1] Harting takes in these measurements the thickness of the wires and the size of the interspaces separately. We do not consider this distinction to be necessary for the test in question, since we have always obtained nearly the same result with different network, in which the proportion of the thickness of the wire to the size of the meshes varied between 1 : 5 and 1 : 2.

[2] "Vergleichende Untersuchungen neuerer Mikroskop-Objective." Berlin, 1873.

a more recent one, which left the factory in the year 1867. The objectives of Hartnack examined date from 1860–64, as also those of Bénèche. On the other hand, Mueller's measurements refer to objectives of Bénèche, which were presumably constructed between 1868–71, while those of Hartnack correspond to about the same year as ours. The only exception to this is No. 14 Imm. which is one of the prize specimens of the Paris Exhibition. The objectives of Gundlach (with the exception of No. 8 Imm., which is of an earlier year,) date from about the years 1868–71; those of Zeiss partly from 1864, partly from 1869–70; and those of Mœller from 1869.

The objectives are arranged according to their focal lengths; those with correction-adjustment are marked "Corr."; the immersion objectives "Imm." The latter were all supplied with the correction-adjustment.

Opticians.	Designation of Objective.	Focal length in mm.	Objective Amplification.	Eye-piece Amplification.	Combined Amplification.	Diameter of Meshes in Mic.
Gundlach	I	22·239	10·2	3·38	34·5	3·03
*Oberhæuser	4	18·6	12	4	48	2·3
Mœller	I	17·909	13	3·74	48·6	2·566
Zeiss (1864)	A	14·368	16·4	3·73	61·3	2·039
Gundlach	II	13·706	17·2	3·82	65·8	1·531
*Bénèche	4	13·44	18·2	4·4	80	1·6
*Plossl	1. 2. 3.	13	22·3	5·1	114	1·5
*Merz	¼″	12·0	12·5	7·0	88	1·8
*Hartnack	4	11·3	20·4	4·7	100	1·6
Gundlach	III	11·049	21·6	3·68	79·5	1·411
Hartnack	4	10·917	21·9	3·69	80·9	1·44
Bénèche	4	10·412	23·0	3·7	85·2	1·239
Zeiss	B	10·040	23·9	3·73	89·1	1·44
Mœller	II	8·117	29·8	3·73	111·1	1·08
Zeiss (1864)	C	7·851	30·8	3·64	114·6	1·023
Bénèche	6	6·964	34·9	3·72	130·3	0·843
Hartnack	5	5·825	41·9	3·7	155·1	0·904
*Merz	⅙″	5·0	30	7	210	0·8
*Bénèche	7	4·7	51·6	4·4	227	0·84
Bénèche	7	3·934	62·6	3·81	238·3	0·609
*Plossl	5. 6. 7.	3·9	73·5	5·1	375	0·78
Hartnack	7	3·894	63·2	3·74	236·7	0·596
Zeiss	D	3·702	66·5	3·58	247·4	0·629
Mœller	III	3·639	67·7	4·03	272·9	0·611
Gundlach	V	3·571	69·0	4·20	289·8	0·539
*Bénèche	9	3·42	72·7	4·4	320	0·62
*Hartnack	7	3·3	72·2	4·7	340	0·62
Bénèche	9	3·209	76·9	3·88	299·1	0·565
*Merz	1/12″	3·0	50	7	350	0·6
*Bénèche	8	2·95	83·6	4·4	368	0·60
Gundlach	6	2·769	89·3	4·12	367·8	0·442
Zeiss	E	2·596	95·3	3·73	354·1	0·526
Mœller	4 Corr.	2·545	97·3	4·08	396·5	0·488

IMAGES OF WIRE-GAUZE AS TEST-OBJECTS.

Opticians.	Designation of Objective.	Focal length in mm.	Objective Amplification.	Eye-piece Amplification.	Combined Amplification.	Diameter of Meshes in Mic.
Gundlach	6 Corr.	2·472	100·2	4·29	429·9	0·488
Bénèche	11	2·339	105·9	3·77	398·9	0·458
*Bénèche	11	2·22	90.9	4·4	400	0·54
*Hartnack	9	2·02	118·1	4·7	555	0·58
Hartnack	10 Corr.	1·966	126·2	3·8	483·8	0·420
Zeiss	F	1·919	129·3	3·8	492·2	0·458
*Hartnack	9 Imm.	1·87	140·0	4·7	558	0·45
Hartnack	10 Imm.	1·793	138·4	4·19	581·0	0·364
Gundlach	7 Imm.	1·771	140·1	3·94	551·6	0·339
Bénèche	10 Imm.	1·750	141·9	4·22	599·2	0·343
Hartnack	10 Imm.	1·734	143·2	4·09	585·5	0·342
Zeiss	F	1·727	143·8	4·16	598·0	0·433
Hartnack	14 Imm.	1·231	202·1	4·44	898·6	0·317
Gundlach	8 Imm.	1·207	206·2	4·34	895·1	0·314
Bénèche	12 Imm.	0·821	303·5	4·41	1336·8	0·348

From this table we arrive at the conclusion, in the first place, that the optical power of the Microscope does not keep pace either with the power of the objectives or with the total magnifying power. Selecting any particular case, we observe that more than twice the amplification is required to reduce to one-half the diameter of the meshes still just distinguishable; and conversely, an amplification less than half as great to render recognizable meshes twice as large. We may, therefore, with justice require that the lower objectives shall develope the greatest optical power relatively, and on the other hand not require too much of the higher and very highest ones. If, with the latter, the increase of the magnifying power up to the limit of the attainable, produces a further decided gain, the optician has accomplished all that can at present be expected of him. Moreover, it is seen that the power of discrimination in the Microscope is regulated more according to the focal length of the objective than to the total magnification. The slight increase of distinctness, which is observed on the application of deeper eye-pieces, is in no proportion to the increase in the size of the image effected by it.

If, in the second place, we select from the figures those which relate to the highest objectives, with which the smaller instruments are usually provided (for instance, Hartnack's No. 7, Bénèche's No. 9, Zeiss's D, Gundlach's V, &c.), their focal lengths in nearly every case vary between 3 and 4 mm., and the magnifying powers between 250 and 350. We may require of such combinations that they shall resolve clearly the meshes of the wire-gauze with a diameter of ·6 to ·7 mic. With the best of them this diameter may be reduced even to ·54 or ·56 mic. Bénèche's No. 8 proved somewhat more powerful than his No. 9, as appeared by the examination of organic test-objects; still, the two numbers always stand so close to one another that they may fairly be regarded as one. A second No. 9 objective that we had an opportunity of comparing, was between the two, its focal length was 3·15 mm., and the combined amplification with eye-piece No. 2 about 340; the optical power was, however, somewhat less than that of the No. 9 of the table.

The most powerful objectives yet produced, among which are the immersions of Hartnack, Gundlach, Zeiss, Bénèche, &c., admit of the resolution of meshes of from ·4 to ·32 mic. with the eye-pieces generally used for observation, and with the most favourable increase in the eye-piece amplification yield a further addition of optical power up to about ·3 mic. in diameter of the meshes. Whether the best English objectives accomplish more, is not known to us; those we have had occasion to examine were at most of equal power.

A third remark which forces itself upon us, in comparing the above table with the results of other testings, relates to the absolute degree of performance. Measurements of images of wire-gauze afford universally, and especially with objectives of large angular aperture, somewhat larger figures for the limit of discrimination than are obtained with organic test-objects. This might have been expected; the difference obviously arises from this, that both the abnormal path of the rays, which the focal images of the air-bubbles cause, as well as the kind of image-formation in the Microscope dependent thereon, are unfavourable for the objective.

Organic Test-objects.

Among the organic bodies, whose finer structural details a good Microscope resolves only with a given amplification, the small scales upon many insects, particularly upon the wings of butterflies, and also the siliceous envelopes of diatoms, present the most different gradations. The former were for a long time employed almost exclusively as test-objects, and some are even now used to a considerable extent; the latter, which were first suggested in England and are specially adapted for the testing of high-power objectives, have only very recently come into general use, but they now form the most usual resource of the optician. Although we do not consider the results obtained with these test-objects strictly and decisively reliable, yet it cannot be denied that they furnish an easy mode of approximately judging of the performance of a Microscope. A knowledge of the ordinary test-objects is therefore always of practical value, and every microscopist may be recommended to take note of the sharpness and clearness of the images produced of them in a good instrument with a definite amplification. We therefore subjoin a brief list of the butterfly-scales and diatom-valves which may be advantageously used as test-objects, with a short description of the more important.

With regard to the scales of butterflies, they usually show distinct *longitudinal striæ* (which run parallel to each other, or somewhat diverging from the base of the scale to its upper edge), with less distinct *transverse striæ*, which cross the others at right angles and come into view with somewhat lower focal adjustment. The series of markings ought to be seen distinctly with the amplifications indicated.

1. *Lepisma saccharinum.*—Longitudinal striæ of the larger scales, distinct with an amplification of 40, of the smaller with 100—150. The larger scales are wedge-shaped, the smaller ones rounder, with paler and closer markings.

2. *Hipparchia Janira.*—(a) *Scales from the wings of the female.*—Longitudinal striæ, according to Schacht, visible with a power of 80, transverse striæ by oblique illumination with Hartnack's No. 5, Bénèche's No. 7, and Zeiss's C objectives, that is, with a power of 120—250. With higher powers the transverse striæ should be clearly seen, even with direct (axial) light, as sharp lines (not knotted or broken). The long transparent scales are always more difficult than the darker coloured ones. (b) *Scales of the upper side of the wings of the male.*—Longitudinal striæ, according to Mohl, visible only with the highest objectives of Amici and with oblique illumination. There is no trace of transverse striæ, which, without doubt, are also present.

3. *Lycæna Argus.*—*Scales of the upper surface of the fore-wing.*—(a) Those which by reflected light appear blue, and by transmitted light yellow. Longitudinal striæ visible with lower powers (50—80 linear), transverse striæ only with about 300. Deserving especial

attention, on account of the great uniformity of the scales. (*b*) Those which by reflected light appear light brown, and by transmitted light grey or dusky. Longitudinal striæ, as in the previous case; transverse striæ, although closer together, somewhat less difficult, though not perceptible with a lower power than 300. (*c*) Scales of oval form, which appear yellow by reflected and transmitted light. With rows of dark points about 2½—3 mic. distant from one another, and which, therefore, ought to be easily seen sharply defined with medium powers.

The markings of the diatom-valves are much more varied than is the case with the butterfly-scales. As test-objects those are generally used which admit of two or three series of lines being observed; in the first case the lines intersect at a right angle, in the latter obliquely. Whether these series of lines are due to ridge-like projections, or to slight elevations or depressions, disposed in rows, or to differences of density, is in most cases undecided; the various views here and there met with, and at times expressed with great confidence, need confirmation. We leave this an open question, and confine ourselves to the characteristics of the microscopic image, which is the sole matter entering into consideration in testing.

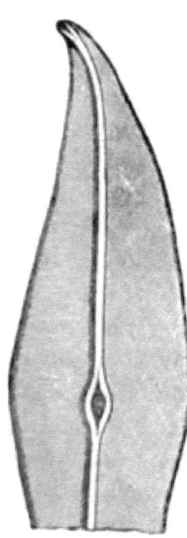

Fig. 82.

4. *Pinnularia viridis.*—A diatom nearly related to Navicula, with transverse striæ of about 1·5 mic. These striæ ought to be resolved even by objectives of 10—15 mm. focal length.

5. *Pleurosigma angulatum* (Fig. 82).—Two series of striæ equally strong and equally inclined towards the median-line, intersect at angles of about 53°—58° (not, as is usually said, at 60°); there is a third series, somewhat less distinct, at right angles to the median-line, forming, with the two oblique series, angles of 61°—63½°. With oblique illumination incident at a right angle, each of these series of lines ought to be distinctly seen alone with medium objectives, such as Hartnack's No. 7, or Bénèche's No. 9—that is, with a magnification of about 250—300 linear[1]—so that on revolving the object, the three series

---

[1] These magnifications necessitate an eye-piece amplification of about 4½ to 5 linear, *e.g.*, Bénèche's No. 2 eye-piece, and Hartnack's No. 3—measured

come into view successively. The higher powers, which give an amplification of 500—600 linear, exhibit the three series of lines simultaneously with direct (axial) light, and resolve them with oblique illumination into bright points, which are arranged in rows in the three directions. With the best objectives it is possible to see these bright points resolved as somewhat irregular hexagons, the two longest sides of which are parallel to the median-line (Fig. 83). The dark lines, which with lower power appear perfectly straight, are therefore in reality zig-zag lines, and the two oblique series consist, as shown in the figure, of elements running alternately lengthwise and obliquely, whilst the transverse lines are formed only by the latter. Since the elements running lengthwise are throughout coarser than the oblique ones, the greater clearness of the oblique series of markings is accounted for. Their apparent difference of level is due merely to an optical effect, which is explained by the unequal position of the oblique striae in relation to the incident light[1],—hence one and the same series of striae, with a given position of the midrib, appears to lie decidedly higher, but when rotated about 90°, just as decidedly lower.

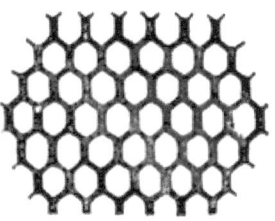

FIG. 83.

This excellent test-object, and the others we have described, can be obtained from Bourgogne, of Paris, and Mœller, of Wedel (Holstein), and also from most of the opticians. The specimens are generally mounted in balsam, and are so numerous that one suitable for observation may be quickly found. The larger ones exhibit, as might be expected, somewhat more distinct markings than the smaller ones. The microscopist ought, however, to be satisfied if the Microscope resolves the largest and finest specimens in the manner above described.

6. *Pleurosigma attenuatum.*—With strong longitudinal striae parallel to the midrib, and fine transverse striae at right angles,

---

according to the power of the objectives; objectives with a focal length of 4 to 5 mm. should show the markings with favourable eye-piece amplification.

[1] For, since the coarser elements, situated lengthwise, belong to the two oblique systems of markings, and therefore act similarly in every position towards the incident light, the difference in the effect can arise only from the obliquely-situated elements.

Longitudinal striæ with direct light appear distinct with a power of 150—200 linear; for instance, with Bénèche's No. 7 objective and No. 1 eye-piece, transverse striæ are shown in places distinctly with direct (axial) light with an amplification of about 300, and with oblique illumination very sharp throughout, with Bénèche's No. 9 and Hartnack's No. 7. With higher power objectives the bright lines are resolved, as occurs always with similar markings, very distinctly into small squares or, more accurately, rectangles, whose angles are more or less truncated, so that the dark division-lines appear thickened into knots.

7. *Pleurosigma balticum.*—With longitudinal striæ, and rather fine transverse striæ; the former visible with a power of 200 linear in oblique light, the latter scarcely distinct with 300. Bénèche's No. 7 shows the longitudinal lines separated only with the deeper eye-pieces, and the transverse lines on the margin; with Hartnack's No. 9 and No. 7 both series are resolved distinctly.

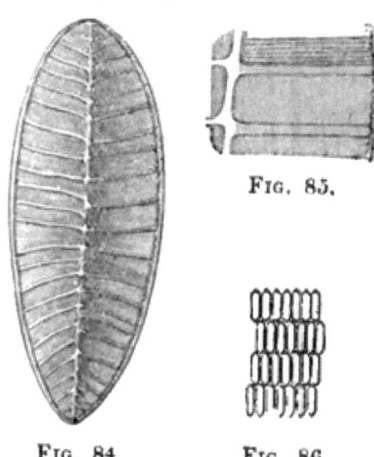

FIG. 84.   FIG. 85.   FIG. 86.

8. *Grammatophora marina.*—Marginal parts with transverse striæ, and two less distinct oblique series of striæ as in *Pleurosigma angulatum*. The three series of striæ can be seen with Bénèche's No. 9, and even the transverse striæ with the usual eye-pieces Nos. 1 and 2, but to see the oblique ones clearly deeper eye-pieces are required. Hartnack's No. 7 gives about the same results. We may therefore insist that a Microscope should at least exhibit the three series of striæ with an amplification of 340—400 linear.

9. *Surirella Gemma* (Fig. 84).—Between the strong cross-bars which appear at irregular distances on both sides of the midrib, and parallel with them, are fine transverse striæ, which are themselves intersected at right angles by exceedingly fine longitudinal striæ (Fig. 85). These latter are distinctly seen only with the higher powers, such as Hartnack's Nos. 9 and 10; the former, however, require merely a magnification of 400—500. With the best objectives that are now manufactured the fine longitudinal

lines ought to be resolved into small apparently hexagonal spaces (Fig. 86). This appearance is stated by Flœgel[1] to be incorrect, on account of the interference phenomena.

10. *Nitzschia sigmoidea.*—This long and narrow valve shows fine close transverse striæ, which, according to Schacht, were first resolved by Hartnack's No. 10 immersion with oblique light— they are, however, visible with the same illumination with the corresponding objectives of Bénèche and Zeiss. These conclusions of Schacht refer to dry specimens, which, it is well known, are less difficult than those mounted in balsam.

11. *Grammatophora subtilissima* (Fig. 87, *1*).—Marginal portion *a* with extremely fine and close transverse striæ (Fig. 87, *2*), which are resolved (if the preparation is in balsam) only by very good objectives. With Hartnack's No. 9 immersion we can scarcely see any indication of this striation. The most powerful immersion objectives of recent construction resolve these striæ distinctly, and also exhibit isolated traces of oblique series of striæ, as in *Pleurosigma angulatum* and in the larger *Grammatophoræ*.

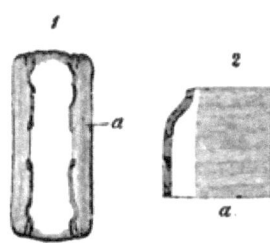

FIG. 87.

12. *Navicula rhomboides.*—With longitudinal and transverse striæ, the former more distinct and not very difficult, the latter exceedingly sharp and fine. Was used as *N. affinis*, as a test-object (balsam mounted) at the London International Exhibition (1862).

In addition to the preceding test-objects several others have been recently proposed which may be of service; for instance, different species of *Navicula, Coscinodiscus, Frustulia saxonica, Hyalodiscus subtilis,* and others, on several of which a striation as fine as 4 to 5 lines in a micromillimetre should be seen. Our list is, however, sufficiently complete, and we are not inclined to attribute importance to the discovery of new test-objects with our present knowledge of practical Optics. We would advise the microscopist to select three or four, and to note most carefully their appearance under different Microscopes. This will always be of more advantage to him in testing an unknown instrument, than the simple fact of whether it resolves a particular test-object or not.

---

[1] "Botanische Zeitung," 1869, p. 757.

We may, however, specially mention the test-plates of Mœller, consisting of twenty diatoms arranged in a row, according to the difficulty of their resolution. The names of these diatoms are:— 1, *Triceratium Favus*; 2, *Pinnularia nobilis*; 3, *Navicula Lyra rar.*; 4, *N. Lyra*; 5, *Pinnularia interrupta rar.*; 6, *Stauroneis Phœnicenteron*; 7, *Grammatophora marina* (more coarsely marked than those of Bourgogne); 8, *Pleurosigma balticum*; 9, *P. acuminatum*; 10, *Nitzschia amphioxys*; 11, *Pleurosigma angulatum*; 12, *Grammatophora oceanica subtilissima (marina)*; 13, *Surirella Gemma* (for transverse lines); 14, *Niztschia sigmoidea*; 15, *Pleurosigma Fasciola rar.*; 16, *Surirella Gemma* (for longitudinal lines); 17, *Cymatopleura elliptica*; 18, *Navicula crassinervis, Frustulia saxonica*; 19, *Nitzschia curvula*; 20, *Amphipleura pellucida*.[1]

For convenient comparison of organic test-objects with the images of wire-gauze and of Nobert's test-plates, we add a table of the distances of the lines according to measurements made by Dippel,[2] Flœgel,[3] and Abbe.

Names of the Objects.	Observer.	Distance of the lines in Micromillimetres.	
Lepisma saccharinum	Dippel	large	2·22
" "	"	small	1·42
Hipparchia Janira	"	transverse striæ	0·99
" "	Abbe	longitudinal striæ	2·00
Lycaena Argus	Dippel	light scales	1·33
" "	"	dark scales	0·87
Pinnularia viridis	"	transverse striæ	1·53
Pleurosigma angulatum	Dippel & A.	" "	0·46 to 0·50
" attenuatum	Dippel	" "	0·69
" balticum	"	longitudinal and transverse striæ	0·74
Grammatophora marina	"	" "	0·41
" subtilissima	"	" "	0·32
Surirella Gemma	Dippel	longitudinal striæ	0·32
" "	"	transverse striæ	0·46
Nitzschia sigmoidea	"	" "	0·33
Navicula rhomboidea	"	" "	0·33
Frustulia saxonica	"	" "	0·29
" "	Abbe	" "	0·25
Amphipleura pellucida	Flœgel	" "	0·28

The two last-named test-objects, *Frustulia saxonica* and *Amphipleuri pellucida*, are clearly amongst the finest and most difficult tests with which we are at present acquainted; but that in

---

[1] Mœller supplies also diatom type-plates with 100 to 400 objects, arranged systematically in rows.

[2] "Das Mikroskop," p. 134.

[3] "Botanische Zeitung," 1869, p. 741, *et seq.*

Amphipleura, as stated by Rabenhorst, transverse striæ are at the rate of 135 in ·001″ (according to which the distance of the markings amounts to only ·187 mic.) is evidently a mistake.

Nobert's Test-plate.

The employment of organic test-objects is always attended by the disadvantage, that the results which the observer obtains are never exactly comparable with those of another observer,—objects of the same kind differing so much from one another in size and distinctness. To overcome this inconvenience, Nobert invented a method of producing an artificial test-object, in the form of a micrometric series of lines upon glass. In 1846 he ruled test-plates with ten bands of lines, in which the distances of the lines were greatest in the first band, and progressively less in each successive band. Subsequently he increased the number of bands to twelve, fifteen, twenty, and, later on, even up to thirty, at the same time altering the distances from those in the older plates, so that the results obtained with the latter cannot be accurately compared with the results obtained with the modern plates. Nobert has recently reduced the number of the bands to nineteen. The distances of the lines in the two most recent[1] test-plates, and their number in a millimetre, are collected in the following tables:—

A.—THIRTY-BAND PLATE.

Band.	Distance in Mic.	Lines to 1 mm.	Band.	Distance in Mic.	Lines to 1 mm.
1	2·256	443	10	0·620	1612
2	1·945	513	11	0·591	1692
3	1·647	607	12	0·565	1772
4	1·399	715	13	0·533	1875
5	1·240	806	14	0·508	1969
6	1·082	924	15	0·451	2216
7	0·902	1108	20	0·377	2653
8	0·789	1267	25	0·323	3098
9	0·676	1478	30	0·282	3544

[1] Still more recently Nobert has produced a test-plate on which are ruled twenty bands of lines; those lines ruled in the tenth band are equivalent in fineness of division to the nineteenth group described in *Table B*; and those in the twentieth group are of twice that fineness of division.—[Ed.]

B.—NINETEEN-BAND PLATE.

Band.	Distance in Mic.	Lines to 1 mm.	Approximate equivalent Test-objects.
1	2·256	443	*Lepisma saccharinum*, large
2	1·504	665	*Pinnularia viridis*
3	1·128	886	
4	0·902	4108	
5	0·752	1329	*Pleurosigma balticum*
6	0·645	1550	,, *attenuatum*
7	0·564	1773	
8	0·502	1992	} *Pleurosigma angulatum*
9	0·451	2216	
10	0·410	2439	*Grammatophora marina*
11	0·376	2653	*Nitzschia linearis*
12	0·347	2870	*Navicula rhomboides*
13	0·322	3105	*Grammatophora subtilissima*
14	0·301	3322	
15	0·282	3546	
16	0·265	3768	*Frustulia saxonica*
17	0·251	3989	*Amphipleura pellucida*
18	0·237	4211	
19	0·226	4433	

It is evident from these tables that the test-plates of Nobert are not inferior in fineness of division to the most difficult organic objects. They agree with them, also, in that the image of the lines is a pure interference image, produced by the diffracted rays. We must, however, leave it undecided, not being ourselves intimate with the subject from personal observation, whether they really offer a more reliable standard than the siliceous envelopes of diatoms. According to Max Schultze[1] the equality of the later plates with nineteen bands leaves very little to be desired. That a complete uniformity of the corresponding bands of lines, especially with regard to the strength of the lines, cannot be attained with any mechanism, may, *à priori*, be asserted with certainty.

2.—RELATIVE POWER OF DISCRIMINATION.

Although in the foregoing the present standpoint of practical Optics is to a certain extent arithmetically defined by the results obtained with images of wire-gauze, in so far as the limit of the performances of high-class Microscopes can be determined in

[1] "Archiv für mikr. Anat." Bd. i. pp. 3-4.

accordance with the numbers given, it is of special interest to compare what has been already attained with that which ought theoretically to be attainable, provided the Microscope were mathematically perfect in every respect. It is geometrically evident, that with such an instrument the image of a wire network should be seen just as distinctly with an amplification of m times, as if viewed by the naked eye in an m-fold proportion. The brightness of the image is not, however, necessarily the same, as it is dependent on the angle of aperture of the objective, and, so far as this circumstance exercises an influence upon the discrimination of the meshes, it is conceivable that even in the most perfect instruments the optical power should not increase in exact ratio with the magnifying power. We will, however, for the sake of simplification, assume provisionally that the brightness in the Microscope either $= 1$, or that the choice of the source of light is so made, that the resulting images possess equal brightness. If, then, D is the diameter of the meshes at the limit of distinct vision with the naked eye, d the diameter of the same on a given amplification m; then (disregarding all influences of a physiological nature) $d = \frac{D}{m}$, or, what comes to the same, $\frac{D}{m\,d} = 1$, if the instrument is perfect; on the other hand $\frac{D}{m\,d} < 1$, if it is not perfect. In the latter case the larger or smaller value of the fraction is of course related to the optical power; it does not give its absolute magnitude, but only its relation to the performance of a perfect instrument. We may therefore also say that $\frac{D}{m\,d}$ is the mathematical expression of the *relative power of discrimination*.

The quantity D may easily be determined, in default of sufficiently large air-bubbles, by placing a suitable objective (for instance Hartnack's No. 7) upon the object-stage with its anterior surface turned upwards, and observing the image which it forms of the wire-gauze through the empty microscope-tube, or, if necessary, with the naked eye after the removal of the tube. In this case the eye must be at the distance from the object at which it most easily accommodates itself, and we determine the magnitude of the image by the micrometer, when it is just clearly recognizable.

We do not, however, obtain in this way (nor in general with similar observations with the naked eye) such harmonious results as in vision through the Microscope. Measurements which we often repeated gave, for instance, with a vision of 200 mm., values for D fluctuating between 100 and 128 mic., increasing respectively to 125 and 160 mic., if we suppose the image to be projected at the conventional distance of 250 mm., for which magnifying powers are usually calculated.

Otto Mueller[1] obtained 103 mic. for his eye as the mean diameter of the meshes, Harting[2] 99 mic., and Helmholtz[3] places the limit of discrimination with a trellis-work even as low as 78 mic.[4]

Under these circumstances a graphic representation is best adapted to exhibit the results of the calculation, since the construction remains the same whatever may be the amplification selected as the starting point. We do not, therefore, give a tabulated view of these quantities, but have drawn the magnifying powers in Fig. 88 according to an entirely optional scale for the abscissæ, and the corresponding values of $\frac{D}{m\,d}$ according to another

---

[1] Cf. "Untersuch. neuerer Mikroskop-Objective," p. 11 (259).

[2] "Mikroskop," 2nd ed. i. p. 81. The diameter of the meshes and the thickness of the wire must, of course, be added, which Harting in another place (i. p. 342) has overlooked. The methods of observation of Harting, moreover, give figures which cannot be compared with those obtained with the Microscope.

[3] "Physiologische Optik," p. 218. The author gives 63·75 seconds, which is about equal to the above value. Helmholtz employed for his observations a grating of black wires, their interspaces being equal to the diameter of the wires, and placed them under a bright sky. The figures thus obtained evidently do not admit of comparison with those obtained by our method.

[4] How little such observations in general agree is shown by the table given by Harting ("Mikroskop," i. p. 72), in which the visual angles of the meshes at the limits of discrimination for five observers are tabulated. We consider the methods of observation upon which it is based (according to which the wire-gauze is viewed direct, and the object-distance is increased to three metres), to be inapplicable in the given case, since we cannot assume that the eye always accommodates itself equally well to different distances. At any rate, measurements of dioptric images are to be preferred for comparison with microscopic observations, if they are made with as favourable a distance of the eye as possible. The minute image on the retina depends, therefore, in the one case, as well as in the other, on approximately equal conditions, and the reduction to the conventional distance of 25 cm. is merely a matter of calculation.

optional scale for the ordinates. The connection of the terminal points of the ordinates corresponding to the objectives enables us to compare them *inter se*, and at the same time to estimate the relative productions of the opticians named. The absolute magnitude of D first comes into account when dealing with the determination of the ordinate corresponding to the naked eye (amplification 1), and it can easily be imagined for every desired value of D. In the figure, the point O.M. represents the limiting

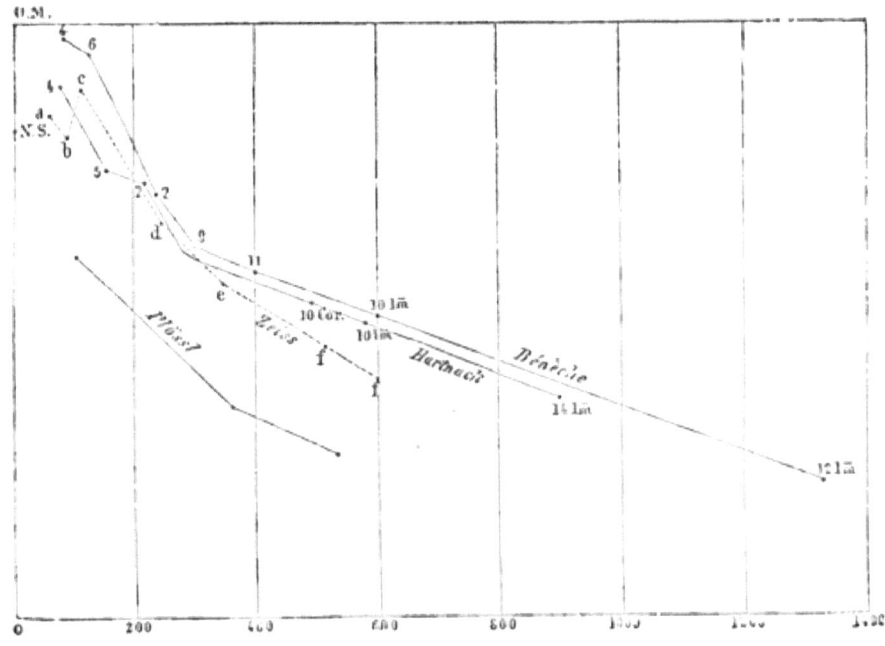

Fig. 88.

value of the diameter of the meshes of 103 mic. given by Otto Mueller, and the point N.S. the minimum observed by us, 125 mic.

What first attracts the eye in the figure is the striking fact, that the highest curves so much exceed the line representing the naked eye, taking the mean value as decisive. The contrary assertions, which have been made with regard to the power of discrimination of the naked eye, afford no grounds for questioning this fact. They may be explained by the unequal constitution of different eyes; but they leave our repeated measurements, in which D sometimes becomes less than $m\,d$, wholly untouched.

It may therefore be considered certain that there are Microscopes which, in spite of their imperfections, accomplish more with low powers in comparison with a healthy eye, than a perfect, ideal instrument should accomplish according to theory. It is evident that this paradoxical phenomenon can be explained only by peculiarities which distinguish microscopic observation—whether the refracting lenses are more or less perfect—from vision with the naked eye. We may mention as such peculiarities: (1) the microscopic production of images due, as regards detail, to interference; (2) the difference of brightness dependent upon the angle of aperture, which has already been discussed; (3) the chromatic aberration of these pencils of light, which it is known can never be quite eliminated; (4) their spherical aberration.

Of these four points—and we know of no others—the first is of preponderating influence, as might be expected. For, as above remarked, the microscopic structure-image is formed through interference, and the image-forming pencils of light which graze the margin of the objective-aperture are so narrow that the aberrations of the eye cannot produce any appreciable error. The sharpness of the image on the retina is therefore necessarily increased to a still higher degree than by diminishing the cones of light reaching the eye by circular diaphragms. These diaphragms, however, have a favourable influence. If, for instance, we look through a small aperture of about 1 mm. in diameter at the page of a book, or its image, the printing appears decidedly clearer, and the letters seem blacker and more sharply defined.

The second of these points, as experimental testing immediately shows, is unimportant, for the phenomenon remains the same even when, by a suitable adjustment of the illumination, the brightness of the microscopic image is so modified that the naked eye has a considerable advantage. Just as little can the spherical aberration, mentioned as point (4), turn the scale, as the eye does not produce a noticeable aberration of the marginal rays of pencils of light, which are incident approximately in the direction of its axis,[1] and consequently the compensation of such an aberration by means of an opposite one of the incident rays cannot take place.

[1] Cf. Fick, "Medizinische Physik," 1st ed. p. 310, and Wagner's "Handwörterbuch der Physiologie," article "Das Sehen," by Volkmann.

As regards the chromatic aberration of the eye—which reaches a very marked amount, occasioning principally the phenomena of *irradiation*—it is conceivable, that it could be eliminated by a rather over-corrected Microscope, and thus an increase of discriminating power could be obtained. This explanation is certainly applicable to Bénèche's No. 4 objective, which is never capable of compensating in the whole middle part of the field of view the opposite chromatic aberration of the eye, although distinctly over-corrected. We will discuss later on the testing of such phenomena, noting the simple fact here, in order to establish without doubt the correctness of the explanation for at least one definite case.

The diagram shows, secondly, that the relative optical power of Microscopes decreases considerably in proportion to the higher and highest amplifications, since the corresponding curves approach the abscissa-axis rather rapidly towards the right, in the most unfavourable cases sinking to even less than half their height. We might hence conclude, that practical constructions are still far from the limit of the attainable; and so far as regards the perfection of the dioptric image merely, this conclusion would be wholly justified. As, however, at the limit of the discriminating power the interference image is exclusively operative, the minuteness of the details depending chiefly upon the extent of the angular aperture, the case is essentially altered. The angle of aperture of the highest objectives, having 180° in air, has already reached the limit which is in general attainable, and the power of discrimination determined by it is raised to about that amount which for physical reasons it is impossible to exceed.[1] On the other hand, there is still required a greater uniformity of correction in the different zones of the objective, as well as more harmony in the combined action of the various factors which determine the optical power of a system,—at all times a very desirable aim, and by no means yet attained in practice.

For the medium amplifications, 300—400, the objectives of the various opticians we have mentioned are nearly equal with regard to the power of discrimination, while with the low-power objectives not unimportant differences appear. The explanation of these differences lies not so much in the impossibility of cancelling them, as in the absence of a desire to do so, because

[1] Cf. upon this point our chapter on the Theory of Microscopic Observation.

in the case of the low powers, which are usually employed only to obtain a larger field, the degree of perfection is of less importance, and consequently the improvement of these powers does not appear to be so remunerative. We must not, however, estimate too lightly the advantages of a good low-power objective; on the contrary, we shall learn always to value them highly for all investigations, where sharpness of image and extensive field have to be combined, as, for instance, in researches upon fibro-vascular bundles.

The testing of Microscopes by means of dioptric images, which are gradually diminished to the limit of discrimination, admits, of course, of various modifications. Instead of wire-gauze we may choose any other objects, spherical, thread-shaped, square, &c., and employ as the standard of comparison not only the power of discrimination, but also the capacity of recognizing form, as, for instance, the angles of a square or hexagon. Two or more minute perforations in an opaque screen form also a useful test-object, inasmuch as the distance of the air-bubble can be so chosen that the single points in the image are seen just separated from each other. It follows, of course, that the results obtained by any one method are comparable only with each other, and not with the figures found in another way. For the recognition of the form of a single square we shall, with a given amplification, obtain limiting values altogether different from those obtained for the discriminations of the square meshes of a wire-gauze, and the same differences between the absolute magnitudes also appear with the naked eye. On the other hand, it cannot well be supposed that the relative magnitudes, as represented by the curves, should be essentially different.

We have made use of several of these modifications, and have invariably found that none afford the same certainty of judgment as the one first mentioned. The limit of recognition, especially of spherical and thread-shaped objects, is by no means easy to determine, and is therefore rather variously estimated by different observers. Further, a condition is reached in which we no longer observe the *sharp* image of the object, but its feeble image formed of dispersion circles, which remains visible somewhat longer because of its greater extent. This is evident from the fact that the calculated size of the retinal image with thread-shaped objects is often found decidedly smaller than the diameter of the sensitive retinal elements;[1] for it cannot be supposed that these latter render qualitatively different excitations appreciable. In the last instance, therefore, the visibility of a dark line upon a bright ground depends chiefly upon whether the eye of the observer is still capable of appreciating a slight difference of light, which is dispersed over a part of the retinal elements. The supposed testing of the Microscope virtually passes into a testing of the eye.

Under these circumstances we refrain from giving a table of our measurements, and refer those who take a special interest in such observations to the detailed work of Harting.

The testing of objectives by means of diatom-valves, or Nobert's test-plates, as regards the relative power of discrimination, leads, naturally, to the same result, *cæteris paribus*, as the comparison of the images of wire-gauze meshes that are still just perceptible, always on the supposition that, to a certain degree, the different series of lines admit of comparison. The *limit* of discrimination usually appears to be slightly displaced, owing to the varying nature of the details and the unequal method of illumination. Thus, for instance, it is a fact that diatoms with faint markings, although their distance from each other is exactly the same, are always more difficult to resolve than those with strongly marked series of striæ. *Amphipleura pellucida* is for this reason more difficult than *Frustulia saxonica*, and *Navicula rhomboides* than *Nitzschia sigmoidea*, &c., although the distances of their markings are approximately equal. This is also the reason why organic test-objects are much less adapted for comparative examinations than images of wire-gauze, or the series of lines on Nobert's plates.

---

[1] The "bacillary layer," which is regarded as the true light-perceiving organ, consists in the yellow spot (the point of clearest vision) merely of so-called "cones," whose diameter amounts to about 5 mic. This magnitude, if the distance of the focal point of the retina is taken at 15 mm., corresponds to a vision of about 68″, and with an object distance of 25 cm. to a diameter of 83 mic. According to Harting ("Mikroskop," p. 298; 2nd ed. i. p. 333), thread-shaped objects, or rather their images formed by air-bubbles, are microscopically perceptible even with an amplification of 50, if their actual thickness equals ·194 mic., and consequently the diameter of the microscopic image equals $50 \times ·194 = 9·7$ mic. The corresponding retinal image would not, therefore, occupy even the eighth part of one retinal element.

3.—DEFINING POWER.

The above-mentioned details of test-objects, which serve as a standard for the resolving power of objectives, belong, as already mentioned, exclusively to the interference image which is produced by the pencils of light diffracted in the object, and reunited in the Microscope. Since these pencils with ordinary illumination pass through the marginal zone of the objective only, they must, of course, be regarded as furnishing unreliable vouchers for the accuracy of its construction in general. It is therefore necessary to test the course of the rays in the objective by other means, which admit chiefly of our judging of the central portion of the objective-aperture. In other words,—in addition to testing the structure-image, a careful testing is also necessary of the contour-image which is produced by the direct rays, especially when the middle part of the objective-aperture comes into play. This testing is most effective on objects which show clearly marked contrasts of light and shade in the same plane, or, what is equivalent, sharp outlines on surface-elements that are not too small. Bright spots of different shapes on a dark ground are specially suitable, and they can readily be made upon an object-slide coated with Indian ink or tin-foil. Fragments of diatom-valves having very sharp edges are excellent test-objects for this purpose; and since, moreover, as Abbe rightly insists, they afford the advantage of testing the structure-image at the same time as its coincidence with the dioptric image, they are unconditionally preferable for a rapid yet sufficiently reliable examination. If these fragments are to satisfy the two-fold purpose, they must first be perfectly flat, and at the same time so thin that outlines and structural details may be regarded as lying in the same plane; and, secondly, so marked that the interference images may be nearly equal in brightness to those formed by the direct rays, which obtains with dry objects to a higher degree than with those mounted in balsam.

As regards the choice of objects, the gradations for the lower and medium objectives are easily found; for the higher powers, however, particularly for immersion objectives, the selection is limited. Fragments of *Pleurosigma angulatum* may be strongly recommended. According to Abbe, this object may be advantageously used even with the highest powers, if fragments of delicate specimens are selected, and if the eye is directed specially to note the appearance of the image near the margin. For testing the highest dry objectives, coarser specimens of the same object, also in fragments, may be used, although the markings are rather too fine for an angular aperture of, say, 100°.

With regard to the method of illumination to be employed in the testing, Abbe recommends principally two fixed positions of the mirror. First, that it should be placed vertically to a series of lines on the object, so that one edge is approximately in the axis of the Microscope; the direct rays thus form a slightly inclined cone of light, which grazes the centre of the objective, and consequently occupies the intermediate portion of the aperture; the diffracted rays, however, pass through the opposite marginal zone of the objective, and produce with the direct rays a structure-image dependent on this zone. Secondly, to move it laterally to produce the most oblique illumination. In this case the two cones of light change places; the direct one passes through the periphery, and the diffracted one through the centre of the objective.

With an accurately constructed objective, the contour-image, in both cases, should not only appear equally sharp throughout, but should coincide with the structure-image, without difference of level and without lateral displacement. If an objective fulfils these requirements, at least in the centre of the field of view, it may be depended upon for producing accurate images. An intermediate position of the mirror would furnish additional proof that the different zones co-operate simultaneously in the production of the images.

II.

TESTING THE SPHERICAL ABERRATION.

Having explained the methods of testing the optical power of the Microscope in general, we purpose examining the different elements or factors which determine that power. Among these factors, according to our experience, spherical aberration is the most influential; its determination, and the discovery of the precise action of the lens-combinations, together form one of the most important problems, especially for the optician.

In the first place, the question arises in practice, how the presence of spherical aberration is manifested in the microscopic image; and, secondly, we have to decide whether it indicates an under-correction or an over-correction. The problem demands still further consideration, for it frequently occurs that the objective is under-corrected for the central rays (say, up to an inclination of 10°—15°), but over-corrected for the marginal. Since, however, the optician judges by the total effect, he considers a lens-system as far as possible aplanatic, if either the larger central part or, as more often occurs, the larger peripheral portion of the incident cone of rays exhibits no marked aberration.

The general action of spherical aberration may be shown graphically. If $a\,b$ (Fig. 89) is a self-luminous object, whose

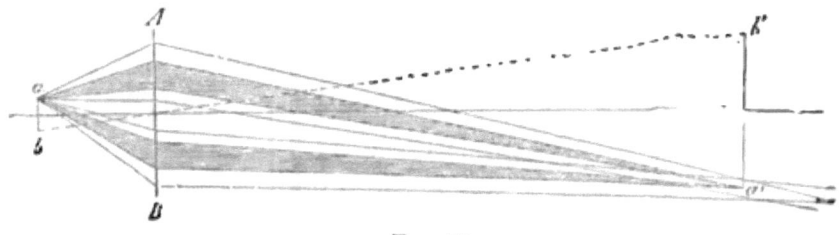

Fig. 89.

cone of rays completely fills the aperture of the objective $A\,B$, and if we assume that the rays emerging from a and b unite after refraction for the most part in the points a' and b' (in the Fig. this is assumed for the rays of medium inclination), while the others converge towards points lying somewhat nearer or further, and which are more or less separated, as shown in the Fig., for a central and a peripheral pair of rays: the real image of the point a appears therefore in a', and similarly, that of b in b'; consequently the images of the other points of the object appear in the intermediate points between a' and b'. An image-plane drawn through the points a' and b' would, as the construction shows, be met beyond these points by other rays, whose focal points, according to the hypothesis, fall in front of or behind this plane. These rays do not originate merely from the extreme marginal points of the object, a and b, but also from the adjacent ones up to a certain distance from the margin, from which point, also, the aberrant rays cut the surface of projection of the image. The observer, therefore, looking through the eye-piece at the image formed by the objective (the plane of the diaphragm representing the image-plane), perceives, outside the margins a' and b' of the illuminated image-surface, a glimmer or fog due to the rays affected with aberration.

The testing of the aberration can therefore be simply effected by viewing under the Microscope a small luminous surface, or a real or virtual image which acts as such a surface, taking care that no light reaches the eye from the surrounding part. The fainter, therefore, the glimmer or fog surrounding the microscopic image, the more perfect is the correction of the aberration.

We know of no better object than that which was employed in principle by Mohl, and later also by Harting. A glass slide is coated with an opaque layer of Chinese ink, in which fine points or lines to admit light are made with the point of a needle; or the layer of ink is allowed to dry, and a number of fine, sharply bordered cracks will be formed, which are preferable for testing to the artificial ones. Better still, but not so easily made, are the small silvered glass discs (as employed by Abbe), upon which bands of coarse and fine lines are ruled. Opaque and transparent parts here lie in exactly the same plane, and the different bands afford the necessary gradations for testing the various objectives.

In practice the condition must be fulfilled, which every test-object ought in general to satisfy, viz., that the cones of rays proceeding from the illuminating points, should occupy the whole aperture of the objective (a fact which Mohl and Harting have apparently disregarded). For this purpose the diaphragm must, with the lower powers, be removed or sufficiently enlarged, and the mirror must be brought nearer; but with the higher powers a condensing lens or an objective of suitable focus must be applied to increase the convergence of the incident rays correspondingly. If the objective to be tested has, say, an aperture of $60°$, the illuminating cone incident upon the glass slide should have an aperture at least as large.

The other objects which have been recommended by microscopists—for instance, the minute images of window-frames formed by an air-bubble or a mercury-globule (on a dark ground)—

appear to us less suitable for testing the aberration than the above-mentioned fissures in an otherwise black field of view. The latter afford, moreover, the essential advantage that a withdrawal of the diaphragm or of the condensing lens renders the marginal zone of the objective inoperative; the observer is thus enabled to confine the testing to a gradually diminishing central portion of the refracting surfaces. It is often observed that the strong appearance of fog surrounding the microscopic image disappears immediately if the extreme marginal rays are excluded, whilst in other cases it decreases almost imperceptibly, and is therefore due, chiefly, to the central rays. The determination of such differences is, of course, a part of the task of testing. If mercury-globules, or air-bubbles, are used for this purpose, suitable diaphragms must be applied to the objective; and there is the further disadvantage that the window-frames or their surroundings are never sufficiently dark to form a black background in a catoptric or dioptric image, and hence the presence of a slight fog can be less easily recognized. Regarding this point, we should still give the preference to a large spherical air-bubble, which forms the image of a small diaphragm, rather than to a mercury-globule, because the field of the image appears perfectly black up to the bright inner ring. The amount of light which an image of this kind produces is always somewhat less than that given by a fissure, because the marginal rays are weakened by partial reflexion at the upper surface of the air-bubble.

The answer to the further question, whether an objective, affected by aberration, is under- or over-corrected, is based upon the phenomena which are produced by the varying distance of the eye-piece, or, what in principle is equivalent, the increase or decrease of the object-distance. If we take the simple case of the aberrations all following the same direction, so that the refraction of the objective gradually increases or decreases from the centre to the periphery, the effect of change in the focal adjustment can readily be understood. If $A\,B$ (Fig. 90) is the objective, p the focal point, or the small focal space occupied by all the central rays up to a certain inclination, and $p\,p'$ the greatest length of the aberration of the marginal rays, for which, therefore, the objective is over-corrected; the real image will evidently be most clearly seen on a screen when it coincides with a plane f_1 drawn through the point p. This image appears, however, bordered by fog caused by the marginal rays incident on the surrounding parts. If the screen is moved further to the left, in the position f_2, it cuts both the central cone and the aberrant peripheral rays in a larger surface; the image must consequently be dim, while the fog increases in extent. If, on the other hand, the displacement takes place in the opposite direction, the cross-section of the solid cone of rays becomes at first smaller, attaining its minimum in f_3, where the peripheral rays intersect with the image-forming rays, and consequently the fog disappears. The screen is therefore illuminated in f_3 by a small sharply bounded circle of light. Further to the right the latter again increases in extent; it retains, however, the same sharp outline, because the peripheral rays are projected within it. In the neighbourhood of f_4 even the marginal zone of the circle of light appears decidedly brighter, because the aberrant rays here approach the upper surface

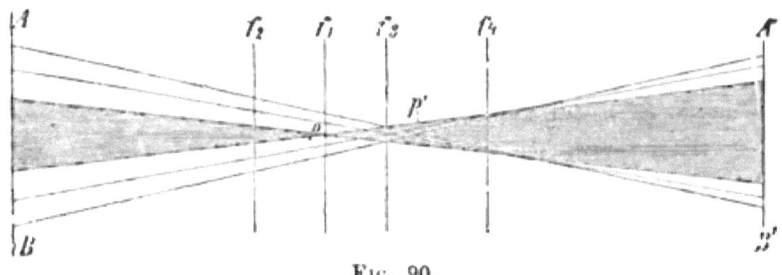

Fig. 90.

of the central cone; it contrasts, therefore, still more decidedly than before with the dark surroundings.

The phenomena are precisely reversed when the spherical aberration is under-corrected for the peripheral rays. In order to apply the same construction to this case, let us suppose in the above figure that the objective is at $A' B'$ instead of $A B$. The path of the rays corresponds to the opposite aberration; and the same movement of the screen, which, as regards $A B$, was an approximation, is now receding with regard to $A' B'$. The projections of the united cones of rays remain obviously the same in the different positions.

If, therefore, the Microscope is focused on the illuminating surface as accurately as possible, the approximation of the eye-piece is accompanied, in an over-corrected instrument, with a stronger appearance of fog; in an under-corrected one, by the formation of a sharply defined disc of light somewhat clearer at the edge; whilst increasing the distance of the eye-piece always produces the opposite appearance. The same effect is also obtained within the limits here considered, if, for the displacements of the eye-piece, are substituted corresponding displacements of the object,—instead of moving the eye-piece nearer or further with respect to the objective, the object is brought nearer or further, or the body-tube is lowered or raised. Where, therefore, an approximation produces more fog, or increasing the distance produces a well-defined circle of light, there is over-correction, and where the converse phenomena appear, under-correction.

With these effects of spherical aberration, which are clearly apparent to the eye, are associated those of chromatic aberration, which, as a rule, further increase the contrasts. In consequence

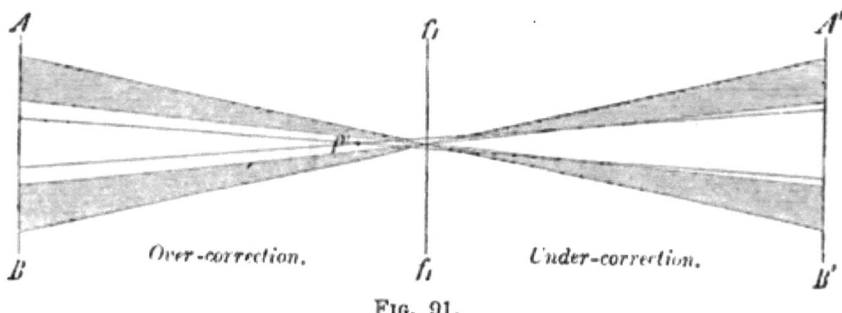

Fig. 91.

of this, the fog appears bluish, and the disc of light has a greenish-yellow border, red at the extreme edge, which appears, up to a certain limit, the more decided the greater its diameter. The distinction between under-corrected and over-corrected aberration is thus still further facilitated.

In certain micrographic works is added, as a further criterion of the two opposite aberrations, the fact that the fine details of the image (for instance, the window-frame, when mercury-globules are used, or the margins of small apertures in an otherwise dark field of view) disappear at once on the approximation of the object, but only slowly on its withdrawal, if the objective is under-corrected; and conversely in an over-corrected instrument. Thus generally stated, this assertion is inaccurate; we have submitted the matter to the test of experiment. In one special case, as shown in Fig. 90, precisely the converse took place. For it is clear that a displacement of the screen in the direction of f_1 towards f_4, influences the distinctness of the image less than the opposite displacement, as all the focal points of the peripheral rays, and consequently the images corresponding to them, lie to the right of f_1, between p and p'. A displacement in this direction corresponds, in an under-corrected objective, to an approximation, and in an over-corrected one to a withdrawal of the object. If the objective is so constructed that the marginal rays of the cone of light, up to a certain minimum of inclination, are the image-forming ones, while the central rays aberr towards the one or the other side (Fig. 91), then, of course, the rule stated by these writers is applicable. This, however, is only *one* of the various combinations imaginable; and even if it frequently occurred, which we doubt, it should not be accepted without further investigation. We must not therefore take the slower or quicker disappearance of the minute image on raising or lowering the focus as a criterion of over- or under-correction, except in so far as its applicability has been otherwise demonstrated.

In the path of the rays delineated in Fig. 91, the position of the screen (or of the eye-piece setting) to the plane $f_1 f_1$, in which the objective-image is formed, necessarily involves a slight fog, due to the central rays; but this disappears immediately on a slight displacement to the right or left. It is immaterial whether the objective is supposed to be in $A B$ or $A' B'$; the circle of light, which is projected upon the screen by different focal adjustment, appears clearly defined in either case. Spherical aberration, therefore, is exhibited only in the plane of the image as fog, which does not appreciably alter on the withdrawal of the diaphragm that limits the aperture of the whole effective cone.

If to the image-forming peripheral rays others are added which, like the central ones, are affected with aberration, the phenomena become more complicated. By the effect which the marginal rays produce with higher or lower focal adjustment, they are seen to be over- or under-corrected, whether the central rays aberr in the same or in opposite directions. In this case it is necessary to diminish or to withdraw the diaphragm until the bluish fog suffers no further alteration with a medium adjustment, and totally vanishes with a lower or higher adjustment. The residual traces of aberration must therefore proceed from the central rays; and the above-mentioned criterion of under- and over-correction, that is, the quicker or slower disappearance of the fine details of the image on raising or lowering the body tube, is applicable—as shown by the construction. In most cases, however, unless the aberration of the central rays is considerable, we shall obtain somewhat uncertain results, because the image-forming rays do not, as we have supposed, intersect exactly in the same point, but are partly affected by the opposite aberration. The outlines of the image do not then disappear suddenly either with high or low adjustment, and it is difficult to say which acts more slowly. For this reason we do not enter upon the explanation of other combinations which may occur with respect to the aberrations of the differently inclined rays. The phenomena which have been theoretically determined for either assumption, appear too complicated in practice to supply more reliable data for estimating the quality of the objective.

The practical processes of testing are consequently confined to the following points, summarizing the preceding remarks:—

(1.) We give to the cone of light proceeding from the mirror an aperture at least as large as that of the objective to be examined. With high-power objectives this is accomplished by the application of a suitable illuminating apparatus, or by placing an objective-lens of 3 to 4 mm. in diameter under the object-stage.

(2.) We focus the Microscope to that plane, in which the details of the image (for instance, the edges of fine fissures in a blackened plate) are most distinctly seen. If then a bluish fog is noticeable in the parts surrounding the image, the aberration is not sufficiently corrected.

(3.) In order to decide which rays of the incident cone are affected by aberration, the illuminating apparatus, or the diaphragm, must be gradually withdrawn. If the bluish fog rapidly decreases or entirely disappears, the aberration is confined to the marginal rays; if it retains the same intensity for some time, the cause of the phenomenon lies in the central rays. In the latter case, which on the whole occurs less often, the fog should disappear if the central part of the objective is covered by a suitable stop.

(4.) We test the phenomena which a higher or lower focal adjustment causes. In doubtful cases an alteration of the position of the eye-piece is to be preferred. The over-correction of the objective, with regard to the marginal rays, is hence recognized by the fact that the bluish fog, with lower adjustment, extends further, but with higher adjustment immediately disappears, so that the microscopic image appears as a sharply bounded circle of light with greenish-yellow border. Under-correction is characterized by the same contrasts in an opposite sequence.

(5.) The aberration of the central rays can only be more accurately ascertained, if the effect of light thereby occasioned is apparent—a case which, according to our experience, hardly ever occurs.

(6.) If the phenomena mentioned in (4) are undecided (the bluish fog, for instance, remaining perceptible for a short time both with higher and with lower adjustment), then both kinds of aberration are present in the peripheral part of the cone, perhaps in consequence of an inaccurate curvature of the refracting surfaces.

An accurate and practically valuable process of testing has recently been expounded by Abbe.[1] It is intended to exhibit the combined action of all the zones of the aperture of the objective in the centre and margin of the field of view, and at the same time to enable the respective images to be clearly distinguished. For this purpose the illumination is regulated by suitable diaphragms having several circular apertures (most conveniently with the apparatus described above on page 110), so that two or three isolated pencils simultaneously traverse the objective in the same number of correspondingly situated zones. For instance, if the objective has an aperture of 6 mm. the track of one of these pencils of light would appear as a bright circular disc of 1 mm. in diameter in the posterior focal plane, near the centre (Fig. 92, middle circle); a second track would be situated on the left, from 1—2 mm. from the centre; the third would be on the right, near the margin. This device shows the sensitive course of the rays, in which all faults of correction are most active. If the Microscope be focused on one of the test-plates mentioned on p. 151, in which transparent and opaque lines lie in the same plane, each pencil of light from one of the bands of lines occupying the field of view forms its own image, and the three partial images at a certain point of focal adjustment coalesce and form one colourless and sharply outlined image only when the objective is perfect. In every other case a

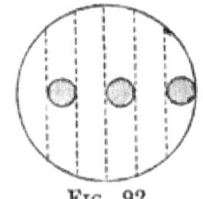

FIG. 92.

[1] "Archiv für mik. Anat." Bd. ix. 434.

complete coincidence either does not take place at all, or only in one part of the field of view, whilst in other places the corresponding lines are seen displaced laterally, or in different planes; further, in consequence of chromatic aberration, coloured borders appear on the bright lines, which vary according to their position in the field of view. Abbe states that "A test-image of this kind exhibits at one and the same time the whole condition of the correction of a Microscope in all its peculiarities before the eye. By means of the explanation which theory gives for the diagnosis of the different errors of delineation, the comparison of the coloured borders of the single partial images suffices to define exactly their mutual displacements laterally, and their differences of plane, in the central part of the field of view and in the four quadrants of the marginal zone, as well as all faults of correction in the last constituent parts according to their nature and magnitude."

III.

TESTING THE CHROMATIC ABERRATION.

THE simplest and most reliable method of testing the chromatic aberration of a system of lenses is to cover one-half of the anterior or posterior surface with black paper or tin-foil, so that only the other half remains optically effective. If we then view a line of light or a small luminous surface—for instance, a minute aperture in a blackened plate—it will appear colourless in an instrument that is perfectly achromatic; but in an over- or under-corrected one it will be encircled on the one side by a blue border, on the other by an orange-red or a yellow one. What arrangement these colours will exhibit, depending on the kind of aberration present, may be found without further discussion from Fig. 93. Let $A\,B$ be the objective to be tested, the right half covered by the diaphragm P. Since the uncovered half acts as a prism, then, if the chromatic aberration is under-corrected, the differently coloured real images of the object $a\,b$ must obviously be so displaced that the violet one ($a'_v\,b'_v$) projects the farthest on the right, and the red ($a'_r\,b'r$) on the left. The microscopic image has, therefore, on the right side a violet or blue border, and on the left a red or orange-coloured one. If, on the other hand, the chromatic aberration is over-corrected, the left side appears violet, and the right red or orange.

Upon this fact is based the method recommended by Mohl and Harting. Let w (Fig. 94) be the self-luminous surface under observation—for instance, the virtual focus of an air-bubble or mercury-globule. The medium focal adjustment gives a sharp and generally colourless image, because the incident cones of light

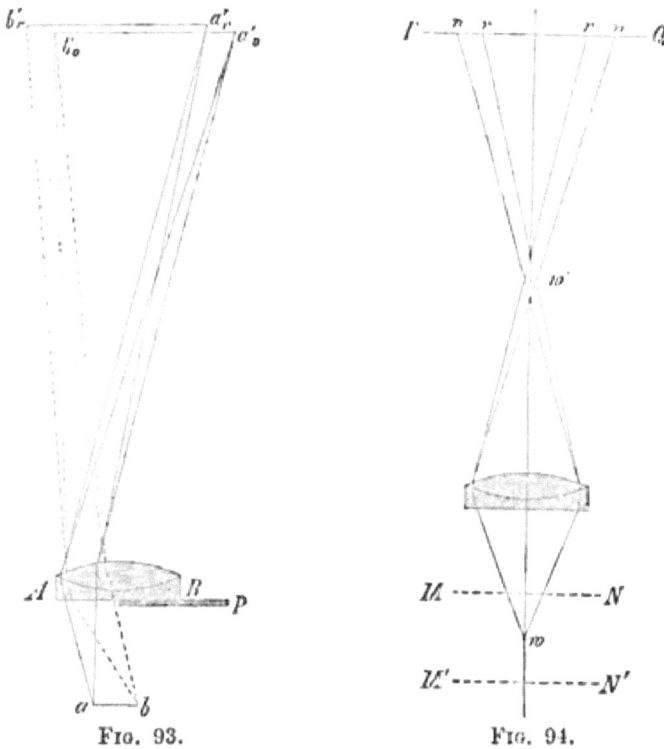

Fig. 93.   Fig. 94.

occupy the whole aperture of the objective, and the opposite aberration-colours of the right and left half unite to form white light. If we raise the focus of the Microscope to the plane $M N$, the conditions of the refraction of the rays are essentially altered. The real image now appears lower down in the body tube at w', and the eye-piece receives only the diverging rays which are projected as a circle of light on the image-plane $P Q$. Each half of the eye-piece, therefore, acts separately; a given sector of the circle of light is formed by a corresponding but opposite sector of the objective, and consequently shows also traces of the aberration peculiar to the latter. If the objective is under-corrected—that is, if the violet rays (v) are more strongly refracted than the red (r)—the circle of light has a blue border, and in the case of over-correction a red one, as shown by the construction.

A precisely analogous proof may be given, that if the Microscope is focused to the lower plane M' N', the opposite colours will appear; a red border indicating under-correction, a blue one, over-correction.

This method of testing is very simple, and it is theoretically accurate. The former method appears to give more definite results where minute traces of aberration are under consideration. To avoid the somewhat delicate operation of covering up portions of the lenses, a minute aperture in an opaque ground may be adjusted in the centre of the field of view; the mirror is then moved laterally out of the axis of the Microscope; the incident cone will then meet only that half of the objective which is opposed to the mirror.

In order to test severely, not merely the halves of the objective, but several of its concentric zones, either singly or variously combined, Abbe's method, mentioned above, is most efficient. A diaphragm having two separate apertures is adjusted so that one of the effective pencils of light in the posterior focal plane of the objective passes centrally through the opening, while the other touches its margin (cf. Fig. 92, right half). The examination of the zones which lie between the centre and the margin is effected by moving the diaphragm from right to left. This method has the additional advantage, that the phenomena of interference produced by small diaphragms are totally eliminated.

If we test various Microscopes by the method just described, we shall discover a decided under- or over-correction in the majority. Objectives which are achromatic for red and violet rays, and which therefore exhibit the colours of the secondary spectrum only, are, according to our experience, somewhat uncommon. In many Microscopes the coloured borders appear far more intense near the margin of the field of view than in the centre; examples are found where the aberration observed in one part of the field changes to the opposite aberration on viewing a more central part or the other side of the field. This obtains in a marked degree in

Bénèche's Nos. 7 and 9 objectives. Others are achromatic for the marginal parts of the field, though under-corrected for the central.

We have tacitly assumed that the coloured borders, which arise from the chromatic aberration of the objective, are seen by the eye unaltered. This supposition was admissible for a preliminary consideration where the question was chiefly relating to qualitative and not quantitative differences in the phenomena; but it is not, however, in any case strictly accurate, inasmuch as the refractions in the eye-piece, and in the eye of the observer, are accompanied by colour-dispersion. These points merit an exhaustive discussion.

As regards the eye-piece, it is a widely-spread error, which has become stereotyped in micrographic works and text-books of Physics, that the combined action of the field-lens and eye-lens compensates, to some extent, the defective achromatism of the objective. According to the usual representation we are directed to construct, by means of the rays of direction, the red and violet image produced by the objective, as well as the corresponding images produced by the field-lens and eye-lens (*i.e.*, of the rays which pass through the optical centre of the lenses), in which case it is, of course, always possible to give to the eye-lens such a position that the final virtual images are seen almost at the same angle, and must therefore coincide. This theory of the Campani eye-piece, as developed, for instance, in Mueller-Pouillet's "Physik," Harting's "Das Mikroskop," &c., is entirely erroneous, as will now be shown. The conclusions are accurate, but the hypotheses are wrong.

It has already been shown that the real image formed by the field-lens appears under altogether peculiar conditions, in consequence of which its construction cannot be found by simple drawing of the rays of direction. This obtains likewise for the virtual image of the eye-lens. Each point of these images is formed by a surface-element of the lens, different from the adjacent one. The further the image-point is from the centre of the field, the more remote also from the centre of the lens is the active part of the refracting surface. Moreover, the incident cones of light are so reduced where they meet the lens that they are hardly 1 mm. in diameter, and are therefore refracted almost as simple rays.

Having established the actual relations, which are ignored in the traditional method of description we will endeavour to explain the action of the Campani eye-piece by a construction, in which all essential factors are taken into account. Let A (Fig. 95) be the objective, B the field-lens, and C the eye-lens. We may take for the object the whole field of view, or a small aperture in an opaque plate, &c.; the condition necessary is that the emergent cones of rays occupy the whole aperture of the objective, or its central part. On this supposition, the differently coloured rays are evidently so refracted that the emergent cones corresponding to a given object-point have a common axis. The lines of direction in which the apices of differently coloured cones lie, are therefore determined; but the plane of these apices is unknown, and is dependent on the chromatic aberration of the objective. In our figure, the two symmetrical cones of light

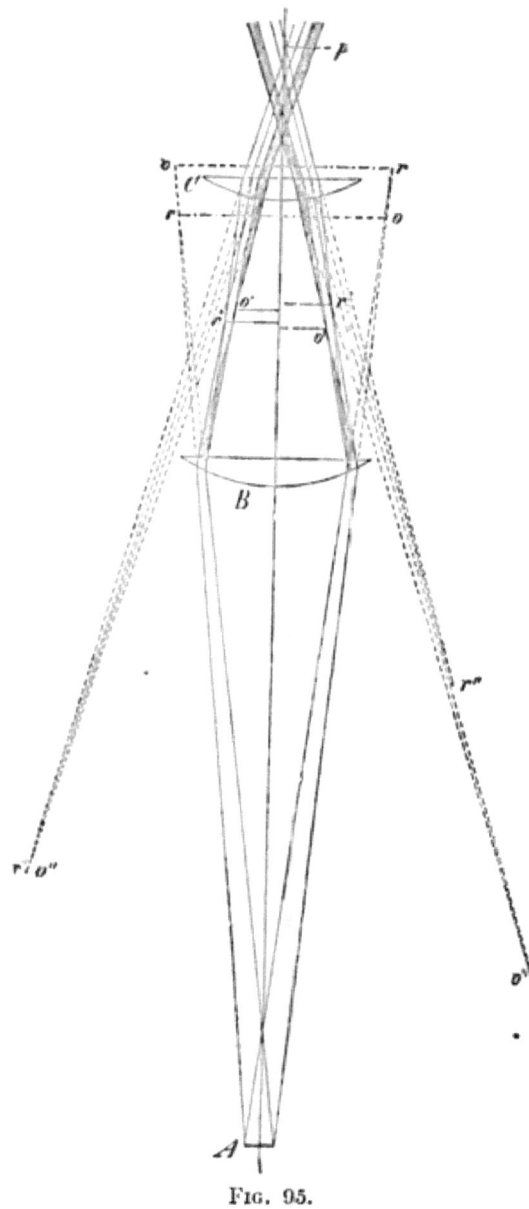

Fig. 95.

to the right and left of the middle line correspond to two opposite marginal points of the object. Since the refractions which they undergo are, of course, symmetrical also, it would have been sufficient to have made the construction on the one side and to have added the image-points to the other at equal distances. For more convenient comparison, two opposite cases have been placed side by side; on the left the path of the rays when the objective is chromatically over-corrected, and on the right when under-corrected to the same extent. The apices of the violet cones of rays are denoted on both sides by v, and those of the red by r; similarly, the corresponding points of the real image by v' and r'.

It is now clear that the red pencils of light undergo, in the field-lens, the same deviation on both sides, as they meet it also at the same angle; this obtains also with the violet pencils. The focal points o and p of similarly indicated pencils therefore, of necessity, lie in the axis. With regard to the position of the real image-point v' and r', it is clear that on the left r', and on the right v', is nearer to the field-lens than the similarly indicated image-point of the other side. The distance in the direction of the axis is, however, necessarily less than that of the two r's or of the two v's in our figure, therefore less than 6·5 mm. Similarly, it is clear that, with equal distance of the virtual objects v and r, the violet image v' will be at somewhat less distance from the field-lens than the red one r'. The real images on both sides must therefore be situated as represented in the figure.

The refraction in the eye-lens, as was above pointed out (Fig. 61 and the preceding), is dependent on its position. If the centre of curvature of the refracting spherical surfaces lies between the focal points o and p, the first refraction of both pencils takes place in the opposite direction, and a relation may be produced, which will lessen its divergence to any degree, or perfect parallelism may possibly be effected; the refraction in the opposite direction is not, however, an indispensable condition. The fact that the red rays meet the eye-lens at a greater distance from the centre than the blue rays, and are therefore also more strongly refracted on account of spherical aberration, is quite enough to cancel these differences of inclination, and might even reverse their direction. If we consider that the differences in deviation due to this are the more important the further the eye-lens is removed from the field-lens, it becomes intelligible that a gradual increase of this distance will at last bring the axes of the emergent pencils to convergence. In this case, however, if produced backwards, they meet the plane of the virtual images so that the red is on the outside, and the violet on the inside, of the field of view. The object has therefore a red or orange-coloured border. Conversely, too great an approximation of the eye-lens to the field-lens causes a blue border, because the divergence of the emergent rays is increased, and the violet pencils meet the field of view at a greater distance on the outside.

From what we have just stated, it follows that for every eye a particular position of the eye-lens may be found, at which the differently coloured emergent pencils, when produced backwards, will cross each other exactly at the distance for which it is adjusted. They then unite in one point of the retina to form white light, and the resulting image is free from coloured borders.

So far the conclusions we have drawn have a general validity, since the over-correction or under-correction of the objective does not here come into account. For complete achromatism it is, however, necessary that the virtual images of r' and v', which are last observed, should be situated at the same distance at which the corresponding cones of light intersect; in other words, the axes of the differently coloured cones of light must converge backwards to the same points, from which also the single rays of a cone appear to come. This condition cannot be fulfilled for the right and left sides of the figure simultaneously, because the distances of the similarly coloured real images from the eye-piece are not the same. In consequence of the chromatic aberration[1] of the eye-piece it is even indispensable that v' should be brought somewhat nearer than r', if the virtual images v'' and r'' are to coincide. The opposite arrangement, which accompanies the under-correction of the objective, is, therefore, incompatible with the conditions of

---

[1] The following figures may be of service for more exact consideration. A plano-convex crown lens, whose radius of curvature is equal to 15 mm., gives for the focal length of the violet rays 27·77 mm., for the red rays 28·84 mm.; consequently a longitudinal chromatic aberration of 1 mm. Similarly, with a radius of curvature of 25 mm., the corresponding focal lengths are 46·3 mm. and 48 mm. If the latter lens is used as the field-lens, and the former as the eye-lens, and if the distance of the virtual object from the field-lens is taken as 50 mm., the violet image of the latter appears at a distance of 24 mm., and the red image is $\frac{1}{2}$ mm. higher still. Since the chromatic aberration of the eye-lens requires exactly the opposite arrangement, the objective must be over-corrected to such a degree that the red rays converge to a point about 6 mm. lower.

achromatism, and to this extent the traditional method of explanation is well grounded. Yet it does not by any means follow that an under-corrected objective-image will receive a red or orange-coloured border by the dispersion in the eye-piece. For since the image-points v'' and r'' are the apices of the emergent cones of light, and therefore approach each other to a certain extent on approximation or withdrawal of the eye-lens (while the distances remain constant), it is always possible to give them such a position that they will coincide for the observing eye. In this case, however, the resulting image appears either perfectly colourless or bordered only by colours of the secondary spectrum.

The elimination of the blue or red coloured borders does not therefore afford any criterion of true achromatism; it is equally possible with under-corrected and achromatic objectives, as with over-corrected ones,—and so far the usual method of explanation is inaccurate.[1]

This explanation becomes quite untenable when the real image is formed by an excentric part of the objective-system, *e.g.*, by one half, which occurs most frequently in microscopical examinations. The defect of achromatism must therefore manifest itself, as was above pointed out, in the displacement of the different colours laterally, so that on the one side red, and on the other violet, preponderates. It is evident that under such circumstances a correction of the objective-image by means of the eye-piece is impossible. The coloured borders may certainly be somewhat modified, but never eliminated on both sides simultaneously.

The chromatic aberration of the objective cannot, as a rule, be eliminated by means of the opposite aberration of the eye-piece, except in certain particular cases. This is also unconditionally true with regard to spherical aberration. For since the incident cones of light in the eye-piece generally meet less than a square millimetre of the refracting surfaces, it can hardly be assumed —indeed it is almost impossible—that, with proportionally weak curvatures, considerable aberration, capable of elimination, could still be present, or that aberration already present could be compensated.

---

[1] The absence of the coloured borders shows only, according to what has been stated, that the field-lens and eye-lens form, in the effective peripheral zone, a combination which is aplanatic for red and blue light. The points of convergence of the differently coloured pencils do not here come into account.

We thus arrive finally at the result, that *the defects of correction, which are rendered apparent by the given method of testing (in which only the half of the objective is optically effective), are to be referred solely to the objective in all cases where they are readily seen.*

The second point we have to consider, namely, the chromatic aberration of the eye, may in principle be regarded as solved. That the eye is affected with a very marked chromatic aberration, may be easily shown by covering up half of the pupil and directing the gaze to any bright object. One margin will then appear distinctly blue, just as in the microscopic image, and the other orange or yellow. The only question remaining is whether this chromatic aberration can increase or compensate that of the Microscope to any considerable degree. To decide this point, it is only necessary to cause the pencils of light proceeding from the

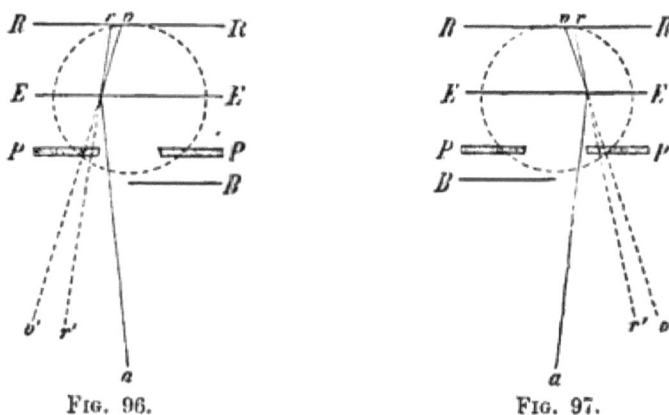

Fig. 96.   Fig. 97.

eye-piece to reach the retina through a part of the pupil as excentric as possible. The eye should be placed at about the height where these pencils of light cut the optic axis (at the so-called eye-point), and moved so far to the right and left that the microscopic image is only just visible. It will then be observed that the red and blue margins appear fainter or more intense, according to the position of the eye, and that the colours are often reversed and even disappear wholly. These phenomena may be explained by taking the action of the naked eye upon white light as the starting-point. If $E\,E$ (Fig. 96 and 97) is the principal plane of the eye, R the retina, and P the pupil, which is half covered in the manner indicated by the diaphragm B; a colourless object a must appear, under the conditions represented in Fig. 96, bordered on the left by blue, and on the right by red, because the violet image r of any object-point appears upon the retina more to the right, and is therefore mentally assigned to an object r' situated more to the left. If, on the other hand, the left half of the pupil be covered up, the arrangement of the coloured borders is reversed, as shown in Fig. 97. When, in a dioptric image, the same coloured borders are present in the same order as in refraction by the eye, this refraction is further intensified; while the converse of course diminishes, and under certain circumstances cancels it. It follows that the aberration of the eye may, in all cases where lateral displacement of the differently coloured image-points is concerned, be just as well attached to an under-corrected as to an over-corrected objective-image, since the arrangement of the colours varies according as the right or left half of the objective, or of the eye, is effective. If, on the other hand, chromatic aberration is due partly or exclusively to differences of plane of the differently coloured images—as, for instance, in the testing of optical power with the meshwork—it is evident that the eye is capable of lessening or compensating only the contrary aberrations. This fact may be worthy of some attention in practical Optics; it is not, however, decisive, because with organic objects the lateral displacements are the most frequent, and do not admit of elimination, since they take place in the most different directions.

The most favourable combination for the retinal image must therefore be a microscopic image free from aberration as the object, and the usual vision through the central part of the pupil; and we conclude from these discussions that the best objective is that in which the spherical and chromatic aberration is corrected as far as possible.

Whether the manufacture of achromatic or slightly over-corrected eye-pieces would compensate for the outlay involved, we will leave undecided. It is evident that, by the most favourable combination of the eye-piece, the microscopic image gains only in certain subordinate points, such as extent of field of view, &c.

IV.

TESTING THE FLATNESS OF THE FIELD OF VIEW.

As the traditional expression, "curvature of the field of view," includes, as we have above shown, two independent notions, viz., (1) the distortion of the image, and (2) the curvature of the image-surface, they must be considered separately in the testing of the Microscope. As regards the distortion of the image, its extent may be determined by viewing a rectangular network (*e.g.*, a glass micrometer divided into rectangular spaces) under the Microscope. If the enlargement increases or decreases from within outwards, the image will appear as in Fig. 24 or Fig. 23 respectively; but if no distortion takes place, the meshes of the image will coincide with those of the object. From the curvature of the lines in Figs. 23 and 24 it also follows that a single straight line, if brought near to the margin of the field of view, will form an excellent test-object.

We may test the behaviour of the images, which the objective forms alone, or with the aid of the field-lens, as follows:—A cover-glass with a straight edge is placed upon the diaphragm in the eye-piece setting (the eye-lens being removed) so that the edge appears in the circular diaphragm as a chord of arc (*m n* in Fig. 98). The real image of another straight line, which is viewed as

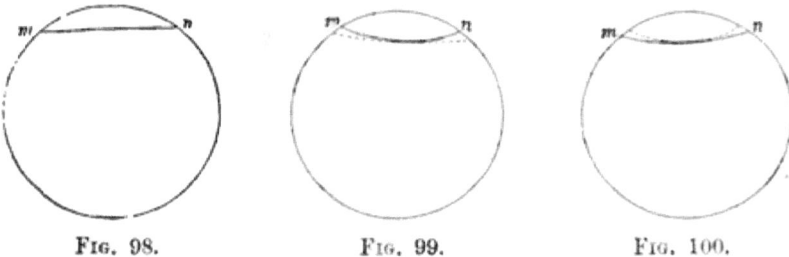

Fig. 98.   Fig. 99.   Fig. 100.

an object, is made to coincide with this chord precisely as the adjustment of a particular division-line to the margin of an object is effected in micrometric measurements. If a complete coincidence takes place—whether or not they appear straight or curved in the virtual image—then the real image is free from distortion; in all other cases it is distorted.

A distortion as in Fig. 23, where the marginal parts appear less enlarged than those in the centre, may be distinguished thus: when the two lines are in contact in the centre, the ends of the straight line *m n*, which is in the eye-piece setting, are bent outwards, while the opposite distortion is indicated by the deviation of these ends inwards (Fig. 100). From the curvature of the straight line *m n* the curvature of the virtual image which the eye-lens forms of a flat surface, under the given circumstances, may be estimated.

The instruments we have examined in the manner indicated (by Oberhaeuser, Hartnack, Bénèche, Merz, and Kellner) showed but slight distortion of the objective-image, when examined either with or without the field-lens. In the latter case the final virtual image was more distorted, because the eye-lens was optically effective in its whole diameter and with far greater aperture, and thus the direction of the incident rays was more conducive to this effect. The same result was also obtained by another method of testing, in which the eye-piece being removed, the objective-images, produced with or without the field-lens, were received at the upper end of the body-tube on a screen of ground glass.

In order to test the *curvature of the image-surface*, we examine fine markings on a plane surface (for instance, the markings produced by pressing the tip of the finger on a glass slip), and observe whether all points of the field of view can be clearly seen at the same time. If we find, as we generally do, that the peripheral points require a lower focal adjustment than the central ones, then the image-surface is evidently convex to the upper side. The curvature is, of course, the greater in proportion to the difference of plane.

The application of a cover-glass with similar markings on the diaphragm of the eye-piece, will also exhibit the curvature of the objective-image according to the same principles which determine the distortion of the image.

V.

TESTING THE CENTERING.

WE have shown that imperfect centering of a system of lenses manifests itself generally in two ways: (1) by producing indistinctness in the images, which first affects the margin of the field of view, and only in the case of greater deviations extends to the centre—in all cases, however, decreasing towards the centre; (2) by the changes of position which any fixed image-point undergoes if the objective or one of its component lenses is rotated on the optic axis. The first point may be tested by determining the optical power of the Microscope for different positions of the field of view, and comparing the results. It must be specially noticed whether the left margin gives decidedly more indistinct images than the right, and whether there is inequality in the distribution of optical power, and in its decrease from the centre towards the periphery. If these differences are not noticeable, or but a slight loss towards the margin can be distinguished, conditioned by the aberrations of the eye-piece and therefore proportional to them, we may be sure the errors of centering are as far as possible corrected, *i.e.*, are reduced to zones which are situated outside the field of view, or project only very slightly over the borders.

Regarding the second point, it is usually stated in micrographic works, that the changes of position in question are apparent on revolution of the objective-lenses by unscrewing the mounting; it is therefore only necessary to adjust the Microscope on any small object, so that the image may touch the margin of the field of view or a line of the eye-piece micrometer; imperfect centering would then be shown, on unscrewing the lenses, by a corresponding circular movement of the microscopic image.

We mentioned this method of testing in our first (German) edition, adding further details with regard to the factors upon which the amount of the displacement is dependent. We then stated that "the movements of the image observed during the revolution of the objective lenses by unscrewing are dependent upon the accuracy of the construction of the mountings and screw-threads quite as much as upon the deviations of the optic axes before the unscrewing"—a proposition the theoretical accuracy of which is unquestionable. Since that date we have become convinced that the process amounts to nothing more than a testing of the mountings. The errors of centering, which still occur in the mounting of the lenses, even in the best workshops, are so small in comparison with the excentricity of the screw-thread, that they play only an insignificant part in the movements of the image. If then the lenses, independent of the mountings, are to be tested for centering, their unscrewing is evidently detrimental to the purpose. The only efficient process of testing the centering is that applied in practical Optics as follows:—

The metal cell of the back combination, which we will assume to be concentric, is attached to a lathe, so that its own axis approximately coincides with the revolving axis. The accurate testing of this coincidence may be effected by means of a so-called "lever of contact," the shorter arm of which rests, during the revolution, upon the free surface of the cell, while the longer one projects freely backwards, and in case of inaccurate centering exhibits the movements of the short arm on an enlarged scale. When the cell is adjusted satisfactorily concentric, the lens is temporarily inserted and moved variously, until the errors of centering are eliminated as completely as possible. These errors are shown by the motion of the small reflected image which the lens forms of a window-frame or lamp-flame. The optician observes this image with a magnifying-glass, and moves the lens in its cell until the image remains perfectly still during the revolution. Then the projecting metal edge of the cell is burred upon the lens, by which it is permanently mounted. The second and third combinations are tested similarly after the cells have been adjusted.

In the testing of completed objectives, we screw the brass-work on the lathe, and commence with the back combination alone, and so on, adding the others. When in each case the reflected image remains motionless during the revolution, the centering is practically perfect.

VI.

DETERMINATION OF THE ANGLE OF APERTURE.

Since the power of discrimination of Microscopes increases and diminishes with the aperture, as was shown above, the latter belongs to the factors which must be taken into consideration in testing instruments. Its determination may be made in very different ways; yet all methods which have been proposed by the microscopists are essentially equivalent. They cannot possibly have any other basis than the easily understood phenomena, which have reference to the known path of rays in the Microscope; one diagram will therefore suffice to verify and elucidate the most different methods of testing.

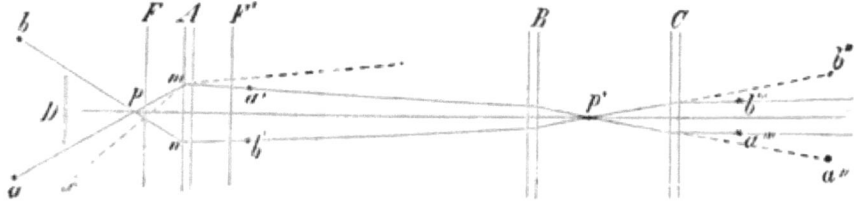

Fig. 101.

Let A, B, and C (Fig. 101) be the three pairs of principal planes of the objective, field-lens, and eye-lens respectively, and let F and F' be the focal planes belonging to A, and p the centre of the field of view. If, then, $m\,p\,n$ is the angle of aperture of the objective, an illuminating cone of equal aperture is, of course, necessary to fill it entirely; it is also evident that any small portion of the cone will illuminate the surface-element p, according to the degree of its intensity. If, therefore, a source of light of slight extent (whether it be a diaphragm illuminated by the mirror, or the illuminating apparatus itself, or the distant flame of a lamp, &c.) is moved in the direction from a to b, or conversely from b to a, the object p appears to be illuminated as long as the source of light is situated wholly or partially within the space limited by the angle $a\,p\,b = m\,p\,n$. Outside this angular space the rays directed towards p do not reach the objective, or are so refracted that they do not reach the eye-piece; the field of view consequently appears dark. (In the figure this case is indicated by the dotted line $x\ m$.) The like result would obtain if a circular diaphragm D, entirely within the illuminating cone, were brought near to the object p, until its edges touched the bounding surface of the cone.

The real image is formed between the field-lens and the eye-lens. From this point the rays again diverge, and are refracted by the eye-lens, so that they appear to come from a point in the optic axis, which is at the distance of distinct vision from the eye. If the eye is adjusted to infinite distance the rays are parallel to each other and to the axis. Conversely, if a pencil of parallel rays is incident on the eye-piece in the direction of the axis, the rays first intersect in p' and then in p, and thence diverge as a solid cone between $p\ a$ and $p\ b$. A screen placed on the left of p at right angles to the axis is therefore illuminated to an extent which is dependent upon the distance from p and the angle of aperture of the objective. If the screen has the form of a circular arc described about p, and is divided into degrees, the angle of aperture may be read off.

If the source of light a, placed at some distance, is regarded as the object, its objective-image will appear somewhat behind F'' in a'. Of this image, the field-lens would form a second one at a''; but the eye-lens takes up the rays before their union, so that the actual image is now formed at a'''. If the light is moved from a to b, the image passes from a''' to b'''; it disappears, however, as soon as the limiting line $b\ p$, or the opposite one $a\ p$, is exceeded. The diminution of the image may readily be calculated for given distances and focal lengths. If, for instance, the distance of the light from the objective = 1 metre, and the focal lengths of the objective, field-lens, and eye-lens respectively, 5, 50, and 25 mm., and the length of the eye-piece setting 50 mm., we get, with an ordinary tube-length of about 200 mm., a diminution of about 800 linear.

On the basis of these theoretical deductions sundry practical methods of measuring apertures have been devised, of which we will describe the most simple and efficient.

(1.) *Lister's Method.*—The Microscope is placed horizontal, and a lamp is adjusted at some distance from the objective in a darkened room. The lamp is then moved to the right and left of the axis, until only half the field of view appears illuminated. The angle at which the extreme positions of the lamp (*a* and *b*, Fig. 101) are seen, from the focal point of the objective, is the angle of aperture sought. It is therefore only necessary to place the Microscope in a suitable position, draw lines to the two extreme points right and left occupied by the lamp, and measure the included angle with the protractor.

The method is still simpler, if the lamp (a few feet distant) remains fixed, and the Microscope is revolved round a vertical axis passing through the focal point, so that the end of the eye-piece describes a horizontal arc. The result will be precisely as in the previous case. The Microscope is turned first to one side and then to the other, in each case until half the field is dark; in these extreme positions lines parallel to the axis are drawn and produced till they meet at the focal point; the angle which they include is equal to the angle of aperture, and may be measured directly by the protractor. The result is practically the same, if the revolution does not take place exactly on the focal point, but merely on the anterior end of the Microscope.

If a special mechanical contrivance be desired for such measurements, Goring's apparatus is good. This consists essentially of a circular brass plate upon which the Microscope is placed, and which rotates upon a fixed plate that is graduated on the circumference; a pointer on the rotating plate enables the observer to read off the angle on the graduated plate. A more detailed description, with illustrations, is given by Mohl in his "Mikrographie," p. 193.

(2.) *Wenham's Method*.[1]—The Microscope is placed horizontal and directed towards a distant source of light, *e.g.*, the flame of a lamp. Instead, however, of observing the field of view in the usual way, the image of the flame is examined with a magnifying-lens above the eye-piece (a''' and b''', Fig. 101). The Microscope is then rotated in a horizontal plane to the right and left, till a certain point of the image—for instance, its centre—just disappears. The arc, which the Microscope thus describes, may be regarded as corresponding to the angle of aperture of the objective.[2]

---

[1] Mr. Wenham ascribes this method to Amici, *vide* "Quart. Journ. Micr. Sc." ii. (1854), p. 209.—[ED.]

[2] The remark of Harting ("Das Mikroskop," 1st ed. p. 263), that this

This method has the great advantage, that not only is the aperture of the objective determined (that is, as far as it admits light), but also the extent of the really effective part, *i.e.*, the part which produces images free from aberration. It is, indeed, frequently observed that the images are indistinct and distorted before they disappear, while they seem otherwise sharply outlined and in their true shape. These indistinct and distorted images are, of course, produced by the marginal rays, for which the objective is no longer sufficiently corrected, and which are therefore better cut off.

The really effective part of the angle of aperture is consequently determined by the limits within which the flame appears sharp and clear.

(3.) *Robinson's Method*.[1]—A pencil of parallel rays is projected in the direction of the axis through the eye-piece, and, after the crossing of the rays in the focal point of the Microscope, is received on a screen.[2] From the diameter d of the circle of light which is here formed, and from the distance l of the screen from the focal point of the Microscope (or the anterior end of it, if

---

method could not yield quite similar results to those of Lister, because by the addition of a magnifying-glass the principal focal point was displaced, is at any rate inaccurately explained. It is, on the contrary, quite immaterial whether the images a''' and b''' are viewed with the naked eye, with a magnifying-glass, or with a second Microscope. Their disappearance in no way depends upon the additional system of lenses. Nevertheless, a slight difference does exist, though the cause is not as Harting supposes. With Lister's method the pencils of rays are refracted in the margin of the Microscope so that they cut the axis in p'; it is only on this condition that the centre of the field of view is seen illuminated. In Wenham's method, on the other hand, it is sufficient that the pencils reach the field-lens after refraction in the objective—a condition which is still fulfilled if their path is at some distance from the axis. The angle of aperture is somewhat larger than it would be if the incident cone of light emanated from the focal point of the objective.

[1] *Vide* "Proceedings of the Royal Irish Academy," v. (1858), pp. 38—47, "On a new method of measuring the angular aperture of the objectives of Microscope," by the Rev. T. R. Robinson, D.D., President of the Royal Irish Acad., communicated Jan. 23, 1854; partly reprinted in "Quart. Journ. Micr. Sc." ii. (1854), p. 295.—[Ed.]

[2] Dr. Robinson stated that for measuring very large apertures he "was obliged to modify the method by receiving the light on a screen made to travel in a cylindric surface concentric with the focal point," which he considered a decided improvement. *Vide* "Quart. Journ. Micr. Sc." iii. (1855), p. 164.—[Ed.]

the distance is proportionally great), we obtain the angle of aperture ω according to the formula

$$\tan \frac{\omega}{2} = \frac{d}{2\,l}.$$

Its determination may also be obtained by construction with sufficient accuracy.

(4.) *Abbe's Method.*—Under the object-stage, at a known distance from the plane of adjustment, a horizontal scale is applied, whose zero point is approximately in the centre of the field of view. Pointers are moved along the scale to the right and left until they touch the margin of the bright circular aperture-image in the upper focal plane of the objective. The positions of the pointers are now read off, and the arithmetical mean taken, by which the possibly somewhat excentric position of the zero point is corrected. We thus obtain the linear distance of the extreme

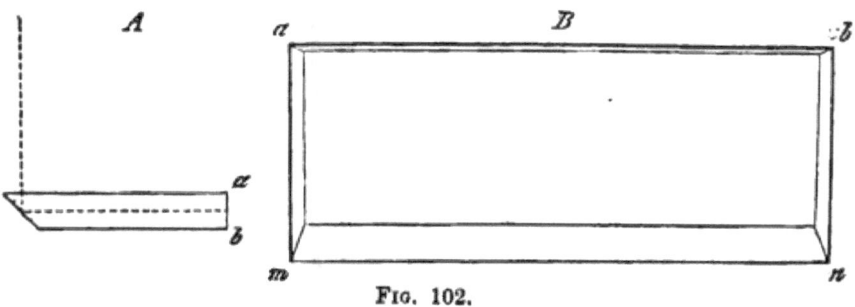

FIG. 102.

marginal points shown in the focal plane of the objective; and since this distance is proportional to the sine of the angle of aperture, the latter can at once be determined by reference to a table, in which the angular values corresponding to the different distances are once for all computed.

In order to obtain a favourable illumination for greater distances, Abbe's scale is applied to the side $a\,b$ (Fig. 102 A and B) and to the two adjacent sides $a\,m$ and $b\,n$ (Fig. 102 B) of a rectangular plate of glass,[1] and is viewed from above by the reflexion at the surface $m\,n$, which is inclined at an angle of 45°. By continuing the graduations of the scale upon the two sides at right angles to $a\,b$,

---

[1] This apparatus is now well known as Abbe's Apertometer, and is made semicircular in shape. *Vide* "Journ. Royal Micr. Soc." i. (1878), p. 19, art. by Mr. C. Zeiss; and iii. (1880), p. 20, art. by Dr. E. Abbe.—[ED.]

the size of the plate is rendered more convenient, and the illumination is stronger.

Whichever method is employed, the determination is but approximately accurate, and with higher powers the possible error may attain two degrees. It follows that light and shadow in the field of view, or on the illuminated screen, do not appear sharply defined on account of the various losses which the marginal rays suffer—in fact, they pass gradually into each other, so that the limiting points must always be somewhat arbitrarily chosen. This disadvantage is not practically important, since it is perfectly immaterial whether the angle of aperture of a Microscope amounts, for instance, to 70° or only 69°. It is absurd, as Harting justly remarks, to record the angle of aperture, as many have done, up to a fraction of a degree. And it is just as absurd as it is unpractical to manufacture objectives with angles of aperture of 160° and upwards, if at least 40° to 50° belong to a totally useless peripheral part of the system, as obtains in certain objectives of English manufacture.

Finally, we remark, that in those objectives which are provided with correction-adjustment, for use with immersion, or to compensate for different thicknesses of the cover-glass, the angle of aperture is necessarily altered if the distance of the lowest lens from the next one is increased or diminished, because an alteration of the focal length of the whole system of lenses is produced thereby. On the other hand, the influence which the eye-piece magnification exercises upon the angle of aperture is by no means so considerable as would be supposed from the representation of Harting. The magnification itself does not come into the consideration, but merely the fact that with shorter eye-pieces the distance of the field-lens from the objective is somewhat greater, which, as a rule, necessitates also a slight increase of the posterior focal length of the latter, and consequently a diminution of the object-distance. The angle of aperture is consequently somewhat increased; but it is equally clear that the refractions, which take place after the formation of the real image, do not alter the angle of aperture, since the path of the rays in the objective is quite independent of them. With a given position of the objective-image it is therefore quite immaterial whether the eye-piece magnifies five or fifty linear.

VII.

DETERMINATION OF THE MAGNIFYING POWER AND FOCAL LENGTH.

It is customary to estimate the power of a Microscope by the total magnification, but that of objectives by their focal lengths. However little the former may be a safe criterion, since it is determined by factors of very unequal values, yet a knowledge of it is indispensable even from a practical point of view. The focal length, on the contrary, stands approximately in inverse ratio to the objective amplification; it may to a certain extent be regarded as an expression of the magnification not dependent upon the tube-length, and for the comparison of different objectives furnishes, therefore, the requisite standard. The determination of the focal length, as well as of the total magnifying power, is therefore essential in a complete testing of a Microscope.

1.—Magnifying Power.

We have already shown that the magnifying power m is proportional to the distance of the virtual image from the second focal point, or, what is nearly equivalent, from the eye-point of the Microscope, so that twice the distance corresponds to double the magnification. If, therefore, the magnifying powers are to be comparable *inter se*, they must logically be referred to equal distances. The distance chosen is entirely immaterial to the comparison. In recent times it has been customary to calculate the magnifying power for a distance of 25 cm., or 10 inches.

The so-called "curvature of the field of view" involves the fact that the amplification of the marginal parts is usually somewhat greater than that of the central ones; hence an accurate determination of the magnification is only possible for a relatively small part of the field of view; from well-known practical considerations it is, as a rule, determined only for the centre, omitting a tolerably broad peripheral zone.

The methods which are employed for the determination of the magnification, all amount to the same thing, viz., the projection of the image of an object of known magnitude to the conventional distance of 25 cm., and the direct comparison of its diameter with that of the image-forming object. The projection may be made by means of Sœmmerring's mirror or a reflecting prism in which the reflecting surface must be situated so that the second focal point of the Microscope shall coincide with the focal point of the rays in the eye of the observer. Let N^* (Fig. 103) be the last refracting surface (the plane surface of the eye-lens), F^* the second focal point, and s a reflecting surface inclined 45°; the path of the rays will thus be altered as if the optic axis ($o\,x$) of the system were horizontal, and the focal point brought to $(F)^*$. The eye must

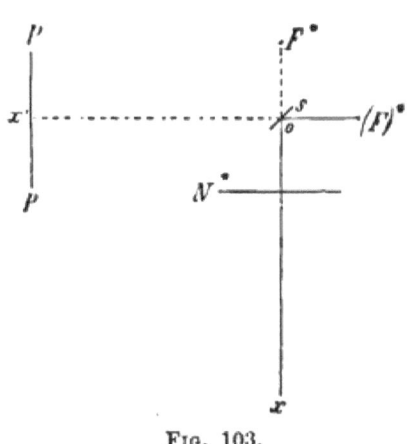

Fig. 103.

therefore be placed at this latter point, if the results of the measurements are to be exact and capable of comparison. Its position corresponds very nearly to that in which the emergent pencils form the smallest circle of light on a screen of ground glass placed at right angles to the axis. It is, moreover, evident that the plane upon which the image is projected ($P\,P$, Fig. 103) must form a right angle with the direction of the optic axis as altered by the reflexion.

In selecting an object, it is essential that its margin, or rather the points whose distance is required, be clearly seen in the image. A glass micrometer is convenient for this purpose; but if the lines appear very indistinct, as generally occurs with the higher magnifications, other objects with sharp margins (for instance, air-bubbles, globules of mercury, crystals, &c., whose diameter can be micrometrically determined) should be employed. The accuracy of the determination is dependent in every case upon the exactness of the divisions of the micrometer.

The measurement of the virtual image on the plane of projection is most easily effected by means of drawing-compasses, which can then be applied directly upon a finely divided scale. The limiting lines can also be drawn repeatedly upon the plane of projection with the fine point of a pencil, and the mean value reckoned as the diameter of the virtual image. If this diameter amounts, for instance, to 26·5 mm., and that of the object to 100 mic. = ·1 mm., the magnifying power $m = 265$.

Where an accurate determination of the magnification is required, all the conditions which theory presents must be satisfied as far as possible. The magnitude of the object should then be determined in a more reliable manner than is possible by means of the micrometric divisions (which are not always exact). For this purpose, Harting proposes to wind a thin metal wire some hundred times round a thicker wire so that each revolution exactly touches the preceding one—which must be verified by the Microscope. We then measure the space which the whole of the revolutions occupy on the thicker wire, and finally count the number of revolutions, by unwinding it on a lathe. The total length of the windings, divided by their number, gives the thickness of the wire with an accuracy which is not attainable by micrometric measurements. Such a wire can either be viewed as the object, or used for testing the exactness of the micrometric divisions.

In practice, however, such extreme care is generally superfluous. It is unimportant whether the magnification differs by a few units, more or less—whether it is, for instance, taken as 360 or 355—for the objects to be measured and represented always differ in magnitude by more than this amount. We need not, therefore, pay much regard to the second focal point of the Microscope, the position of the eye, the accuracy of the divisions of the micrometer, &c. ; in general it will suffice if we measure the distance of distinct vision starting from the middle of the eye-piece, and bestow the usual amount of care upon the determination of the relative magnitudes.

The magnifications, which we obtain by the different objectives with the same eye-piece, are, of course, proportional to the linear dimensions of the *real* objective-images. If, for instance, the image of one objective covers ten divisions of a micrometer in the eye-piece, and that of another objective fifteen divisions, the magnifications are in the ratio of 10 : 15. If, conversely, a micrometer-division is viewed as the object, the number of the divisions which are seen within a given space (for instance, the field of view contracted by a small diaphragm) is inversely as the magnification, and the ratio is again proportional. These relations afford us a means of determining by calculation the values for different objective-systems in connection with the same eye-piece, if this value is known for a single one of them. If, for instance, the numbers of the divisions of a stage-micrometer, which are seen in the diminished field of view with four different objectives, are respectively, 60, 45, 20, and 12, and if the total magnification m obtained with the first $= 50$, with the other three it will amount to $\frac{60}{45}$, $\frac{60}{20}$, and $\frac{60}{12}$ times 50, that is $66\frac{2}{3}$, 150, and 250.

It is in many cases advantageous to know not only the magnifying power, but also the exact diameter of the field of view. Its determination is effected simply by measuring the apparent diameter on the plane of projection at the distance of distinct vision, and dividing by the magnification. The measurement may also be effected directly with a micrometer, which occupies the whole of the diaphragm in the eye-piece, or which is movable laterally.

Similarly, we may require to know the objective-amplifications. The field-lens is removed from the eye-piece micrometer, which is then adjusted until the given objective-image is clearly seen. As, however, the divisions of the micrometer are at a known distance, generally ·1 mm., the diameter of the real image may be measured thereby. The ratio of the latter to the diameter of the object evidently gives the objective-amplification.

2.—Focal Length.

In the determination of the focal length, special attention is required in the application of the mathematical formulæ given in the Introduction, in order to obtain a result as accurate as possible. Since the principal points of the objective-system are unknown, the direct measurement of the focal lengths or of the conjugate foci is impossible. The formula $\frac{1}{p} + \frac{1}{p^*} = \frac{1}{f}$, in which p and p^* denote the conjugate foci, and f the principal focus, cannot be immediately applied. We may assume that we know the relative magnitude of image and object, and one of the two foci, provided that it is taken so large that merely an approximate measurement is sufficient. From these data, however, the focal length may be very easily calculated. Let D and d, where $D > d$, be the diameters of the image and the object, or conversely, and p^* the longer of the two foci; then we obtain, from the proportion $D : d = p^* : p$, and from the above equation $\frac{1}{p} + \frac{1}{p^*} = \frac{1}{f}$, the relation $\frac{1}{f} = \frac{1}{p^*} \cdot \frac{D+d}{d}$, and hence the simple formula

$$f = p^* \cdot \frac{d}{D+d},$$

or, if d is infinitely small as compared with D,

$$f = p^* \cdot \frac{d}{D}.$$

The determination of p^*, d, and D can be accomplished as follows:—

(1.) A suitable object of known size (for instance, a micrometer-division, or an air-bubble, whose diameter can be accurately measured) is adjusted in the middle of the field of view, and then the objective-image is received on a screen of ground glass, which is placed on the body-tube after the removal of the eye-piece. If the correct adjustment is found at which the image appears sharply defined, its magnitude can be measured with sufficient accuracy with the compasses. The measurement is more accurate if we observe the objective-image through an eye-piece micrometer from which the field-lens is removed.[1] Since, however, the divisions of the eye-piece micrometer are, as above mentioned, at a known distance, generally of ·1 mm., the diameter of the real image can be determined within a small fraction of a millimetre. Finally, we measure the distance p^* of the plane of projection from the objective (really from its second principal point) with the compasses or scale, in which an error of 2 to 3 mm. is of no importance.

If, for instance, $d = 100$ mic. $= \cdot 1$ mm., $D = 7\cdot 8$ mm., and $p^* = 210$, then $\frac{d}{D}$ will equal $\frac{1}{78}$; consequently $f = \frac{210}{79} = 2\cdot 658$. If p^* is taken as 207, this value falls to $2\cdot 620$, differing, therefore, only by ·038, or about $\frac{1}{26}$ mm.

(2). The objective is placed upon the stage with its plane

---

[1] We assume here also an ordinary Campani eye-piece. In a Ramsden eye-piece it is not, of course, necessary to remove the field-lens.

anterior surface upwards, a suitable object (for instance, a slip of glass blackened over a candle-flame) is adjusted on the work-table or on the window at a distance of ·2 to 1 metre or more. The illuminating mirror, which must be plane, is adjusted to reflect the image of the object in the optic axis of the objective to be tested, which consequently forms a small dioptric image above; the diameter of this image is directly measured with suitable amplification. The magnitudes D and d are now known; the distance p^* is then measured with the scale, to which the distances from the object to the mirror and thence to the objective must be added. If necessary, a horizontal position can be given to the Microscope, and the object may be viewed directly.

If the focal length of one objective has been accurately determined, that of another can be readily found by comparison. From the formula for the magnification

$m = 1 - \dfrac{p^*}{f}$ we obtain $f = -\dfrac{p^*}{m-1}$, or, if m is regarded as positive, $f = \dfrac{p^*}{m+1}$, that is, the focal lengths are in inverse ratio to the objective amplifications increased by 1. Since the latter are proportional to the linear dimensions which the images of a given object occupy in the eye-piece micrometer, we have only to read off the divisions from one end to the other, in order to determine the ratio of the known focal lengths to one or more unknown, which may thence be calculated.

The focal lengths of single lenses may, of course, be determined similarly. Where, however, the curvatures are very shallow, and the focal lengths amount to several centimetres, it is generally sufficient to measure the distance of the image of the sun from the lens, or one of the simple methods described in the text-books of Physics may be employed.

VIII.

DETERMINATION OF THE CARDINAL POINTS.

LASTLY, the problem remains to be solved of examining a given objective with regard to the position of the cardinal points, both of the separate double-lenses and also of the whole system, *i.e.*, to discover empirically the distances of the principal and focal planes from each other and from the refracting surfaces in a manner analogous to that which we have above obtained, under definite assumptions, by calculation. In order to solve this problem with approximate accuracy, the following method may be adopted:—

(1.) The focal lengths of the separate double-lenses, then those of the two posterior lenses of the system, and, lastly, that of the whole objective, are determined. For greater accuracy it is advisable to make repeated measurements with higher powers, and with a different scale, &c.

(2.) We determine the distances of the first and second focal points from the anterior and posterior surfaces of the double-lens, or of the system. For this purpose we place it upon the stage with the surface in question turned upwards, and adjust the plane mirror of the Microscope so that the image of a distant object (*e.g.*, of a tree, or a cloud, &c.) is visible in the focal point of the combination. The Microscope being focused as sharply as possible, we mark a fine line on the body-tube at the upper or lower edge of the socket in which it slides, and then lower the focus until the surface of the system to be investigated (which is turned upwards) appears in the field of view; to aid the focusing, the surface may be slightly marked with the tip of the finger. This position of the body-tube is registered by marking a fine line. The distance of the two lines, which is obviously the magnitude required, can be micrometrically determined, either by the compasses, or, if very small, by a second Microscope of low power.

(3.) We measure the thickness of the double-lenses or of the system—that is, the distance between the first and last refracting surfaces. Since the anterior surfaces of the double-lenses generally lie exactly in the same plane with the edge of the cell, the measurement is most easily made in the manner above described by placing the lens with its anterior surface upon the stage, and focusing the Microscope successively upon the surfaces. Where the cell projects so much that it cannot be neglected, it is determined subsequently, and subtracted.

From the data which these measurements furnish, the position of the principal and focal planes can be calculated by simple addition and subtraction. If we denote by N_0, N_1, N_2, N_3, N_4, N_5, the successive surfaces of the three double-lenses; by ϕ_0, ϕ_1, ϕ_2, their focal lengths; and e^0 i^0, e' i', e'' i'', their pairs of principal planes; and further by (f) the focal lengths of the two posterior lenses of the system; and by (E), $(E)^*$, (F), $(F)^*$ the principal and focal planes of this system; similarly, by f the focal length of the whole objective; and by $E\ E^*$, $F\ F^*$, its principal and focal planes; and finally, by d_1 and d_2, in any given combination, the distance of the corresponding focal points from the anterior and posterior surfaces: the abscissæ of the different principal planes, reckoned from the front backwards, are given by the following equations:—

$$e^0 = N_0 + \phi_0 - d_1 \qquad i^0 = N_1 + d_2 - \phi_0$$
$$e' = N_2 + \phi_1 - d_1 \qquad i' = N_3 + d_2 - \phi_1$$
$$e'' = N_4 + \phi_2 - d_1 \qquad i'' = N_5 + d_2 - \phi_2$$
$$(E) = N_2 + (f) - d_1 \qquad (E)^* = N_5 + d_2 - (f)$$
$$E = N_0 + f - d_1 \qquad E^* = N_5 + d_2 - f$$

and in like manner we obtain

$$E^* - E = N_5 - N_0 + d_2 + d_1 - 2f$$
$$N_0 - F = f - (E - N_0).$$

The relations, determined at the commencement, between the cardinal points of two systems intended to be combined, and those of the resulting systems, serve here as a check. We obtain, by means of evident substitutions and transpositions,

$$(f) = \frac{\phi_1\,\phi_2}{\phi_1 + \phi_2 - (t)},$$

in which $\qquad (t) = e'' - i' = \phi_1 + \phi_2 - \dfrac{\phi_1\,\phi_2}{(f)};$

and similarly, for the whole objective-system,

$$f = \frac{\phi_0\,(f)}{\phi_0 + (f) - t},$$

in which $\qquad t = (E) - i^0 = \phi_0 + (f) - \dfrac{\phi_0\,(f)}{f}.$

The magnitudes $N_0 - F$ and $E^* - E$ are determined by the equations

$$e^0 - F = f\left[1 - \frac{t}{(f)}\right],$$

$$E^* - E = \left[(E)^* - e^0\right] - ft\left[\frac{1}{\phi^0} + \frac{1}{(f)}\right],$$

which are generally applicable for any two systems of cardinal points by substituting the corresponding values.

As an example of a determination of cardinal points, we give the following numerical data, which we obtained for a No. 9 objective of Bénèche. The focal lengths, focal distances, and surface-distances, expressed in millimetres, were as follows:—

(1.) *Focal Lengths.*

$\phi_0 = 2 \cdot 6 \qquad \phi_1 = 12 \qquad \phi_2 = 11 \cdot 4$
$(f) = 6 \cdot 24$
$f = 3 \cdot 15$

(2.) *Focal Distances.*

For the first lens ... ... $d_1 = 1 \cdot 75 ; d_2 = 2 \cdot 8$
„ second „ ... ... $d_1 = 9 \cdot 5 \ ; d_2 = 13 \cdot 5$
„ third „ ... ... $d_1 = 9 \cdot 4 \ ; d_2 = 11 \cdot 4$
„ second and third lenses $d_1 = 2 \cdot 86 ; d_2 = 5 \cdot 8$
„ whole objective ... $d_1 = \cdot 43 ; d_2 = -1 \cdot 05$.

(3.) *Surface-Distances.*

$N_1 - N_0 = 1 \cdot 37 ; N_3 - N_2 = 2 ; N_5 - N_4 = 2 ;$
$N_5 - N_2 = 5 \cdot 15 ; N_5 - N_0 = 7 ;$

and therefore

$N_2 - N_1 = \cdot 48 ; N_4 - N_3 = 1 \cdot 15.$

From these quantities we obtain, in the first place, for the principal planes

$e^0 = N_0 + \cdot 85 \qquad\qquad i^0 = N_1 + \cdot 1$
$e' = N_2 + 2 \cdot 5 \qquad\qquad i' = N_3 + 1 \cdot 5$
$e'' = N_4 + 1 \cdot 9 \qquad\qquad i'' = N_5 + 0$
$(E) = N_2 + 3 \cdot 38 \qquad\qquad (E)^* = N_5 + \cdot 44$
$E = N_0 + 2 \cdot 72 \qquad\qquad E^* = N_0 + 2 \cdot 8$
$i'' - e' = 2 \cdot 65$
$(E)^* - e^0 = 5 \cdot 71.$

The further deductions are contained in the following table

(column "Observed"), which we give for comparison with the values calculated according to the above formulæ.[1]

	Observed.	Calculated.
$e'' - i'$	1·55	1·5
$(E) - i^0$	3·76	3·66
$(E)^* - (E)$	1·33	1·05
$\Lambda_2 - (F)$	2·86	2·92
$E^* - E$	·08	·57
$\Lambda_0 - F$	·43	·45

The most striking peculiarity of the objective examined, and to which we draw special attention, is the fact that the anterior lens is stronger than the whole system. The equation of condition for this case is an immediate result from the formula for the combined focal length, taking

$$\phi_0 \left[\frac{(f)}{\phi_0 + (f) - t} \right] > \phi_0$$

and then dividing both sides by ϕ_0, we obtain

$$t > \phi_0.$$

Of the medium-power systems we have examined, the No. 9 of Bénèche is the only one possessing this peculiarity; neither of the corresponding objectives of Oberhæuser or Plœssl has it. On the other hand, it occurs somewhat frequently in low-powers, especially those composed of two double-lenses separated considerably. We observed it, for instance, in the No. 4 of Bénèche, and the No. 4 of Hartnack. In the No. 9 immersion of Hartnack, the anterior lens is also somewhat stronger than the whole combination; and this may possibly be the case in the highest

[1] The differences in this table between the observed and calculated values might be explained by the fact that the determination of d_2 is impossible for the single lenses with sufficient accuracy, on account of the indistinctness of the image in the corresponding position. Moreover, the supposition of equal focal lengths in both positions is, strictly speaking, only justifiable for central rays, and becomes untenable for marginal ones, because the amount of the spherical aberration, and consequently also the distance of the focus on reversal of the lenses, is slightly altered. Such an alteration of distance is noticeable even in deeper lenses as soon as the conjugate foci are in essentially different ratio. For this reason the values, for instance, of the focal lengths found according to different methods, never exactly coincide.

powers of other opticians, especially as the front lens is usually a single crown.

We add a few more data with regard to the principal and focal planes in various other objective systems. The last refracting surface is denoted by N^*, all the other quantities as above.

	No.	f	N_0-F	N^*-F^*	N^*-N_0	E^*-E
Hartnack	7	3·3	·5	+ ·45	7·1	+ ·55
,,	9	1·87	·35	+ 1·45	5·1	+ ·27
Bénèche	4	12·5	3·7	+ 3·0	16·5	− 7·8
,,	7	4·7	1·5	− 2·0	7·25	+ 1·35
,,	11	2·22	·25	+ 3·2	5·2	− 2·19
Kellner	3	3·87	·58	0·	8·2	+ 1·04

PART IV.

THEORY OF MICROSCOPIC OBSERVATION.

Microscopic observation differs in many respects from ordinary observation with the naked eye. It demands, therefore, long practice, in order to arrive at that certainty of judgment in regard to the sensations received from the light which alone assures the correctness of the observation. To see through the Microscope, and interpret the form and nature of an object in accordance with what has been thus seen, is an art requiring special training.

Amongst the peculiarities of microscopic vision we may primarily remark, that the image which we observe does not exactly correspond to the object itself, but only to a particular sectional plane to which we happen to adjust the focus. Whatever lies above or below is indistinct to the eye, or invisible. The Microscope therefore gives us direct information only of the dimensional relations of the different sectional planes, but not of their distances in the direction of the axis of the instrument. The latter can in many cases be approximately estimated; for instance, we may distinguish at once whether a given object is spherical, flattened, or compressed at its sides, but a more accurate determination of its form can only be attained by turning it on its axis —that is, by the combination of different views.

This is not so in ordinary vision. A mere glance is in this case often sufficient to inform us of the form and grouping of closely adjacent bodies, because the limits within which the eye sees distinctly at one view lie somewhat distant from each other. We therefore see the bodies in their stereometric form, and in their positions relative to those which are contiguous to them. Whether we observe with two eyes or one is immaterial, precisely as with the Microscope.

Secondly, the illumination in microscopic observation is entirely different from that of ordinary vision. Through the Microscope we see the objects, as a rule, by transmitted light; with the naked eye, by reflected light. Any given point of the Microscope image therefore appears the brighter the greater the number of the rays which before their entrance into the objective, if produced backwards, are directed towards the corresponding point of the object. If these rays form a perfect cone of light whose base fully occupies the optically effective part of the anterior surface of the objective, the point appears of maximum brightness. If the base of the cone is only half as large, the point appears in the image only half as luminous—and so on. The Microscope forms therefore, to a certain extent, a shadow-image, in which the opaque parts of the object appear dark, and the transparent ones more or less bright, provided the inequalities and differences in density of the object, in consequence of the refraction they produce, contribute essentially to the distribution of light in the image.

This obtains also in vision with the naked eye, if transparent bodies are observed under the same conditions. Nevertheless, the image which we see in this case is essentially different from that formed in observation through the Microscope; and contrary assertions which are met with here and there in micrographic works[1] must be regarded as erroneous. They are based upon the tacit assumption that the unequally large apertures of the incident cones of light directed towards the eye by the Microscope have no influence upon the resulting effect of light. A mathematical discussion of this question proves that the distribution of light and shade in the microscopic image is dependent upon the angle of aperture of the instrument, and is not therefore the same even for different amplifications, much less for the naked eye.

There can be no doubt that objects are seen in the Microscope differently than with the naked eye. The observer with the Microscope begins by viewing images which are new and unusual to him; he must first learn to interpret them correctly, precisely as with the words of a foreign language. Although practice is most essential and cannot be replaced by theory, still a theoretical foundation of microscopic vision would not be superfluous even to the experienced observer, since it presents, in many cases, useful data for further deductions. Hence we have submitted a series

[1] Cf. for instance, Harting, "Das Mikroskop," 1st ed. p. 339; 2nd ed. ii. p. 26.

of cases to a searching discussion based upon mathematical developments.[1] The phenomena of refraction and reflexion which determine the microscopic image of given objects are first discussed; then the effects of light depending upon interference, which were formerly called, as far as they were known, diffraction phenomena, the origin of which may be of a very different kind. Finally, a short consideration is devoted to oblique illumination and to the appearances presented by objects in motion.

I.

SPHERICAL AND CYLINDRICAL OBJECTS.

1.—AIR-BUBBLES IN WATER.

LET $A\,B$ (Fig. 104) be the vertical section of an air-bubble freely floating in water, and therefore spherical, and $M\,N$ the plane of adjustment of the Microscope; then any point P in this plane is illuminated in the microscopic image by the rays which, after their passage through the air-bubble, seem to come from this point, and are directed towards the eye by the refraction in the Microscope. The inclination of these rays to the perpendicular cannot be greater than half the angle of aperture of the objective. If this latter is assumed to be 60°, then the ray reaches its maximum at 30°. Similarly, the

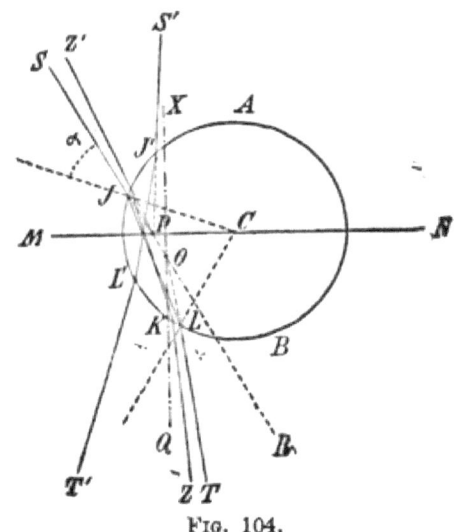

FIG. 104.

[1] Harting ("Das Mikroskop," 2nd ed. ii. p. 31) expresses the opinion with regard to these developments, that it is possible to explain the real point in question less tediously. We do not for a moment say that Harting is alone in holding this opinion. But in our view, and others certainly agree with us, it is important here as elsewhere to understand the phenomena we observe, and to

inclination of the rays coming from below, if the mirror and source of light are of sufficient extent, is determined by the size of the diaphragm. The angle under which the latter is seen from the centre of the air-bubble is the same as that which is found by the peripheral rays of the incident cone of light. If this angle $= 30°$, then also the angle of deviation, which the emergent rays form with the corresponding incident ones, is determined for marginal rays, such as $T\,S$, which before and after the refraction lie on the surface of the cones of light.

If we draw through the point of intersection O, of the incident and emergent rays, the perpendicular $X\,Q$, and if the angles of aperture taken at 60° and 30° are denoted by ω and δ, we get

$$L\,S\,O\,T = 180° - L\,T\,O\,R;$$

or, since
$$L\,T\,O\,R = L\,Q\,O\,R - L\,Q\,O\,T = \frac{\omega - \delta}{2},$$

$$L\,S\,O\,T = 180° - \frac{\omega - \delta}{2};$$

therefore, in the given case,
$$L\,S\,O\,T = 180° - 15° = 165°.$$

Now, since a radius drawn through O bisects this angle of deviation and crosses the direction of the ray in the air-bubble at right angles, if the angle of incidence is denoted by a, the angle of refraction by a', and half the angle of deviation by ρ, we obtain the further relation

$$a' - a = 90° - \rho = 7\tfrac{1}{2}°.$$

If we assume the mean refractive index of water to be 1·3356, this equation will be satisfied if $a = 20°\,45'$, while a' will therefore $= 28°\,15'$.

The position of the point P is thus determined: the triangle $C\,P\,J$ gives the relation

$$C\,P : r = \sin a : \sin\left[180° - \left(90° - \frac{\omega}{2}\right)\right];$$

consequently,

$$C\,P = r \cdot \frac{\sin a}{\sin\left(90° + \frac{\omega}{2}\right)} = \frac{\sin 20°\,45'}{\sin 60°} \cdot r = \cdot 64838 \cdot r\ .$$

interpret them correctly in detail. We can at any rate affirm that Harting's statement does not explain the microscopic image of the air-bubble, hollow cylinder, &c. The reader wishing to follow out these points should not shrink from mathematical explanations.

If we now suppose a second adjacent ray $Z\,K$, which is less inclined to the perpendicular and which also appears after refraction to come from the point P, it is evident that at some point it intersects the first ray, and therefore meets the objective somewhat nearer the centre. For if we follow the two rays in the opposite direction, from above downwards, their point of convergence, P, moves slightly to the left, in consequence of the refraction at the surface in J, assumed to be infinitely small, and hence lies in a different level; an intersection takes place invariably within the air-bubble. This applies also to each succeeding ray with reference to the preceding one. The further we proceed in the emergent cone of light from left to right, the further is the corresponding incident ray moved from right to left, its inclination to the perpendicular becomes gradually less, and then passes into the opposite inclination, whose maximum also amounts to 15°.

It is also evident that this maximum deviation must occur before the emergent rays on the right have reached the limit of 30°, as the two refractions always cause a deviation to the left. Calculation shows that in the given case the marginal ray $T'L'$ is inclined, after its passage through the air-bubble, slightly more than 18° to the left.

These discussions lead first to the conclusion, that all rays of the incident cone of light[1] lying in the plane of the paper contribute to the illumination of the point P. Moreover, of the rays not lying in this plane none are lost. For since, after refraction, they all appear to proceed from the point P, and furthermore lie in the same plane with the radius drawn to the point of emergence, consequently cutting the plane of the paper in the line $M\,N$, we shall exhaust all possible positions thereof, if, while retaining the point of convergence, all rays between $S\,J$ and $S'J'$ (Fig. 104) are raised above the plane of the paper, and allowed to diverge so far upwards and downwards, that the corresponding incident rays graze the margin of the diaphragm. That the inclination to the plane of the paper of the rays thus raised can at the most amount to 15°, in the given case, and for the two marginal rays $= 0$, is evident without further remark. There is, moreover, no difficulty

---

[1] The expression cone of light is not to be taken with strict mathematical accuracy for the incident rays, inasmuch as they have no common point of convergence.

in coming to the conclusion that under these conditions the whole are incident upon the effective portion of the objective.

Hence the point P, since the entire cone of the incident rays contributes to its image, appears equally bright with any other point of the field of view. The like holds good, of course, with respect to all the points that lie nearer to the centre, except that for these the refraction is less, so that the emergent cones of light meet a more central part of the anterior surface of the objective. The centre itself is illuminated by rays which traverse it without refraction.

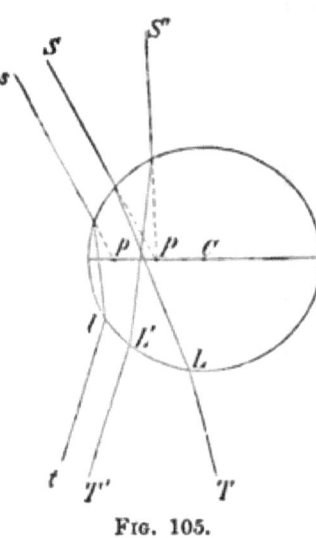

Fig. 105.

On the other hand, at the points further from the centre, as, for instance, p (Fig. 105), a diminution of the light is perceptible. That is to say, for them the angle of emergence (= that of incidence) a, as a glance at the figure demonstrates, is greater, and consequently $\rho \ (= 90° - [a' - a])$ is less. Hence a marginal ray parallel to $T\,L$ (Figs. 104 and 105) will be so refracted that it deviates more than 30° from the perpendicular, and is therefore lost to view. This diminution of course takes place as well for the point immediately adjacent, and as we get nearer and nearer to the periphery, it becomes also extended to the rays of the cone of light situated more towards the left, until at last only the marginal ray $t\,l$, parallel to TL', reaches the eye. All rays not lying in the plane of the paper are ineffective for this limiting position.

The determination of these limiting points is made in exactly the same manner as that of the point P. If ρ is half the angle of deviation, the equation is

$$\rho = 90° - \frac{\omega + \delta}{4};$$

therefore in the given case

$$\rho = 90° - \frac{60° + 30°}{4} = 67\tfrac{1}{2}°.$$

From the relation above obtained $a' - a = 90° - \rho$ we get

$$a' - a = 22\tfrac{1}{2}°.$$

This latter difference occurs for the assumed index of refraction, if $a = 43°$; therefore $a' = 65° \, 37'$. The distance of the limiting point p, in Fig. 105, is given by the formula

$$C\,p = \frac{\sin 43°}{\sin 60°} \cdot r = \cdot 7875 \cdot r.$$

Under a Microscope with the assumed angle of aperture, the air-bubble will therefore exhibit a black band, the breadth of which $= (1 - \cdot 7875)\, r = \cdot 2125 \cdot r$. The central part up to P would appear just as bright as the field of view, and the zone between P and p would obstruct the light. An alteration in the angle of aperture, by the employment of different objectives and diaphragms, would only modify the numerical relations, but not the general distribution of the light. With greater values for ω and δ the black band would become narrower, and with smaller values broader. Oblique illumination would produce an excentric position of the bright circle.[1]

In reality the microscopic image of the air-bubble presents an essentially different appearance. The bright inner part appears encircled by a dark zone, which gradually merges into a perfectly black one; this zone does not extend to the margin, but is encircled by a bright ring, and again by a somewhat uniform penumbra to the periphery. With more careful observation we recognize, a little to the outside of this bright ring, a second one less distinct; indeed, the whole penumbra proves to be made up of alternately bright and dark concentric circles.

These phenomena cannot be explained by refraction alone; they are due to rays, which have in addition undergone one or two internal reflexions. Let TL (Fig. 106) be such a ray, which is reflected at R and refracted a second time in J; then, as is evident from the figure, JBC is the half angle of deviation, for which we

---

[1] Harting ("Das Mikroskop," 2nd ed. ii. pp. 30 and 32, note) does not agree with this theory, according to which the angle of aperture ω has no influence upon the breadth of the dark zone. He relies upon his observations and measurements, according to which the diameter of the bright central part is simply and solely determined by the angular magnitude of the source of light, dependent upon the aperture of the diaphragm. That is, however, erroneous, and we cannot understand how he could obtain such a result. If we place a piece of paper or tin-foil, with a minute perforation (a needle-hole), as a diaphragm between the objective-lenses, and then observe the image of a spherical air-bubble, it will not be necessary to resort to measurement in order to establish the influence of the angle of aperture.

will again employ the letter ρ. Since ρ is an exterior angle of the triangle JBR, we get $\rho = LBJR + LBRJ$, or, if a denotes the angle of incidence, and a' the angle of refraction,

$$\rho = (a' - a) + a' = 2a' - a.$$

It would be easy to determine from this equation the limits between which the air-bubble appears to be illuminated in consequence of the internal reflexion. The question, however, would not even then be settled. It appears more to the purpose to take a particular case, which leads directly to the solution of our problem—the case, namely, where the incident and emergent rays lie in the same straight line, and consequently ρ becomes a right

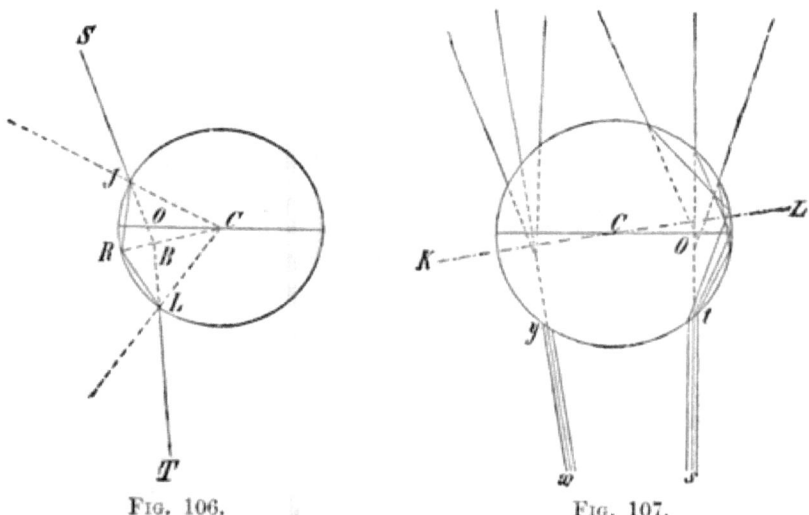

Fig. 106.   Fig. 107.

angle. The difference $2a' - a$ attains the same magnitude; from which there results for a a value of $43\frac{1}{2}°$. If we now suppose, instead of the incident ray, a parallel pencil of rays $s\,t$ (Fig. 107) to be taken, then only that ray whose a has exactly the stated value proceeds in the original direction. All the other rays of the pencil are deflected more or less; and indeed, as calculation shows, to the left of the perpendicular if $a < 43\frac{1}{2}°$, to the right if $a > 43\frac{1}{2}°$. As examples a few calculated values are collected in the following table. The first column contains the incident angle a; the second the corresponding deviation, ϕ, from the perpendicular to the right or to the left; and in the third, the distance of the points in which the emergent rays intersect the plane of adjustment.

Angle of Incidence α.	Deviation from the Perpendicular φ.	Distance of the Points of Intersection from the Centre.
37°	40° to the left	·7856 . r
39°	29° 12' ,, ,,	·7209 . r
40°	23° 24' ,, ,,	·7004 . r
42°	10° 40' ,, ,,	·6809 . r
43½°	0°	·6883 . r
45°	13° 12' ,, right	·7263 . r
47°	36° 32' ,, ,,	·9314 . r

Parallel rays, whose angles of incidence vary between 40° and about 44°, accordingly receive, by the internal reflexion combined with the refraction at the surface, such a direction that they all appear to come from points whose distance from the centre amounts to ·68 — ·7 . r, and therefore differs at most by $\frac{1}{50}$ of the radius. In other words, to an incident pencil of light composed of parallel rays there corresponds an emergent cone of light, whose virtual point of convergence O falls somewhat below the plane of adjustment, though almost on the line which represents the ray emerging without deviation. At this point, therefore, is concentrated for the eye the whole effect of light which the incident pencil is capable of producing in this way. In addition to this, every other pencil of rays $x\,y$ (Fig. 107), which with equal angles of incidence deviates by a few degrees to the right or to the left (drawn on the opposite side in order to avoid confusion), behaves, with reference to the diameter-plane $K\,L$ drawn at right angles to it, in such a manner that the brightness of the point O is further increased by an infinite number of others, which lie somewhat higher or lower, and form with it an uninterrupted line of light.

We thus explain the fact that a small space at a distance of $C\,O = ·69 . r$ from the centre, in the middle of the marginal shadow, appears so luminous, and when viewed from above appears as a bright ring. It is intelligible also that the breadth of this ring, at any rate within the usual limits, increases and decreases with the angles of aperture of the objective and of the diaphragm, and that a higher or lower focal adjustment must cause an obliteration of its outline.

Applying the same reasoning to rays that have undergone two internal reflexions, we can also explain the presence of a less distinct outer ring. If $S\,T$ (Fig. 108) is such a ray, which is reflected at R and Q, and refracted for the second time in P, the

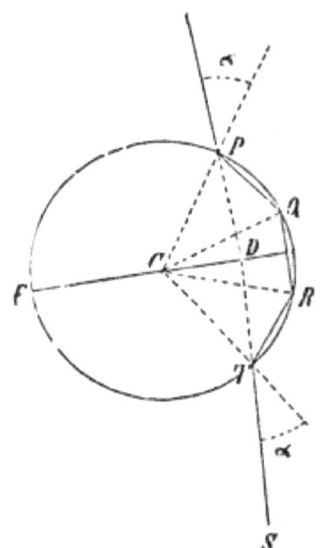

Fig. 108.

triangles $C\,P\,Q$, $C\,Q\,R$, and $C\,R\,T$ are similar, and their angles adjacent to the chords $P\,Q$, $Q\,R$, and $R\,T$ are $= a'$. Consequently the three angles $P\,C\,Q$, $Q\,C\,R$, and $R\,C\,T$ at the centre are together $= 3\,(180° - 2\,a')$. On the other hand, the triangle $P\,C\,D$ gives the relation
$$L\,P\,D\,C = \rho = F\,C\,P - C\,P\,D$$
$$= F\,C\,P - a,$$
or, since
$$F\,C\,P = \frac{360° - P\,C\,T}{2},$$
$$\rho = \frac{360° - 3\,(180° - 2\,a')}{2} - a$$
$$= 3\,a' - a - 90°.$$

For the case when the incident ray emerges without deviation, therefore $\rho = 90°$, the condition accordingly is $3\,a' - a = 180°$. This condition is satisfied if $a = 46°\,28'$, from which the distance of the point D from the centre is calculated at $\cdot 725\,.\,r$. The following table, in which the same quantities have been inserted as in the previous one, shows that parallel rays whose angle of incidence is slightly greater or less than $46°\,28'$, form, as in the previous case, a virtual focus whose smallest diameter lies somewhat above the plane of adjustment and very near to the line of the ray emerging without deviation.

Angle of Incidence a.	Deviation from the Perpendicular ϕ.		Distance of the points of intersection in the plane of adjustment from the Centre.
$43\frac{1}{2}°$	$46°$	to the left	$\cdot 995\,.\,r$
$44°$	$39\frac{1}{2}°$	,, ,,	$\cdot 900\,.\,r$
$45°$	$25°\,12'$,,	,,	$\cdot 781\,.\,r$
$46°$	$8°\,36'$,,	,,	$\cdot 727\,.\,r$
$46°\,28'$	$0°$		$\cdot 725\,.\,r$
$47°$	$11°\,48'$,,	right	$\cdot 747\,.\,r$
$48°$	$42°$,,	,,	$1\cdot 000\,.\,r$

The outer bright ring is consequently at a distance of $\cdot 725\,.\,r$ from the centre; and from the inner one $\cdot 0366\,.\,r$, or approximately the twenty-seventh part of the half-diameter; it appears most sharply outlined at a somewhat higher level. Its lower intensity of light is intelligible, partly by the loss caused by the two reflexions, partly by the narrow limits of the angles of incidence (about $45\tfrac{1}{2}°$—$47°$).

We may hence conclude that further bright rings are formed by pencils of rays which have undergone internal reflexion three, four, or more times. It will be sufficient for these cases to collect the conditions under which the emergent rays show no deviation, and thence determine the points in which (if produced backwards) they cut the plane of adjustment. The distances of these points may be identified with those of the corresponding rings without further explanation.

No. of Internal Reflexions.	Equations of Condition for the ray emerging without Deviation.	Angle of Incidence a.	Angle of Refraction a'.	Distance of the Rings from the Centre.
3	$4a' - a = 270°$	$47°\ 23'$	$79°\ 23'$	$·7359 . r$
4	$5a' - a = 360°$	$47°\ 47'$	$81°\ 33'$	$·7406 . r$
5	$6a' - a = 450°$	$48°$	$83°$	$·7431 . r$
6	$7a' - a = 540°$	$48°\ 7'$	$84°\ 3'$	$·7447 . r$

It is evident that this series terminates as soon as the angle of incidence has reached its limiting value, which amounts for water and air to $48°\ 29'$. The ring corresponding to this limiting ray, which, however, is no longer perceptible, would be $·7487 . r$ from the centre, that is, approximately, $\tfrac{3}{4}$ of the half-diameter. Nevertheless, the remaining portion of the margin may still be illuminated by the internal reflexion. From the table we might rather conclude that all the rays here under consideration, which do not contribute to the formation of the rings, are deflected so that they appear to come from points which lie without the corresponding ring. A feeble illumination of the margin must therefore result, while the umbra appears, within the rings, perfectly black.

Since the distances of the rings are dependent upon quantities which are determined by the index of refraction of the surrounding medium, we might infer that liquids of greater density (as oil, glycerine, &c.) would modify the numerical relations above given. In oil of refractive index $1·5$ we obtain for the innermost bright ring a distance of $·5957 . r$ from the centre, which agrees perfectly with observation. Hence the rings always lie somewhat nearer to the centre in blue light than in red ; and, consequently, when viewed with white light they appear encircled on the inside with a blue fringe and on the outside with a red one.

The bright rings, and at the same time the fringes of colour, appear very distinct at the sides facing the adjacent air-bubbles. The faint half-light, which gives a grey tone to the marginal shadow, becomes here intensified to a band of light, which towards the interior admits of the recognition of distinctly indicated lines corresponding to the rings, while on the other hand a homogeneous glimmer is perceived in the outward direction. How such a reinforcement of the luminous effect is brought about by the action of

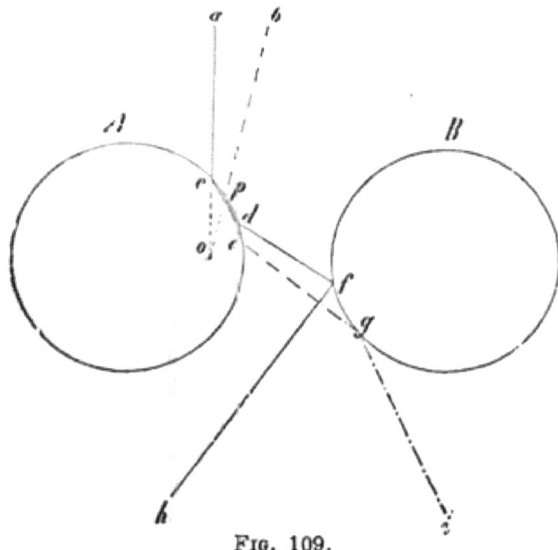

Fig. 109.

one air-bubble upon another one, may be demonstrated in a simple manner by tracing the rays of the emergent cone of light backwards, and, at the points where they are reflected, allowing them to pass into water by refraction, whereby they are partially reflected downwards towards the diaphragm by the adjacent air-bubble. In this manner, for instance, the rays $a\,c$ and $b\,p$ (Fig. 109), which converge to the point o in the second ring of the air-bubble A and after a two-fold refraction proceed in the directions $d f$ and $e g$, reach h and i by reflexion on the air-bubble B. If $a\,c$ and $b\,p$ are the marginal rays of a cone of light incident from above, then $f\,h$ and $g\,i$ are the marginal rays of the corresponding emergent cone.

Conversely, therefore, if the movement of the light takes place from below upwards, the rays of the very oblique cone $h f g i$ must unite to form the acute pencil $a c p b$, and hence illuminate the point o. If $h f$ and $i g$ approach nearer to the perpendicular, which would be equivalent to diminishing the diaphragm, then $a c$ and $b p$ also approach each other. The pencil of light reaching the objective is thereby weakened, but remains, afterwards as before, inclined somewhat to the right. This inclination passes into a contrary one if the air-bubbles are considerably more distant from each other. The lines of direction $d f$ and $e g$ then approach more to the horizontal; $a c$ and $b p$ are consequently removed further to the left, as would occur if the air-bubble A revolved in the same direction round its axis. The emergent pencil of light consequently changes its direction to the axis of the Microscope in accordance with the distance of the two air-bubbles.

Similarly may be demonstrated the strengthening of the other rings, and the intenser illumination of the margin. We have only to construct the corresponding cone of light and to trace backwards single rays, as shown in Fig. 109, in order to account for the different phenomena.

If we consider the air-bubble as the refracting apparatus, without reference to the plane of adjustment, it acts essentially as a bi-concave lens. Its focal length, f, which is of course negative, is determined by the formula

$$f = - \frac{r}{2(n-1)},$$

in which r is the radius, and n the refractive index of the surrounding medium. Since the two principal points, as in every sphere, coincide with the centre, the above expression is also equal to the distance of the focal point from the centre. In oil with the refractive index 1·5, f is equal to $-r$, in water to approximately $-\frac{3}{2} r$, which values are reduced more or less (in water to about $·2 . r$) through the aberration of the marginal rays.[1]

[1] An incident cone of light is always refracted by an achromatic system of lenses as if it were composed of rays of about the mean inclination, *i.e.*, it acts as a conical shell of small aperture. Hence, it is evident that a cover-glass upon the object produces an apparent approximation by a *definite* quantity, although the mathematical expression for this approximation is dependent upon the angle of inclination of the rays of light, and therefore yields no fixed value. A cover-glass of 227 mic. in thickness causes, for instance, with Nos. 7 and 9 objectives of Bénèche, an elevation of the object-point of 80 mic., whence the

If we lower the focus of the Microscope to the level of the focal point, the images of distant objects reflected in the mirror, or which are within the incident cone of light, are seen. The outlines are the more sharply defined the more we exclude rays incident obliquely to the axis by the arrangement of the diaphragm or of the illumination, and the less the marginal rays of the cone of light reaching the objective deviate from the perpendicular. A parallel pencil of light incident vertically from below gives, with a moderate amplification, such a sharp image, that it immediately becomes indistinct on somewhat higher or lower focal adjustment.

The unequal refrangibility of the differently-coloured rays causes the emergent cone to receive only red rays in the centre and violet at the periphery. We might hence be tempted to explain the fact that the bright colourless circle, which the image of the diaphragm presents, is, on slightly higher focal adjustment, not only considerably larger, but also shows a reddish centre and a bluish fringe. These colour-phenomena, as with real images, do not originate in the object itself, but are due to the chromatic aberration of the Microscope, which is over-corrected by the higher adjustment and under-corrected by the lower. This is demonstrated by the simple experiment, which is usually employed in testing the aberration, viz., by viewing the image of a window reflected in a globule of mercury. Raising or lowering the focus produces, therefore, the same colour-phenomena as those exhibited by the images of air-bubbles formed by refraction.

2.—GLOBULES OF OIL IN WATER.

We will consider globules of oil in water as representing in general a circular vertical section of any given object immersed in a medium of lower refractive index. In order to determine

limiting inclination is found to be about 18°, if the refractive index of the glass is 1·5. This inclination determines also the position of the focal point of an air-bubble. It produces, moreover (if the objective is not immersed in water), a further diminution of the focal length, which is due to the fact that an object in water appears the higher the deeper it lies. The focal point is therefore more raised than the centre of the air-bubble, and is hence apparently moved nearer to the latter. This phenomenon, which must not be neglected in direct measurements of differences of level, will be fully discussed later on.

the distribution of light in a given sectional plane to which the Microscope is focused, we will again apply the principle of the reciprocity of the path of the rays, and will produce the pencils backwards from the objective to any desired points of the plane of adjustment, and then trace them to the diaphragm. If again ω and δ are the angles of aperture of the objective and of the diaphragm, MN (Fig. 110) the plane of adjustment, and FL a ray directed towards the point P, which after two refractions proceeds in the direction TS, and if, finally, a and a' are the angles of incidence and refraction; then the half angle of deviation is

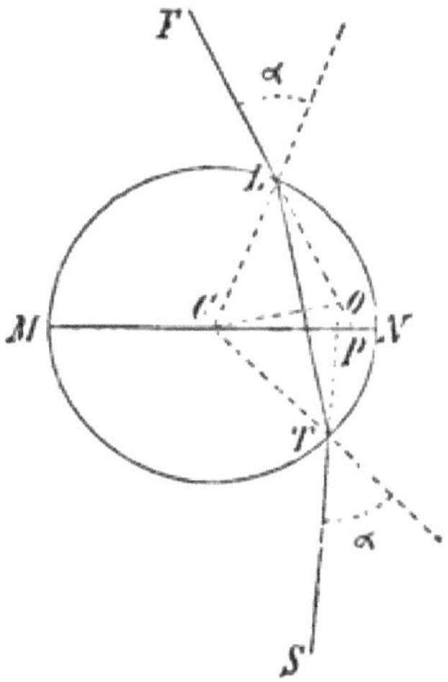

$$\rho = \tfrac{1}{2} L O T = L O C;$$

Fig. 110.

and we get from the right-angled triangle, whose hypotenuse is LO, the relation

$$\rho = 90° - (a - a').$$

For those points in the plane of adjustment, whose brightness is equal to that of the field of view, the relations and deductions above derived also hold good.

The constructions there given are applicable here also, if we suppose the cones of rays transposed from the left half of the circle to the right half, so that the points where the refractions take place again fall in the periphery. We get, therefore,

$$\rho \gtreqless 90° - \frac{\omega - \delta}{4};$$

and hence, by combination with the above equation,

$$a - a' \lesseqgtr \frac{\omega - \delta}{4}.$$

Similarly, there results for the limiting points, which are illuminated only by the outermost marginal rays,

$$\rho = 90° - \frac{\omega + \delta}{4};$$

and consequently

$$a - a' = \frac{\omega + \delta}{4}.$$

By the help of these equations the limits of the umbra and penumbra can easily be determined in any given case. On account of the great difference in the numerical relations which may occur in practice, the discussion of a particular example would be of no special value; a few of the more general consequences may, however, be noted.

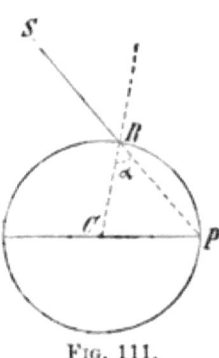

Fig. 111.

Let $S R$ (Fig. 111) be a marginal ray of the emergent cone of light, which (if produced backwards) cuts the plane of adjustment (drawn through C) in the point P. Then, as a glance at the figure shows, its angle of incidence a is equal to the angle $R P C = 90° - \frac{\omega}{2}$. If for this magnitude of the angle of incidence $a - a' < \frac{\omega + \delta}{4}$ or even less than $\frac{\omega - \delta}{4}$, then in the former case the umbra will disappear, and in the latter the penumbra. The sphere will therefore appear, under certain circumstances, uniformly illuminated from one margin to the other. This occurs if its refractive index $= 1·5$, $\omega = 60°$, and $\delta \leq 22°$; whence $a - a' = 9\frac{1}{2}°$, and $\frac{\omega - \delta}{4} \geq 9\frac{1}{2}°$. Globules of oil, spherical starch-grains, cylindrical hairs, &c., whose refractive index does not differ much from that just assumed, will therefore exhibit no marginal shadow with higher amplifications, while the lower-power objectives show it the more distinctly the smaller their angles of aperture. It would, of course, appear broadest when observed with the naked eye, since ω must then be regarded as infinitely small.

If, when ω remains constant, the angle δ varies, the distribution of the light is altered as follows:—If $\delta = 0$, that is, if the incident rays are parallel, the equations for the umbra and penumbra will be identical, *i.e.*, the limits of both will coincide, or, what is equivalent, the penumbra vanishes. If δ gradually increases, the umbra will become narrower and the penumbra broader. The inner limit of the latter finally reaches the centre when $\delta = \omega$, and then no part of the sphere will appear so bright as the field of view. If $\delta > \omega$, the limiting line of the penumbra will again move outwards, and similarly that of the umbra; and both will finally reach the upper surface of the sphere, depending on the magnitude of ω.

If these theoretical deductions had to be somewhat modified by the inequality of the loss, which the differently-inclined rays undergo in consequence of the reflexion at the upper surface of the lens, as well as by the aberrations of the object and of the Microscope, yet the experiment teaches that the calculated limits between light and shadow for every combination of ω and δ will agree approximately with those observed. If, for instance, with the ordinary illumination, we observe through the body-tube (after the removal of the objective and eye-piece) a glass sphere or a cylindrical glass rod in the incident cone of light, the illuminated middle part will appear almost as a point or as a line, and the broad marginal shadow deep black. If, however, we hold the same objects against the open sky, that is, where δ is very large, they will appear bright from one margin to the other.

The focal length f of a sphere or a cylinder is determined for parallel rays by the formula

$$f = \frac{n' r}{2 (n' - n^0)},$$

where n' and n^0 denote, respectively, the refractive indices of the sphere and the surrounding medium, and r is the radius of the sphere. If, for instance, $n' = 1\cdot 5$ and $n^0 = 1$, then the distance of the focal point from the centre is $\frac{3}{2} r$; hence, if we adjust the Microscope to the corresponding level, the real image of more distant objects, from which direct light reaches the object through the diaphragm, will be seen there. This is exhibited also by globules of oil and spherical starch-grains, as well as by cylindrical hairs, bast-cells, &c., with striking clearness by withdrawing the diaphragm some distance, or if the incident rays are otherwise rendered parallel. Under favourable circumstances, it is even possible, from the known distance and size of the object, and the accurate measurement of the image, to determine to the second decimal place the refractive index of the substance. The formula to be applied in this case can easily be developed from the one just found for the focal length,[1] and the known relation of the conjugate

---

[1] The formula for the focal length needs, moreover, a slight correction on account of the spherical aberration of the marginal rays. Our observations regarding the focal length of the air-bubble (*vide* note on p. 201) are applicable here also.

foci $\left(\dfrac{1}{p} + \dfrac{1}{p^*} = \dfrac{1}{f}\right)$, which are proportional to these magnitudes. If M is the diameter of the diaphragm, p its distance from the object, and m the diameter of the image, and finally n^0 and n' the refractive indices; then we find

$$n' = -\frac{2\,p\,n^0}{2\,p - \left(1 + \dfrac{M}{m}\right)r}, \text{ or, approximately, } n' = \frac{2\,p\,n^0}{2\,p - \dfrac{M}{m}r},$$

where it is assumed in the latter formula that p is very large in comparison with r.

In applying these formulæ to spheres with negative foci, the second term of the denominator must be taken positive. Thus we should have

$$n' = \frac{2\,p\,n^0}{2\,p + \dfrac{M}{m}r}; \text{ and hence } n^0 = \left(1 + \frac{M\,r}{2\,p\,m}\right)n'.$$

3.—Hollow Spheres and Hollow Cylinders.

The following consideration is generally applicable to tubular cells, globules of oil, and nuclei with vacuoles, starch-grains with spherical cavities, &c. For the sake of brevity, we will limit our discussion to the case of hollow cylinders. Of the pencils of light, which influence the microscopic image with an object of this kind, we have, (1) marginal rays which pass through the walls of the cylindrical tubes without reaching the lumen; (2) marginal rays which meet the inner surface of the cylinder very obliquely, and are there reflected; (3) rays which enter the lumen, are reflected on the walls, and then after two refractions reach the objective; (4) rays which traverse the lumen in a direct line, and undergo in consequence a four-fold refraction.

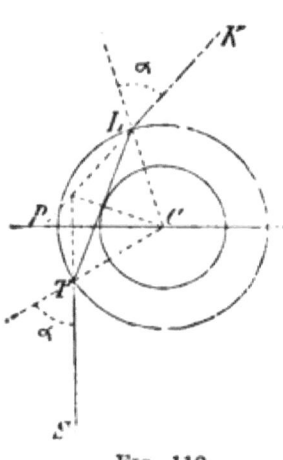

Fig. 112.

The marginal rays, which are only refracted twice, behave exactly in the same way as in the solid cylinder. It here depends not merely upon the angles of aperture ω and δ, but also upon the thickness of the walls, whether and how much they contribute to the illumination of the latter. If r is the short and R the long radius of the hollow cylinder, the thickness of the wall being therefore $R - r$, and $S\ T\ L\ K$ (Fig. 112) is a limiting ray which forms a tangent to the inner wall after the first refraction, and if a and a' are its angles of incidence and refraction, and n the refractive index; we get $\sin a' = r$, $\sin a = \dfrac{n \sin a'}{R}$, and consequently $\sin a = \dfrac{nr}{R}$. For the marginal rays the condition $\sin a > \dfrac{nr}{R}$ will therefore in general hold good; since, moreover, if they are to contribute to the illumination of the walls with medium focal adjustment, they must satisfy the general equation $a - a' \lessgtr \dfrac{\omega + \delta}{4}$, they will be lost in the microscopic image as soon as $\dfrac{nr}{R}$ reaches a certain magnitude. They come into account therefore with the given refractive indices only with thicker walls.

If, for instance, $n = \dfrac{1\cdot 649}{1\cdot 335}$ (the refractive ratio between flint-glass and water), and $\dfrac{r}{R} = \cdot 8$, we find $a \gtreqless 81°$, $a' \gtreqless 53° \ 8'$, $a - a'$ is therefore nearly 28° at the minimum. The sum of the angles of aperture $\omega + \delta$ would therefore amount to approximately $4 \times 28 = 112°$, if the innermost rays only contributed to the illumination of the cylinder-wall. If no other rays illuminated this cylinder-wall it would appear perfectly black even with a somewhat high amplification (ordinary illumination being assumed).

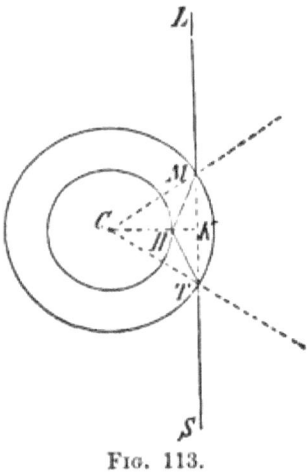

Fig. 113.

The marginal rays mentioned in (2), which at the inner surface of the walls are reflected to the outer walls, and there undergo the second refraction, form somewhat above the centre a virtual focus, which—as in the air-bubble—nearly coincides with the point in which the line of direction of the ray passing through unrefracted cuts the plane of adjustment drawn through the centre.

To determine this point, let $S\ T$ (Fig. 113) be a ray incident vertically from below, which is reflected at H and refracted for the second time in M; further, let a and a' be the angles of incidence and refraction, r the short and R the long radius; and let $\rho = \angle MKH$ and $\eta = \angle MHK$: then the triangle MHK gives

$$\rho = 180° - [(a - a') + \eta];$$

consequently

$$\sin \rho = \sin (a - a' + \eta) = \sin (a - a') \cos \eta + \cos (a - a') \sin \eta.$$

In the triangle CHM we get the proportion $CM : CH = \sin \eta : \sin a'$; and hence $\sin \eta = \dfrac{CM}{CH} \sin a' = \dfrac{R}{r} \sin a'$. The above formula therefore passes into the form

$$\sin \rho = \sin (a - a') \sqrt{1 - \left(\frac{R}{r} \sin a'\right)^2} + \cos (a - a') \frac{R}{r} \sin a'.$$

By the help of this equation the direction of the emergent ray may be determined for any given angle of incidence. In the case where $\rho = 90°$, η is the complementary angle $a - a'$. We get therefore

$$\cos (a - a') = \sin \eta = \frac{R}{r} \sin a',$$

and hence

$$\frac{r}{R} = \frac{\sin a'}{\cos (a - a')}.$$

From the last expression we find that with any angle of incidence a proportion between the radii may be imagined that will give to the emergent ray the direction of the incident one. Since, then, the line of direction of the ray emerging without deviation, and with it the virtual focus, moves the further inwards the smaller a is, it is not without practical interest to compare the position of the focus, which appears in the microscopic image as a bright line with the ratio between R and r, and to investigate the general relations which here exist. We have collected below a few values of $r : R$ with the corresponding distances of the virtual focus from the centre (denoted by F in the table), and have added the corresponding angles of incidence and refraction. The refractive index is taken as $\frac{1\cdot 5}{1\cdot 3356}$, and R as unity. The quantities F and r are therefore expressed in fractions of half the longer diameter.

a	a'	r	F
20°	17° 44'	·304	·3420
25°	22° 6'	·3768	·4226
30°	26° 26'	·4460	·500
35°	30° 43'	·5121	·5735
40°	34° 55'	·5746	·6427
45°	39° 1'	·6331	·7070
50°	43°	·6872	·7660
55°	46° 50'	·7368	·8191
60°	50° 27'	·7801	·8660
65°	53° 48'	·822	·9063
70°	56° 48'	·8594	·9397
75°	59° 19'	·8933	·9659
80°	61° 16'	·9259	·9848

The virtual focus is therefore always further from the centre than the inner bounding surface of the cylinder-wall. Since then, as the calculation shows, all rays, whose angles of incidence are with equal radii somewhat larger or smaller than those above denoted, invariably appear to come from points which lie nearer to the periphery, it follows that the innermost part of the wall will fall for a breadth of $F - r$ in the umbra, while the peripheral part receives (just as the margin outside the rings in the air-bubble) a faint illumination, which is, however, intensified under the above conditions by the marginal rays as mentioned in (1). If the surrounding medium is water, and if ω and δ are taken tolerably large, the resulting brightness will therefore in many cases equal that of the field of view, and thus the bright line will appear in the direction of the margin without definite limits, only slightly increasing the effect of light.

The assumption, usually tacitly made in practice, that the lumen of a cylindrical cell reaches as far as the marginal shadow, is therefore, from what we have adduced, essentially incorrect. The error is the greater in proportion to the refractive power of the substance; it is infinitesimal only when the object lies in a medium of approximately equal density. This, of course, holds good also for the cavities in starch-grains, nuclei, globules of oil, &c.—they all appear somewhat larger than they actually are.

Since with thin walls the reflexion is more complete and in many cases *total*, the bright line becomes in this case also of greater intensity. It is most brilliant in glass tubes filled with air and immersed in water.

The unequal refrangibility of the differently-coloured rays produces here, as with the air-bubble, narrow coloured fringes on the bright line, the outer fringe appearing blue and the inner one red, as is evident from the refractive indices.

Another table may be serviceable here, exhibiting the influence of the refractive index n on the distance of the bright line for different values of r. The numbers have been partly determined by interpolation, but are accurate to three places of decimals. As unity we have taken the longer radius R, as above.

Radius of the Hollow-space r.	Distance of the bright line from the Centre.		
	$n = \dfrac{1\cdot 5}{1\cdot 2356}$	$n = \dfrac{1\cdot 4}{1\cdot 3356}$	$n = \dfrac{1\cdot 35}{1\cdot 3356}$
·5	·5598	·5236	·5054
·6	·6707	·6284	·6064
·7	·7798	·7329	·7075
·8	·8842	·8366	·8085
·9	·9698	·9378	·9095

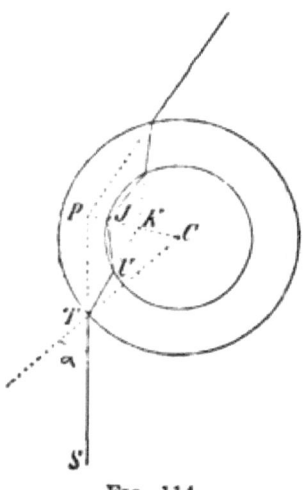

Fig. 114.

We have, in the third place, to trace the rays which enter the hollow space and are reflected at the walls so that they reach the objective only after a four-fold refraction and a single reflexion. This course of the rays involves, as in the corresponding case of the air-bubble, the formation of a second bright line, which always corresponds to the point in which the ray emerging without deviation (if produced backwards) cuts the plane of adjustment. For the determination of this point let us take the ray $S\,T$ (Fig. 114) incident vertically from below, whose angles of incidence and refraction shall be designated by a and a', as before. The angles which for the inner bounding surface of the wall are the angles of incidence and refraction, we will denote by a'' and a''', and the half-angle of deviation $C\,P\,T$ by ρ. Then $\rho = \angle C\,K\,T - (a-a')$, or, since $C\,K\,T$ is an exterior angle to the triangle $U\,K\,J$

$$= (a''' - a'') + a''' = 2\,a''' - a'',$$
$$\rho = 2\,a''' + a' - (a'' + a),$$

where $\sin a'' = \dfrac{R}{r}\sin a' = \dfrac{R}{nr}\sin a$, and $\sin a''' = n\sin a'' = \dfrac{R}{r}\sin a$, n being the relative index of refraction of the substance of the cylinder. As the ray emerges without deviation we obtain the condition $2\,a''' + a' - (a'' + a) = 90°$. If the surrounding as well as the enclosed medium is water, and the refractive index of the hollow cylinder $n = \dfrac{1\cdot 5}{1\cdot 3356}$, we obtain for the following ratios between r and R the values of a subjoined, and thence the distances F of the bright line from the centre.

$\dfrac{r}{R}$	a	a'	a''	a'''	F
·8	51° 48′	44° 24′	61°	79° 13′	·7859
·8904	61° 30′	51° 30′	61° 30′	80° 45′	·8788

In the second series $\dfrac{r}{R}$ was assumed $=\dfrac{1}{n}$, by which $\sin a'' = \sin a$, and therefore $a'' = a$.

The comparison of the last column with the first shows that the values of F are somewhat smaller than those of r, or, in other words, that the inner bright line falls in the hollow space. It is evident that its distance from the wall will increase and decrease with the refractive index; for since a''' is not dependent upon n then, if we bring the equation of condition to the form

$$2\,a''' - [a + (a''-a')] = 90°,$$

the expression within the square brackets will be smaller when n increases, because in that case a'' and a', and at the same time their differences, become smaller also. We must therefore assume a larger in this case to satisfy the equation; and the bright line is moved further inwards. Similarly may be explained the movement outwards of the line when n decreases. These displacements are, however, as calculation shows, so small that they may in general be neglected. We have tabulated below a few values of n and F. The assumed ratio of the radii is $r = \cdot 8\, R$.

Values of n	1·1231	1·2345	1·649
Corresponding values of F.	·7859	·7799	·7694

If the refractive index remains constant, while the ratio of the radii is altered, the difference between r and F will increase until r becomes about $\frac{1}{2} R$, and afterwards becomes less again when $r < \frac{1}{2} R$. But here also the changes are so slight that they may, in most cases, be disregarded. For $r = \cdot 8$ to $r = \cdot 5$ they scarcely reach $\cdot 007 \cdot R$. For comparison is also appended a table, in which the quantity $r - F$ (the distance of the bright line from the wall) is given for different values of r. We have taken the refractive index $n = 1·649$.

Values of r	·8	·6	·5	·4	·2
Corresponding values of $r - F$	·0306	·0365	·0372	·0345	·0211

If the cylinder is filled with air and surrounded by water, the inner bright line takes almost the same position as in an air-bubble; it is brought only very slightly nearer the centre by the influence of the cylinder-wall. If, for instance, the absolute refractive index of the cylinder-wall $= 1·4$, and that of water $= 1·3356$; then we obtain for the distances F, of the line from the centre, the values in the first series below, to which have been added, as a second series, the somewhat higher ones, obtained with equal value of r in the air-cylinder. The numerical relations are expressed in fractions of R, as before.

	Distance of the bright line from the Centre.		
	$r = \cdot 8$	$r = \cdot 7$	$r = \cdot 5$
In the hollow cylinder .	·5485	·4792	·3415
In the air cylinder . .	·5501	·48785	·3442
Difference	·0022	·0026	·0027

If the refractive index of the substance of the cylinder rises to 1·5, the bright line is moved somewhat further inwards. Its distance is reduced therefore to ·5458 R, if $r = ·8\,R$; the difference rises in this case to ·0054. The influence of the cylinder-wall upon the distances in question is therefore in all cases limited to the third decimal, as the preceding examples prove; consequently the displacement of the line produced by it amounts to less than 1 mic., when R is less than 100 mic., and may reasonably be neglected. In this case we can determine, from the position of the line in the enclosed air, the radius r of the lumen, and therefore also the thickness of the wall $(R - r)$. We therefore get, in accordance with what has preceded for the air-bubble, $F = ·68836\,.\,r$, and consequently for the hollow cylinder, taking as mean displacement ·00636 . r, $F = ·683\,.\,r$, whence $r = \dfrac{F}{·683} = 1·464\,F$. The quantity F is most accurately determined by direct measurement of $2\,F$, that is, of the distance of the bright lines on either side.

The inner bright line lies nearer to the centre than any other point which is illuminated in the same way, precisely as the inner ring in the air-bubble. Consequently there appears within it (with reference to this particular path of rays only) a perfectly dark umbra, but further outwards a faintly luminous penumbra. Under favourable circumstances other bright lines appear in the inner part of the penumbra, their origin being explained by repeated internal reflexions.

We have finally to consider the rays which penetrate the lumen without internal reflexion, and which undergo, in consequence, only a fourfold refraction. It is clear from what has already been said, that they will illuminate the central portion of the hollow cylinder, but disappear towards the edges. The points where the penumbra begins, and where it passes into the umbra, alter their position in accordance with the magnitude

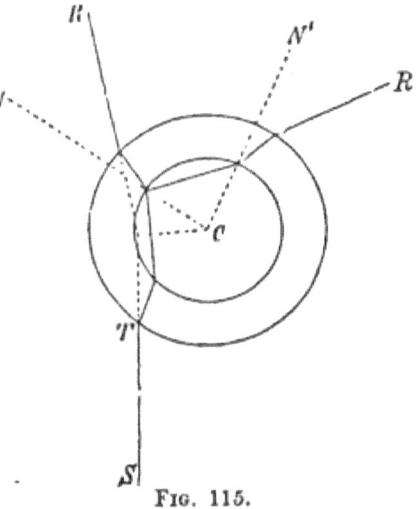

Fig. 115.

of the angles of aperture ω and δ, the thickness of the wall, and the refractive power of the different media. For the determination of the latter by a formula which will be generally valid, the following consideration may be utilized. If the ray $S\,T$ (Fig. 115), incident from below, were, as in the previous case, reflected at the inner wall, the angle of deviation of the ray emerging towards R' would be given by the equation

$$2\rho = 4a''' + 2a' - 2(a'' + a).$$

As there is no reflexion, the path of the rays undergoes the same modification as if, in the above figure, the line $C\,N'$ and the reflected ray corresponding with it were turned, like the hand of a watch, round the point C, but to the left, till $C\,N'$ coincides with $C\,N$. If we mentally follow this movement it will be seen that the angle of deviation 2ρ is increased by the angle $N''\,C\,N = 180° - 2a'''$. Adding this value to the one above obtained, we get the angle of deviation 2ρ for the ray which has been refracted four times, and hence

$$\rho = 90° + a''' + a' - (a'' + a).$$

The direction of the emergent ray is determined from this for any given inclination of the incident ray. If, therefore, the angles of aperture of the objective (ω) and of the diaphragm (δ) are given, the limits of the umbra and penumbra in the plane of adjustment can be calculated. For the distances of these limiting lines from the centre, if $R = 1$, $n = \dfrac{1\cdot 5}{1\cdot 3356}$, $\omega = 60°$, and $\delta = 12°$, we obtain the values given in the following table:—

r.	Limits of the penumbra.	Limits of the umbra.
·8	·862	·947
·5	·557	·57

In reality the limits of the shadow always lie somewhat further inwards, if the above suppositions are approximately accurate. It is evidently in consequence of this, that the intensity of the marginal rays, for which a''', for instance, may be 80° and upwards, is considerably weakened through the repeated reflexions at the refracting surface, and will therefore become *nil* for the observing eye before it should do so theoretically. Hence it is intelligible that the inner bright line, even with tolerably high amplifications, will still fall in the marginal shadow of the hollow space.

To demonstrate the action of the four different systems of rays, we have represented in Fig. 116 the distribution of light and

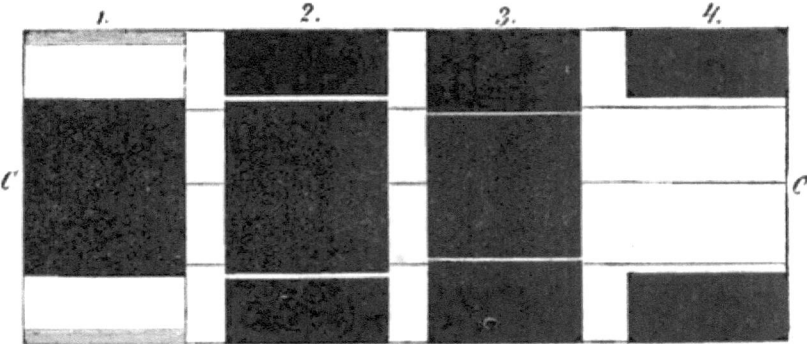

Fig. 116.

shadow for each separate path of the rays where $R = 2\,r$. No. 1 shows the image of the twice-refracted rays which traverse the walls in a straight line; No. 2 represents the outer bright lines which proceed from the rays which are twice refracted and once reflected at the inner surface of the wall; No. 3, the fine inner lines which correspond to the rays which have entered the cavity and been reflected once at the inner surface of the wall; finally, No. 4 represents the marginal shadow which the rays traversing the lumen, and consequently refracted four times, would of themselves produce. Since umbra and penumbra nearly coincide, we have throughout, for the sake of simplicity, drawn only the limiting lines of the umbra, and indicated for clearer determination the outer and inner walls of the cylinder and the axis $C\,C$ by lines through the figure.

If we regard the hollow cylinder, without reference to the plane of adjustment, simply as a refracting apparatus, it acts upon the marginal rays, which do not enter the lumen, as a sphere of equal density; but on the central rays, which traverse the hollow space, as a concave lens. It forms, however, real and virtual images of objects which are reflected in the mirror, both of which become appreciable to the eye as soon as the instrument is focused to a suitable level. The position of the real focus is, of course, approximately the same as with the sphere; that of the virtual focus may be determined from the above formulæ for the optical constants of the refracting media. If we denote the negative focal length reckoned from the axis of the cylinder by F, the absolute refractive index by n, and the radii as before, by R and r, we obtain, under the suppositions given below, the following values:—

Medium.		Values of F when $n = 1\cdot 5$.	Values of F when $n = 1\cdot 6$.
$r = \cdot 5$	In water	$3\cdot 314\,R = 6\cdot 628\,.\,r$	$2\cdot 189\,R = 4\cdot 378\,.\,r$
	In air	$1\cdot 5\,R = 3\,.\,r$	$1\cdot 333\,R = 2\cdot 666\,.\,r$
	Filled with air, surrounded by water	$\cdot 571\,R = 1\cdot 142\,.\,r$	$\cdot 45\,R = \cdot 9\,.\,r$
$r = \cdot 1$	In water	$\cdot 363\,R = 3\cdot 63\,.\,r$	$\cdot 238\,R = 2\cdot 38\,.\,r$
	In air	$\cdot 166\,.\,R = 1\cdot 66\,.\,r$	$\cdot 148\,R = 1\cdot 48\,.\,r$

This table affords, of course, only a superficial criterion of the position of the focal point in hollow cylinders, and is intended to show, by means of examples, the influence which the ratio of the radii and the surrounding media in general exercise upon the focal length. It is easily understood that if r is very small as compared with R, the curvature of the outer surface of the cylinder may be neglected; the optical effect is then approximately equivalent to that produced by a hollow space of equal radius in a homogeneous substance of the density of the cylinder, and with plane bounding surfaces above and below. Similarly, it is evident that if the ratio $r : R$ approaches unity, the focal length will become greater, and at last infinite. The virtual image here remains microscopically observable only up to a certain limit, with extremely small values of r and R, although still perceptible in tubes with somewhat thin walls.

If the focal lengths are determined for the above cases experimentally, by measuring the displacement of the body-tube with a second Microscope placed horizontally, we obtain throughout smaller values than those calculated, because the observed focal lengths, as was above (p. 201) stated, always refer to marginal rays only, which are inclined towards the axis more or less (*e.g.*, about 15° or 20°) according to the power and speciality of the objective.

Since the image of a hollow cylinder, as well as every real or virtual image, may be regarded as a source of light, the focal adjustment to its plane will in every achromatic instrument give a colourless microscopic image; a higher or lower adjustment, on the other hand, producing the well-known red and blue fringes. Consequently, small hollow spaces, which enclose a medium of less refractive power (as the lumina of bast-fibres, the nuclei of starch-grains, &c.), if viewed with a chromatically under-corrected Microscope, will appear colourless and bright with low adjustment, and reddish and finally dark with higher adjustment. The true mean focal adjustment to the centre of the hollow space lies, under all circumstances, between these two extremes, and gives, therefore, if the dimensions are sufficiently small, a reddish image.

II.

OBJECTS OF IRREGULAR FORM.

1.—Membranes with Small Depressions or Holes.

It is evident that small concave depressions or furrows, as represented in Fig. 117, will act as concave lenses, and will consequently give a virtual image of the diaphragm on a suitable lowering of the body-tube. And indeed this virtual image must appear as distinct to the naked eye as if viewed in the Microscope, notwithstanding its minuteness.

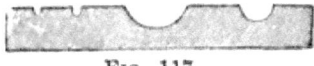

Fig. 117.

As soon, however, as the form of the depression is essentially altered, if, for instance, it becomes prismatic or cylindrical, and has a flat base, the image will be lost to the naked eye, while it remains perceptible in the Microscope by reason of the difference in the angle of aperture. We may prove this by spreading out a solution of common salt, sulphate of magnesia, &c., upon the object-slide, and allowing it to dry in the air. In certain places are formed large homogeneous lamellæ or extended incrustations, with an infinite number of small porous holes or depressions, cracks, furrows, &c., of the most different forms—which are sometimes very beautiful on the addition of diluted alcohol. In most of them the image of the diaphragm, window-frame, &c., appears with tolerable sharpness if we lower the focus slightly below the medium plane of adjustment, and often as distinctly as with a somewhat flattened air-bubble.

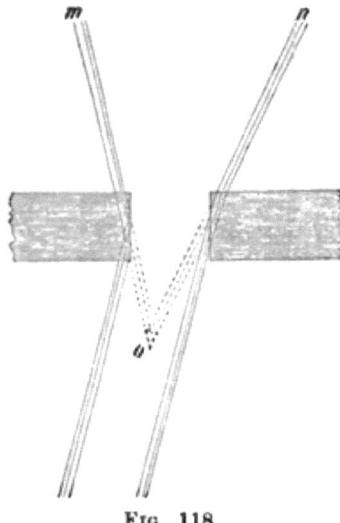

Fig. 118.

The formation of these images is evident from Fig. 118. The edge of the aperture or depression causes on the one side a deviation of the rays emanating from a distant point (therefore incident nearly parallel) towards n, and on the other side a deviation by total reflexion towards m. Both the pencils, after their passage through the object, appear to proceed from a small space o, in which they intersect when produced backwards; they therefore form in o the virtual image of the distant source of light. If the latter is moved further to the right, its image also moves in the same direction, as obtains with virtual images.

The formation of the image is not dependent upon the regularity of the refracting surfaces, as has been assumed in our figure. It is unimportant whether the refracted rays are exactly parallel, or whether they diverge or converge slightly, since the space in which their productions intersect becomes thereby somewhat larger or smaller. It always remains so small that, compared with the deviations which the cover-glass produces, it may be neglected, and to the observing eye will appear as a point.

Though the incident pencils reach the eye through one edge only of the aperture—*i.e.*, do not intersect if produced backwards—an indistinct image is seen precisely as with rays which pass through a minute aperture. It does not, however, in this case assume any definite position, but appears the greater the lower the focal adjustment. With favourable adjustment even those rays which traverse the aperture without refraction must contribute to the formation of this image.

In whatever manner the image is formed, whether it appears more or less indistinct or sharply outlined, the focal adjustment to a higher or lower level will always produce phenomena similar to those observed in the virtual image of the air-bubble. On raising the tube the central part will appear red, and the edge bluish, whilst lowering it below the virtual focus will produce the opposite arrangement of the colours, assuming in both cases that the objective is under-corrected. Since this is generally the case with Microscopes, minute pores, cracks, furrows, &c., will always appear reddish, if we focus the instrument to about their centre, and indeed more or less so according to the peculiarity of the instrument. The bluish fringe, too, though of very slight breadth, is in many cases perceptible.

2.—Membranes Bounded by One Plane and One Undulating Surface.

The elevated portions of membranes of this kind (Fig. 119) evidently act as convex lenses, and their depressions as concave lenses. The distribution of light and colour must therefore in general be as follows: with the highest focal adjustment ($m\ n$) to the real

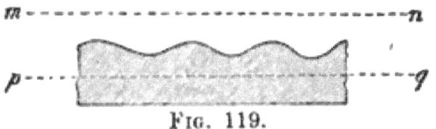

Fig. 119.

images of the projections, the latter will appear bright, and the depressions dark or reddish, according to the curvature; if we lower the tube, the projections will have a bluish tint, and the depressions will remain red, until we reach the plane ($p\ q$) of their virtual images. With still lower focal adjustment, the latter become bluish; the reddish fringes, which they surround, then take the place of the elevations—blended in pairs—which now appear as the depressions did when the focal adjustment was higher. What colours will alternate with a given adjustment depends, of course, upon the curvature of the upper surface.

If the elevations project very much, and rest upon a somewhat narrow base, in addition to the phenomenon just mentioned reflexion occurs, as observed in minute corpuscles, globules of mercury, &c., which we shall more fully discuss hereafter. This obtains, for instance, with many spiral and annular vessels, with branch-like crystalline films (which act in the same way as a thickening of the object-slide), the siliceous envelopes of diatoms, &c. In these objects it frequently occurs that single fibres, even if quite isolated, appear reddish on lowering the focal adjustment.

The optical action of elevations and depressions remains approximately the same in the Microscope, if the undulating form of the bounding surface passes into acute serrations (Fig. 120); for edges projecting outwards always produce real images, and those indented virtual ones. Such forms seldom occur in vegetable organisms; they are more often observed in crystalline films, *e.g.*, in those chains of (imperfect) crystals of common salt, resembling the plan of a fortress.

Fig. 120.

3.—Membranes Bounded by Parallel Undulated Surfaces.

A wave-like membrane, as represented in section in Fig. 121, may be divided into groove-shaped portions of hollow cylinders which turn their convex and concave sides upwards alternately. This division is exhibited in the figure by straight lines, which unite the centres of curvature, as well as by different shading of the portions of the membrane. Since the hollow cylinder acts optically as a concave lens, such portions also act dispersively upon the incident light. The two principal points coincide with the centre of curvature, if the surrounding medium is air, but become separated, and at the same time approach the refracting surfaces, if the surrounding medium is water or other liquid. In regard to this point and the magnitude of the focal lengths, the following table exhibits a few examples which have been calculated. By e and e' are denoted the distances of the principal points from the centres of curvature, by ϕ the focal length, by r the radius of the stronger curvature, and by n the absolute refractive index. The radius of the lesser curvature (R) is taken as standard of unity, and the surrounding medium water with the index $\frac{4}{3}$.

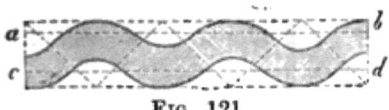

Fig. 121.

r	n	e	e'	ϕ
$\frac{1}{2}$	1·5	$\frac{1}{4}$	$\frac{1}{4}$	6·75
	1·6	$\frac{1}{4}$	$\frac{1}{4}$	4·5
$\frac{1}{10}$	1·5	$\frac{1}{4}$	$\frac{1}{10}$	·75
	1·6	$\frac{1}{4}$	$\frac{1}{10}$	·50

The principal planes of the groove-shaped portions of the membrane, from which the virtual images are at equal distances, do not therefore lie in the same level. For instance, under the relations represented in Fig. 121 for the grooves concave upwards, they coincide with the line $a\,b$, and for those alternating with them with $c\,d$. The virtual images themselves lie therefore alternately higher and lower, and since they act as microscopically small sources of light, the different colours which are observed with higher and lower focal adjustment may hence be explained. As an example, if the instrument is focused to a plane that lies higher than the lower, and lower than the upper virtual images, the latter will appear bluish, the former reddish.

4.—Alternate Solid and Aqueous Layers.

Aqueous layers which alternate with solid ones, when viewed vertically, act as attenuated places or crevices in a homogeneous substance. Equivalent effects are obtained if the ratios of refraction between the solid and the aqueous layers are the same as between the homogeneous substance and the fluid occupying the crevices. The microscopic image exhibits, therefore, just as with fibrous thickenings, alternately bluish and reddish lines, corresponding to the virtual and real images of the portions which are not exactly in the plane of adjustment. The difference in brightness, which is produced by alteration of the object-distance at definite image-points, as well as the intensity of the shadows, offer to the practised eye important data for estimating approximately the differences in density. To distinguish without further investigation, when in a given case actual crevices or merely aqueous layers are present, is theoretically impossible.

5.—Elevations and Depressions as Opposed to Dense and Loose Layers.

Since elevations, as already pointed out, act optically exactly as corresponding thickenings of the substance, the solution of the question, whether fine markings (such as occur in diatoms, striated cell-membranes, &c.) are due to the differences of density or of form, is not possible by mere examination of the objects in water. In most cases, however, means are available which answer our purpose. If we place the object in a medium that refracts the light more strongly than the densest parts of the object—for instance, sulphide of carbon—the marking will remain essentially unaltered if it is caused by differences of density; on the other hand, if the cause lies in the unevenness of the surface the marking changes its character, so that the light and shadow are reversed, as in a photographic negative. The pores of a membrane then appear like papillary elevations, fibrous thickenings, or crevices, and are reversed; we might regard the whole image also as a negative.

The explanation of these phenomena is evident. If the object is bounded by plane surfaces, so is also the surrounding sulphide of carbon. The latter therefore acts just as a plate of glass of corresponding thickness—it raises the object points without altering their images. If there are, on the other hand, depressions or elevations on the upper or lower surface of the object, at the surface of contact of the surrounding medium the opposite relief occurs, and since this latter is denser it will determine the distribution of light in the microscopic image. The object itself acts optically as a cavity of similar shape in a refracting substance.

In experimenting with this process, we must not of course allow the object to dry. The loose and dense layers must be saturated with water, otherwise the latter will project outwards, in consequence of the greater loss in the aqueous portions, and hence interfere with the optical action. In a perfectly dry state no divisions into separate layers will appear.

6.—Vision through Stereoscopic Binocular Microscopes.

The remarkable illusion which stereoscopic contrivances of all kinds produce, is in general a purely psychological phenomenon which we need not discuss. We confine ourselves to the physical question, how the two images which the binocular Microscope presents to the eyes are to be distinguished from one another; whether their mutual relations with reference to the distribution of light and shadow are similar to those which obtain in stereoscopic photographs. If this is really the case, the perception may be classed with other known phenomena, and we may leave to Physiology any further consideration of the point.

If we recall once more the dioptric and the catoptric division of the pencils of rays (cf. Figs. 14—16), it is evident that one eye receives an image which is formed by the right half of the objective, whilst the other eye receives an image formed by the left half of the objective. The image-forming halves of the objective always lie, with the dioptric method, on the same side of the median plane as the observing eye, and with the catoptric method, under certain circumstances, on the opposite side—for instance, if the equilateral prisms of Nachet are employed. The impression which the fusion of the two images renders perceptible will of course differ according to the method of division of the pencils. The dissimilarity of the images formed by the right and left halves of the objective may be determined by calculation, or it may be observed experimentally by successively covering up the halves of the objective. Viewing, for instance, a spherical starch-grain, or a minute globule of oil, the marginal shadow, which by direct (axial) illumination appears in the usual image equally broad, will preponderate by covering up one half of the objective on the corresponding side, so that the spot of light in the image lies on the same side of the median plane as the effective part of the objective. The same phenomenon will occur in general with every other object which acts as a convex lens. On the other hand, if the object is an air-bubble, or a cavity, or if it acts as a concave lens, the spot of light moves under the same suppositions to the side also, but in an opposite direction. The two A images which the binocular Microscope forms, exhibit a distribution of light and shadow as represented in A and B (Fig. 122). The presentation of the one or the other image to the right eye or to the left, depends, with a given object, upon the optical arrangement by which the division of the pencils is effected. B The grouping of A and B to the right or to the left clearly produces, however, the stereoscopic effect.[1]

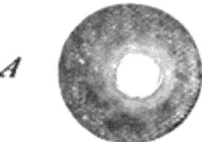

Fig. 122

We may have a combination which represents elevations and

[1] We must not forget here that a half reversal by a single reflexion produces the same effect as the changing of the images.

depressions as such, for instance, the binocular stereoscopic Microscopes of Nachet, but those binoculars in which the images are seen in reversed order ($B\ A$ instead of $A\ B$) will produce the opposite impression—exhibiting therefore the elevations as depressions, and conversely. This pseudoscopic appearance was indeed, as Harting reminds us, observed in 1853 both by Riddell in his catoptric, as well as by Wenham in his dioptric binocular Microscope.[1] The conditions which produce the pseudoscopic effect, as well as the means for its elimination, have now become generally known; many objects are, however, wholly unadapted for binocular observations, since they produce under all circumstances bizarre images.

Among the contrivances by which the true stereoscopic image can be changed to a pseudoscopic one as desired, that devised by Nachet is simple and convenient. Nachet has, for several years, made stereoscopic Microscopes with a movable prism, mounted so that the changing of the images is effected by a sliding motion. The prism is constructed as shown in Fig. 123, and is placed over the objective, so that the horizontal projection of the oblique surface P (45° to the horizon) covers half the aperture. According as the oblique face of the prism is adjusted to reflect one or the other half of the image pencils, the same eye will see the image formed by the right or the left half of the objective. By the movement of the prism the stereoscopic effect is altered (supposing the object is adapted for observations of this kind), and what in the one case appeared as an elevation, is represented in the other as a depression.[2] This arrangement is specially recommended for the production of stereoscopic photomicrographs, because it is possible to employ one and the same tube, without alteration of adjustment, for the reception of the two images.

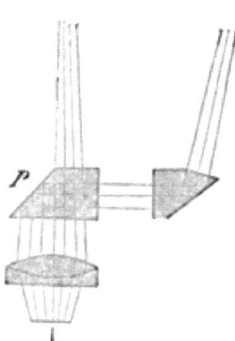

Fig. 123.

[1] Harting: "Das Mikroskop," 1st ed. pp. 195 and 775; 2nd ed. i. p. 199, iii. p. 240.

[2] Numerous examples of this are supplied by Valentin: "Archiv für mikr. Anat." vi. (1870) p. 581. They belong, however, rather to the theory of the stereoscope than to the theory of microscopic observation. The "great future" which Valentin prophesies for the binocular Microscope, especially for the more accurate examination of crystals, we have not at present much faith in.

To what extent the depth of the field of view modifies the stereoscopic effect we leave undecided. Without comparative observations of objects with or without depth (*e.g.* photographs on glass, &c.) nothing decisive can be said on this point. In Harting's first edition it was stated that this factor is of special importance in stereoscopic observation, which we considered quite erroneous. In the second edition, however, he merely states that "in forming a judgment on the performances of the binocular Microscope" the depth of the field of view *also* comes into account (*loc. cit.* i. p. 213). It is not however proved, but merely indicated, that the field of view possesses a certain depth, which no one has doubted. So much is at any rate certain, that the influence in question is of slight importance, and in no case necessary for supplementing the stereoscopic effect. For as in ordinary stereoscopes two plane pictures are combined to form one stereoscopic image, the images of the binocular Microscope would also produce the impression of solidity, even if the depth of the field of view were *nil*.[1]

That the differences in the distribution of light and shadow determine the stereoscopic effect is proved also by the above-mentioned photo-micrographic process, but especially by the use of the so-called stereoscopic "see-saw." Instead of bringing into action the right and left halves of the objective successively, in order to observe the object equally from different points of view, the object-carrier is tilted to the right and left, so that the object appears to be seen under a different angle. The shifting is effected between the two operations, and without alteration of the focal adjustment.[2] The angle through which it is turned amounts with low amplifications to $12°$, with medium to about $7°$ or $8°$, and with high powers (*e.g.* with Nachet's No. 5 objective) to about $4°$ or $5°$.

---

[1] Helmholtz ("Handb. der Physiolog. Optik," p. 682) also explains the origin of the stereoscopic effect by the position of the dispersion circles which the points of the object situated in front of or behind the plane of adjustment produce, according to which, therefore, a certain depth of the field of view would be an important condition. But Helmholtz entirely disregards the fact that the images formed by the two halves of the objective are actually different even without the supposed dispersion circles, and would therefore of their own accord produce a stereoscopic effect.

[2] Cf. on this point Benecke: "Die Photographie als Hülfsmittel mikroskopischer Forschung." Braunschweig, 1868.

Finally, we will refer to a point which we think has hitherto been overlooked. Elevations and depressions may be always replaced in transmitted light by differences of density, which produce exactly similar optical effects. If we suppose a membrane of equal thickness throughout, but with dense and loose parts, the former must appear convex and the latter concave, in the stereoscopic image, and consequently, produce an illusive impression on the mind in the endeavour to combine them. Hence we consider it in all cases preferable to conduct scientific researches with the ordinary monocular Microscope.

III.

INTERFERENCE PHENOMENA.

A. In the Microscope.

1.—Delineation of the Fine Structure of Objects by Interference.

According to the researches of Abbe,[1] which we have already several times referred to, the microscopic image of the object is formed by the superposition of *two* images, of which the one contains more especially the contours, and is generated in the usual dioptric method; while the other reproduces the fine details of structure, and is due to the pencils deflected in the object and which *interfere* in the Microscope. We have now to explain the origin of this interference image and to establish the conditions which influence the power of distinguishing fine details in the object.

The theory of diffraction shows, as is well known, a characteristic change which light undergoes in its passage through objects composed of fine structural details. In general terms the change consists in this, viz., that every incident parallel pencil of rays is resolved into a divergent group of rays with large angular extension, in which a regular increase and decrease of intensity is apparent as a necessary consequence of the difference of phase.

[1] "Archiv für mikr. Anat." Bd. ix. p. 413.

If AB (Fig. 124) is an object with alternately transparent and opaque striæ, the figure representing a transparent part ab and the adjoining opaque parts Aa and bB; then, of the rays incident at right angles on the left side, part only continue in the same straight line after their passage through ab, another part being deflected and forming divergent luminous pencils which completely occupy a given angular space. In the figure one of these pencils is represented with a deviation of 30° to the direction of the incident rays. Assum-

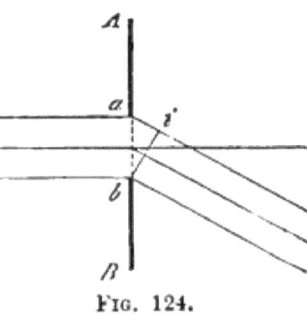

Fig. 124.

ing now the breadth of the transparent portion $ab = 1$ mic., and the wave-length of the incident light (considered to be homogeneous) $= \cdot 5$ mic., then we get for the quantity ai the proportion
$$ai : ab = \sin 30° : r;$$
therefore $\quad ai = \dfrac{\sin 30°}{r} \cdot ab = \cdot 5$ mic. $= 1$ wave-length.

If we divide the deflected pencil into two equal portions, by drawing from the middle of the aperture a line parallel to the two marginal lines, it is at once evident that every ray of the one half, as compared with the corresponding ray of the other half, will appear to be displaced by half a wave-length. The combined action therefore of the two halves must necessarily produce darkness, if the rays are all united by appropriate refractions. If, for example, we adjust the objective of a Microscope to a given object, a dark line will necessarily appear in the focal plane where the rays of the deflected pencil unite. It is the first dark line which appears by the side of the bright line which the incident rays produce in the focus of the objective. Let fe,

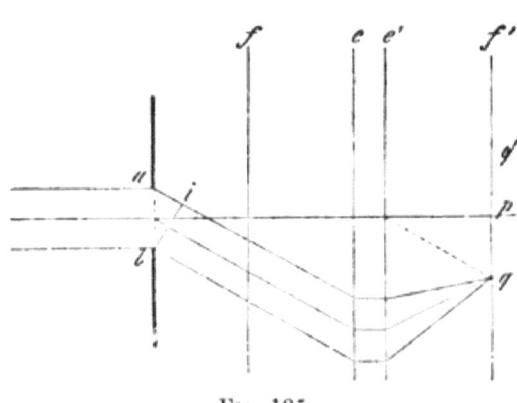

Fig. 125.

$e'f'$ (Fig. 125) be the principal focal planes of the objective (through want of space the focal lengths are represented much too short in proportion to the aperture $a\,b$); then, as the construction shows, the dark line will appear at q, and the bright line at p. Similarly, rays deflected towards the opposite side will form a dark line at q'. The distance $pq = pq'$ is given in the figure by the trigonometrical tangent of the angle $a = 30°$, drawn to a radius equal to the focal length; with a focal length say of 3 mm., this would be $= \dfrac{\tan 30°}{r} \times 3 = 1\cdot732$ mm. In reality, however, these distances are proportional to the sine of the angle of deflexion, as has been shown by Abbe (*see* page 25).

If we increase the width of the opening $a\,b$ from 1 mic. to 1·5 mic., then, *cæteris paribus*, $a\,i = \cdot75$ mic. $= 3$ half wave-lengths. The deflected pencils may therefore be divided into three parts whose marginal rays will be displaced by half a wave-length. Each pair will therefore annul each other, as in the preceding case; the third ray, however, will remain unaffected, and will produce at q a *bright* line instead of a dark one. This line is the first bright line which in the diffraction image follows the focal line of the direct rays. Between this line and the point q the dark line already mentioned appears, for which $a\,i$ is equal to a whole wave-length; for, obviously, a smaller inclination of the deflected rays corresponds to this value for the supposed *larger* aperture. We get $\sin a = \dfrac{\text{wave-length}}{a\,b} = \dfrac{1}{3}$, from which $a = 19° 28'$. The change just mentioned takes place, of course, on the other side of the optic axis; a bright line appears at q', and a dark one between p and q'.

If the angle of aperture of the objective is large enough to admit other pencils, for which $a\,i = \dfrac{5}{2}$ wave-lengths, then another bright line will appear below q (and, similarly, above q'), which again will be separated from the preceding ones by a dark one. For this new line we get $\sin a = \dfrac{5}{2} \cdot \dfrac{\text{wave-length}}{a\,b} = \dfrac{5}{6}$; consequently, $a = 56° 26'$, which supposes an angle of aperture $2a = 112° 52'$. Any further deflected pencils cannot be admitted under the given conditions.[1]

[1] If λ represents the wave-length, and b the width of the opening, then the

If, instead of a single transparent opening, several come into action, the effect is merely intensified, since all the diffraction pencils, which are inclined to the same extent, necessarily intersect in the same point q of the focal plane, and consequently add to the effect. There would now be an end to the question if we could assume that these pencils did not give rise to any new interferences. In reality, however, this is not the case. The diffraction pencils proceeding from adjacent parts are by no means without influence upon one another. For, in addition to the above-mentioned diffraction lines, new lines are formed in the focal plane which originate through the interference of adjacent pencils. The detailed explanation of the phenomena appertaining to this question, especially for any given number of openings, would lead us too far. We must, therefore, refer to the text-books of Physics,[1] or to the detailed exposition of Schwerd,[2] and must confine ourselves here to a brief statement of the facts in question.

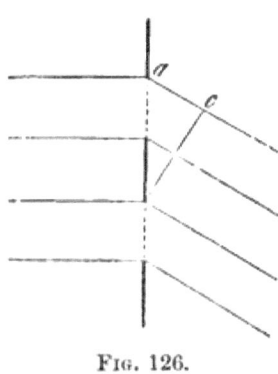

Fig. 126.

If the transparent openings are numerous, and their distances, measured from centre to centre, $= d$, then the pencils of deflected rays give maxima of brightness for all inclinations at which the difference of path $a\ c$ (Fig. 126) is equal to one wavelength, or to a multiple of a wave-length. The bright interference lines (spectra of the second order) will therefore appear in those points for which

$$\sin a = \frac{\lambda}{d}, \frac{2\lambda}{d}, \frac{3\lambda}{d}, \frac{4\lambda}{d}, \&c.,$$

where λ again denotes the wave-length and a the angle of deflexion.

deflected rays which proceed from a single opening will yield *bright* interference lines for the angles of inclination, whose sines are $\frac{3}{2} \times \frac{\lambda}{b}$; $\frac{5}{2} \times \frac{\lambda}{b}$; $\frac{7}{2} \times \frac{\lambda}{b}$ . . . . These sines are, therefore, in the ratio of the odd numbers 3, 5, 7, 9, &c.

[1] *Vide* Wüllner's "Lehrb. der Physik," 2nd ed. In the first edition this matter is erroneously explained. *Vide* also Verdet's "Leçons d'Optique Physique," 2 vols. Paris,—an excellent work.

[2] Schwerd's "Die Beugungserscheinungen," Mannheim, 1835,—specially §§ 148 and 195, where this is illustrated by diagrams.

When the wave-length $\lambda = \cdot 5$ mic., calculation gives the following as the relative values of a and d:—

Values of d in mic.	Sin a, when $\lambda = \cdot 5$ mic.	Values of the aperture a.
·5	1	90°
1·0	$\frac{1}{2}$, 1	30°, 90°
1·5	$\frac{1}{3}$, $\frac{2}{3}$, 1	19° 28′, 41° 46′, 90°
2·0	$\frac{1}{4}$, $\frac{2}{4}$, $\frac{3}{4}$, 1	14° 30′, 30°, 48° 35′, 90°
3·0	$\frac{1}{6}$, $\frac{2}{6}$, $\frac{3}{6}$, $\frac{4}{6}$, $\frac{5}{6}$, 1	9° 35′, 19° 28′, 30°, 41° 46′, 56° 26′, 90°
5·0	$\frac{1}{10}$, $\frac{2}{10}$, $\frac{3}{10}$, $\frac{4}{10}$, $\frac{5}{10}$	5° 44′, 11° 32′, 17° 27′, 23° 35′, 30° ..
10·0	$\frac{1}{20}$, $\frac{2}{20}$, $\frac{3}{20}$, $\frac{4}{20}$, $\frac{5}{20}$	2° 52′, 5° 44′, 8° 38′, 11° 32′, 14° 30′ ..

If $d = \lambda$, a will, under all circumstances, $= 90°$. Hence we find that no objectives can, with direct (axial) illumination, take up even the first image-forming diffraction pencil, if the distances of the openings are less than 1 wave-length $= \cdot 5$ mic.

The relations are different when the incident rays are oblique to the axis (Fig. 127). Besides the difference of phase $b\,n$ of the deflected rays, we have also that of the incident rays $m\,b$. If, then, a is the angle of inclination of the deflected and δ that of the incident rays, the difference of phase in the section $a\,n$ is equal to $\dfrac{\sin a + \sin \delta}{a\,b}$.

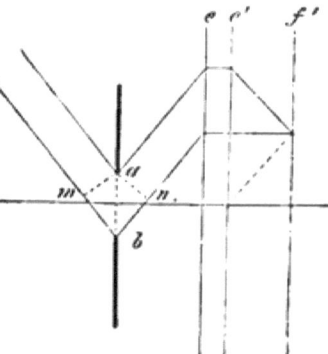

Fig. 127.

The resulting effects can easily be deduced from this. We will confine ourselves to the case where $a = \delta$, therefore $\sin a + \sin \delta = 2 \sin a$, and will consider the first diffraction pencil only. The condition above established for the bright lines with any number of openings, according to which $\sin a = \dfrac{\lambda}{d}, \dfrac{2\lambda}{d}, \dfrac{3\lambda}{d}$, &c.; becomes $2 \sin a = \dfrac{\lambda}{d}, \dfrac{2\lambda}{d}, \dfrac{3\lambda}{d}$, &c.; from which $\sin a$ for the first bright line $= \dfrac{\lambda}{2d}$ (instead of as above $\dfrac{\lambda}{d}$). By suitable oblique illumination it is possible, by means of the corresponding diffraction pencils, to bring still further lines into view, whose distance is only half that which is the limit of resolution with direct (axial) light. For the extreme cases, where a and δ are each equal to 90° and where $\frac{\lambda}{2d} = 1$, the minimum distance d is equal to half a wave-length.

For the further consideration of the effects of diffraction the results which we have above obtained may be conveniently expressed in a somewhat different form. We have hitherto assumed the incident pencils to consist of parallel rays. We will still retain this condition, only diminishing as much as possible the effective source of light, by means of a small diaphragm. The direct (axial) rays will then produce in the focus of the objective a dioptric image of the source of light, that is, of the small diaphragm, which may be seen in the body-tube after removing the eye-piece. But the deflected rays, which make with the optic axis the angle a, will also produce in q and q' (Fig. 125), as well as in every other point where interference produces light, similar images, which, by homogeneous light, agree in form and size with those produced direct. Any object moved between the mirror and the diaphragm is clearly seen in these interference images. In white light the effect is indeed so far essentially different, that the diffraction images will appear coloured, because, as is well known, the maxima of brightness are dependent upon the wave-length. A circular diaphragm, therefore, appears in the diffraction image, which is formed in the focal plane of the objective, to be drawn out radially, red outwards and blue within. We mention this complication incidentally, since it is apparent with the usual illumination. We will confine ourselves to the case of homogeneous light, assuming the incident rays to be strictly parallel, although the diffraction images are distinctly perceptible with ordinary (*i.e.*, not parallel) illumination.

For the demonstration of the diffraction phenomena simple lined objects, such as the small scales of *Lepisma saccharinum*, are the most suitable. The Microscope should be focused on one of the scales, and a small diaphragm used; the eye-piece is then removed, and the effect observed in the upper focal plane of the objective. The direct image of the diaphragm appears symmetrical, and on either side are seen the coloured diffraction images at right angles to the markings of the object. If the angle of aperture of the objective = 60°, the larger scales will exhibit several pairs of diffraction images; the smaller ones, however, are so finely striated, that only the two nearest will fall within the aperture of the objective.

We now come to our proper task, viz., to establish the effect which these diffraction phenomena produce in the plane of the real image. This may be most simply done, if we consider the aperture-images in the upper focal plane, the direct one as well as those due to interference, as so many (secondary) sources of light, whose rays interfere, as in Fresnel's experiment with the mirror. For, since these sources of light are point for point the optical images of the same primary source of light, there is no difference of phase between them.

In Fig. 128, let $A B$ be the optic axis, a the direct image of

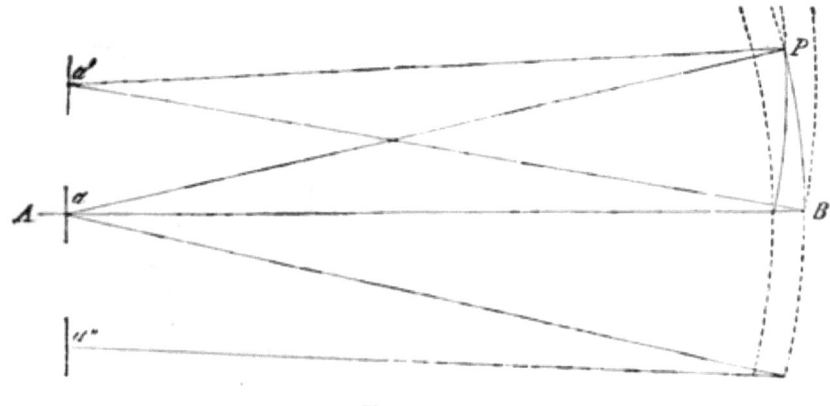

Fig. 128.

the diaphragm, and a' the nearest diffraction image. From corresponding points in these two sources of light (*e.g.* from the centre) let an arc of a circle be drawn through B (where we suppose the plan of the real image), and a second one parallel to the former at a distance of a wave-length; then the point of intersection P will be the point where the first bright diffraction line will be found; for $a P$ is evidently an entire wave-length greater than $a' P$, and therefore the interference reaches the maximum of brightness. In order to determine the distance $P B$, we may regard the two intersecting arcs as straight lines, of which one is at right angles to $a B$, and the other to $a' B$. Consequently, the small triangle, whose vertex is in P, and whose base is equal

to a whole wave-length ($=\lambda$), is similar to the triangle $a\,a'\,B$, and therefore
$$B\,P : \lambda = a\,B : a\,a';$$
whence
$$B\,P = \frac{a\,B \cdot \lambda}{a\,a'}.$$

But $a\,B$ is equal to the posterior focal length p^* less the focal f, that is, $p^* - f$. Similarly, the distance $a\,a'$, between the corresponding points of the aperture-images, is obtained by means of $\sin \boldsymbol{a} \cdot f$, as above explained. Since, then, for the first diffraction pencil $\sin \boldsymbol{a} = \frac{\lambda}{d}$, if d denotes the distance of the lines in the object, this formula becomes
$$B\,P = \frac{(p^* - f) \cdot \lambda}{\frac{\lambda}{d} \cdot f} = \frac{p^* - f}{f} \cdot d\,;$$
or, since $\frac{p^* - f}{f}$ equals the magnifying power m, $B\,P = m\,d$, that is, the distance between the interference lines in the real objective-image is m-times the distance of the bright lines in the object. This implies that the number of interference lines agrees with those of the striæ.

The same action produces, of course, the diffraction image formed at a'', if it is combined with the direct one. On the other hand, if the direct image be stopped off, and the two diffraction images allowed to interfere (in the dark field of view), the denominator $a\,a'$ in the above formula will be doubled, and the distance of the lines therefore reduced to half. Similarly, by means of suitable stops, a reduction of $\frac{1}{3}$, $\frac{1}{4}$, &c., of the actual distances may be effected so far as additional diffraction images come into play. The striæ will therefore appear in this case three or four times as fine as they would be in an image true to nature. In general, there is no difficulty in determining beforehand the effect of interference, as soon as the number and position of the diffraction images in the focal plane of the objective are known.

The method of stopping off the aperture-images in the upper focal plane of the objective, the use of which for the study of diffraction phenomena cannot be too much insisted upon, furnishes unquestionable evidence of the production of the structure-image through interference. As soon as the aperture-images due to diffraction are stopped off in such a manner that the direct image alone is perceptible on looking down the tube, the microscopic image of the object will appear to be entirely blank,—all fine markings have disappeared precisely as though the instrument had suddenly lost its whole optical power. If one of the diffraction images be now brought into action as a second source of light, the markings will immediately appear at right angles to the line of union of the two effective sources of light, whilst they remain as invisible as before in directions parallel to the line of union. If we view *Pleurosigma attenuatum*, for instance, with an objective of about 3 mm. focal length and about 90° aperture, there will be seen in the upper focal plane, outside the direct aperture-image, four others due to diffraction, of which each two *diametrically* opposed images correspond to a single system of striae. According as the one or the other pair is stopped off, the transverse or longitudinal striae will disappear in the image; and if all four are stopped off, absolutely nothing is seen of the striae.

For the sake of simplicity we have throughout regarded the light as homogeneous, and have taken the wave-length at ·5 mic. As a rule, the differently-coloured rays act together, and since the distances of the lines in the diffraction image above the objective are proportional to the wave-lengths, the differently-coloured images appear displaced laterally, and form a true spectrum in which with favourable illumination separate Fraunhofer's lines are apparent. Since *all* the diffraction pencils, which proceed from the same surface-element of the object, nevertheless—so far as the spherical and chromatic aberrations can be disregarded—intersect in the same point, they produce at that point a colourless, and at the same time sufficiently sharp, interference image. The limit of discrimination may therefore in general be determined, because *all* the colours take part in the production of the image that agree with the rays of mean refrangibility, whose wave-lengths are actually about ·5 mic. It may, however, occur in certain cases that rays of greater refrangibility (*e.g.* green or blue) are the determining ones; and in photo-micrography, as is well known, wave-lengths come into action which produce no visible effect on the naked eye. It would therefore not be without interest to tabulate the wave-lengths of the different rays, and the limits of discrimination deduced from them. As above shown, the values of the limits in question correspond to half a wave-length, with illumination which we assumed to be as oblique as possible.

Wave-length in Mic.		Limit of Discrimination in Mic.
Red	·68	·34
Yellow	·58	·29
Green	·52	·26
Blue	·43	·22
Violet	·39	·20
Line M	·37	·19
Line N	·36	·18

The figures given for the lines M and N are taken from the paper of Draper.[1] The so-called chemical rays therefore approximately coincide with the most refrangible rays. In photography, consequently, the conditions are about the same as would be given by direct vision through the Microscope by a structure coarser in the proportion of 2 to 3.

The proof that the delineation of fine structures is not brought about dioptrically, but through the interference of diffracted rays, is of great import as regards the interpretation of the microscopic image. For while a dioptric image, as produced by homofocal pencils of rays, is point for point similar to the object situated in the plane of adjustment, and admits therefore of a wholly reliable conclusion as to the composition of the object, if a correct stereometric interpretation is made of what is seen on the surface, the interference images produced by the process of diffraction have no constant relation to the nature of the corresponding object. For instance, rows of points produce the same image as actual lines, and where two such series of lines intersect at right angles two further lines appear in the interference image in the direction of the diagonals. Lines which lie in the same level are often rendered visible or invisible by varying the adjustment. It may occur that a simple series of lines is delineated as of doubled or trebled fineness. All these phenomena may at once be placed upon a theoretical basis; they belong to the known class of diffraction phenomena produced by gratings, which are treated of in every text-book of Physics. This implies that the structural indications of the composition of the object, as formed in the microscopic image, occasionally conform to the original; as a rule, however, they do not so conform. It is clear, under these circumstances, that

[1] Poggendorff's "Annalen," Bd. cli. (1874), p. 337.

every attempt to discover the structure of finely organized objects—as, for instance, diatom-valves—by the mere observation of their microscopic images, must be characterized as wholly mistaken.

☞ NOTE.—In the above the deviation of the rays caused by fine structures was regarded in the usual way as diffraction, as Abbe also assumes. Nevertheless, it is worthy of mention that a membrane consisting of a substance throughout homogeneous—that is, equally transparent in every part—if it also has fine striated thickenings, will exhibit the same interference images as a real grating, although in such a membrane refraction only and not interference of the rays takes place. The like holds good for the striations which appear in adjacent aqueous particles of various size. Every part projecting outwards, or optically denser, acts as a cylindrical lens; it forms a real image of the source of light, which acts as a self-luminous line in front of the objective. Such objects are principally met with in diatoms, scales of butterflies, cell-membranes, &c.

2.—Reflexion of Light by Small Spheres, Granules, Fine Threads, &c., and the Interference Phenomena thereby Produced.

If $a\,b$ (Fig. 129) is a small mercury-globule, say 20 mic. in diameter, by medium focal adjustment, if we disregard the reflexion of the objective, it will appear as a black disc encircled by a distinct bright ring—the so-called "diffraction ring." This is due to the rays which are reflected at the margins of the sphere; these margins may therefore be regarded as luminous points. Expressed more accurately, the bright line is the virtual image of the diaphragm, which is formed by the zone of the sphere passing through the points a and b. An object crossing the incident cone of light (a window-frame, pencil, &c.) is represented in this reflected image as a dark line, which forms a shorter or longer arc on two opposite sides of the circumference.

It is evident that if the Microscope is focused to the reflecting margin of the mercury-globule, it will appear in the objective-image as a bright border, since all rays, which contribute to the formation of the image, emerge from points of the plane of adjustment. The result is different if the Microscope is focused to a lower or higher level. The same phenomena will therefore appear, with regard to the luminous edge of the sphere, as those produced by the approximation or withdrawal of a source of light of small extent. But in the case we are considering there is added the special limitation, that the pencils emerging from the source of light, as a glance at the figure shows, always pass through the corresponding marginal portion only of the objective, so that, for a point b of the margin of the sphere, situated to the right, the left half of the objective is ineffective. Under these conditions an elevation of the luminous point b will act as a displacement of it towards the left, *i.e.* the inclination of the pencil reaching the objective, or of its axis, is increased by the elevation, but that of the refracted one diminished (Fig. 130). The image-point b' therefore moves to the right, and for the like reason a' moves just as much to the left. The image of the luminous margin therefore becomes successively smaller; the bright ring contracts more and more, until at length, if the elevation is continued, it will appear only as a luminous point occupying the middle of the dark shadow-image of the sphere. It is assumed that the image-points a' and b', notwithstanding this displacement, do not essentially alter their original distances from the eye-lens, otherwise a distinct image would not be perceptible. In most objectives the image vanishes so quickly, that the last stage, the bright point in the centre, is clearly observed only with extremely small spheres, that is, those which require only a very slight displacement. This is the case, for instance, with the higher powers of Oberhaeuser and Hartnack, with the No. 7 of Bénèche and Wasserlein, &c., whilst the No. 9 of the last-named optician shows, when small spheres of 25 mic. in diameter are used, not only the central point, but on still lower focal adjustment (when the image-points alter their relative

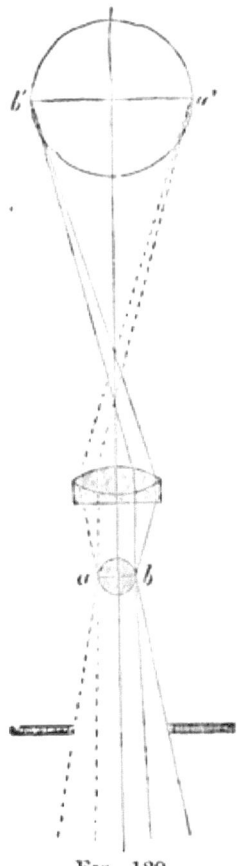

Fig. 129.

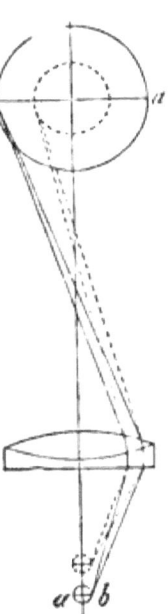

Fig. 130.

positions, b' lying to the right and a' to the left) exhibits a bright circle.

Together with these phenomena, which are referable to the reunion of homofocal cones of light, interference lines are also seen, which are clearly due to rays that have been deflected at the margin of the object. When these rays emerge from the plane of adjustment, their points of intersection of course coincide with the dioptric images of the marginal parts. If, on the other hand, the object is brought nearer to the focal plane, the common point of convergence is moved beyond the level of the objective-image, and we see at this level merely the cross-section of the image-forming cone of light, taken at a greater or less distance from the apex. In consequence of this, the rays are here disposed according to their inclination to the axis; those more inclined, which pass by the edge of the objective, necessarily also pass through the peripheral part of the sectional surface. Of the deflected pencils of light all those annul each other in the focal plane of the objective—just as in the production of the structure-image—whose difference of phase amounts to half a wave-length, or to an odd multiple thereof. The sectional surface of the cone of light consequently appears to be composed of alternately bright and dark circles, which, with white light, show more or less distinct colours. The more intense the deflected pencils, the more clearly of course will these concentric interference lines appear; they are most sharply defined when direct sunlight is employed.

If we focus the Microscope to the plane in which the virtual reflected image formed by the margin of the sphere appears as a small central circle—neglecting the indistinctness and the lesser brightness—it will act precisely like the virtual image of an air-bubble. Window-frames, reflected in the mirror, will therefore appear as straight lines; a slight elevation of the tube causes a red colouring of the centre and a bluish margin, whilst lowering it causes a bluish centre with a red margin. These colour-phenomena are worthy of notice, since they also occur with the smallest spheres of 2—3 mic. in diameter, and obviously augment the difficulty of distinguishing them from small cavities; both indeed appear reddish if focused slightly from the medium focus; with a somewhat lower focus they appear bright, and dark with a higher focus. Only where the minute spheres form a distinct real image, which therefore in every case lies above the red virtual one, is there given a theoretical criterion of discrimination, and in that case a mistake is hardly possible. But in the majority of cases where even the practised microscopist is in doubt this criterion fails us; each separate case therefore requires a special consideration, and can be explained only by those who have made the observation.

It is generally advisable to observe optical appearances of this kind with different Microscopes, since the various appearances of the microscopic image, even if they do not always inform us accurately, save us at least from premature conclusions.

3.—Interference Lines caused by the Withdrawal of a Source of Light of Small Extent.

We have above shown that the approximation of a luminous point to the focal plane of the Microscope produces a displacement of its objective-image, and in certain circumstances the formation of interference lines. A movement of the source of light in the opposite direction, or, what is equivalent, the elevation of the body-tube, produces, of course, an analogous effect; the image-point in the microscopic image is likewise displaced, but in the opposite direction. The border of light which the reflecting edge of a mercury-globule produces is moved in this case outwards. It may therefore, according to the peculiarity of the objective, resolve itself into distinct interference rings, or very rapidly become indistinct. Indistinctness will always be met with if the opposite displacement causes rings, and the formation of rings if it produces indistinctness.

The example adduced is evidently not the most favourable for the observation of this phenomenon, since the dilating border of light lies in the illuminated field of view. More suitable objects are formed by fine lines or points in a blackened glass plate (coated with Indian ink or smoked over a candle-flame), or the minute reflected images of a small mercury-globule on a dark ground. The interference lines, if they appear at all, are then extremely sharp, and the peripheral ones exhibit distinct prismatic colours. If we cover up one half of the objective or adjust the mirror laterally, we may easily convince ourselves that the displacement occurs as we have indicated; by this method also the effect of the elevation and lowering of the body-tube may be conveniently studied. All real or virtual images, whose surroundings appear only faintly illuminated or dark, act of course like the sources of light we have mentioned—for instance, the bright lines in the air-bubble and the hollow cylinder, the focal lines of cylindrical threads and tubes, &c.

B. Interferences in the Plane of Adjustment.

The interference lines observed on the edges of dark bodies, air-bubbles, &c., were formerly explained as diffraction phenomena. It was not considered that though the conditions of diffraction obtained, which in most cases is not the fact, the resulting interferences are always imperceptible. For since the incident rays do not proceed parallel, but form a cone of light of larger or smaller aperture, the dark lines which correspond to particular inclinations are always illuminated by rays of a different inclination also, and the result to the eye is a uniform illumination. In all cases where interference lines are perceptible, it would be useless labour to exhibit the diffractive action by a construction based on measurement; on the contrary, we should always be assured that factors, which have a modifying effect upon the diffraction phenomena, exercise here either no influence at all, or an entirely different one—that the observed distances of the dark lines do not agree with those found by construction or calculation, &c. There is, however, no difficulty in referring the different phenomena in this case to interferences of another kind.

1.—Interference of Direct with Reflected Light.

If we view under the Microscope a plane reflecting surface $A\ O$ (Fig. 131), for instance, a thin cover-glass, or the edge of a piece of metal, and allow light to fall on it through a narrow slit O placed somewhat laterally, there are formed at the edge of the reflecting surface, and parallel with it, alternately bright and dark lines, which with sufficient magnification exhibit coloured fringes. These arise from interference of the direct with the reflected light. The crevice O and its reflected image O' are therefore to be regarded as two sources of light, whose wave-systems strengthen each other at certain points and neutralize at others, according to known laws. Any point P of the plane of adjustment, which here represents the intercepting screen, will appear bright if the rays coming from both the sources of light have at this point the same phase, but dark if the undulations are opposed to each other. Since the reflected ray is retarded by the reflexion itself by half a wave-length, then the first condition is equivalent to an arithmetical progression of 1, 3, 5, 7 . . . . , and the latter to one of 2, 4, 6 . . . . half wave-lengths. The distance of the first dark line from the reflecting surface corresponds, therefore, to such a position of the point P that its distance from the slit O is shorter by a wave-length than from the reflected image O'.

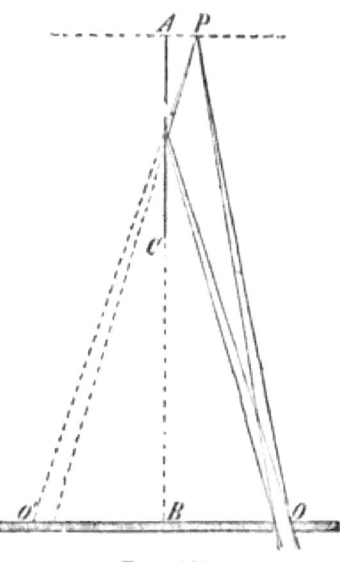

Fig. 131.

If we denote the distance AB of the plane of adjustment from the slit by a, the distance BO of the latter from the plane of the mirror by d, and the angle BAO by ϕ; if we further draw from the points O and O' concentric circles (Fig. 132), which correspond to the waves of light so that the dotted arcs represent wave-depressions, and the lined arcs wave-elevations (in which the retardation caused by the reflexion is to be considered as half a wave-length), we obtain the following relations: —The triangle AQP, whose vertex P corresponds to the first dark line (which may be regarded, on account of the smallness of the arcs AP and PQ, as a straight line), is similar to the triangle $AO'O$. Therefore, if we draw a perpendicular PR, which gives the distance of the point P from the reflecting surface, the triangle RQP is also similar to the triangle ABO, and we get $PR : RQ = a : d$; consequently $PR = RQ \cdot \dfrac{a}{d}$, or, since $PR = \dfrac{\lambda}{2} \cdot \dfrac{1}{\cos \phi}$, if λ denotes one wave-length,

$$PR = \frac{a}{d} \cdot \frac{\lambda}{2 \cos \phi}.$$

This expression admits in all cases, where it is a question of micrometric measurements, of the further simplification, that cos ϕ may be taken as equal to 1, without appreciable error. We then get

$$PR = \frac{a}{2d}\lambda.$$

The other points of intersection, corresponding to the second, third, fourth, &c., dark lines, are at precisely the same distance from the one immediately preceding them as the first is from the reflecting surface. They all lie in an arc drawn from B through P, which, on account of the minuteness of the distances, is to be regarded as a straight line, and is therefore entirely in the plane of adjustment. The interference lines, if observed by homogeneous light, accordingly appear equidistant from one another by the magnitude $\frac{a}{2d}\lambda$. They therefore approach the nearer, the greater ϕ is, $i.e.$, the further the slit is from the reflecting-plane, and, when $d = \frac{1}{2}a$, have a distance of one wave-length only.

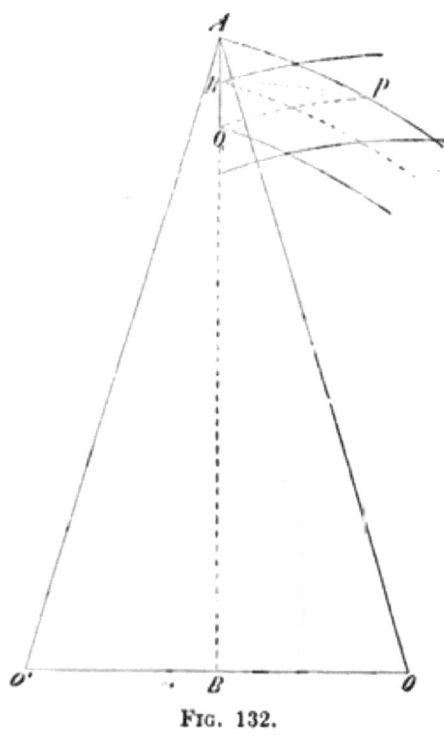

Fig. 132.

With these theoretical deductions measurements agree perfectly. In an actual case, for instance, d was equal to 2·75 mm., $a = 30$ mm., and the distance of the lines in white light 3·1 mic.; calculation gave the wave-length $\lambda = $ ·567 mic., which value corresponds nearly to the mean rays.

The prismatic colours of the bright interference lines are, in reality, arranged in such a manner, that for the first maximum of intensity the violet is turned to the reflecting surface, and the red away from it. In accordance with the laws of the fusion of colours

INTERFERENCE OF DIRECT WITH REFLECTED LIGHT.

the inner margin consequently appears bluish, and the outer one yellowish red. For the other maxima, of course, a less noticeable superposition of the different colours takes place; the combined effect may be determined by reflected light, according to simple rules, precisely as with Fresnel's mirror, or Newton's rings, upon which points further details may be found in the text-books of Physics. Under the Microscope the resulting colours appear unaltered even for the first bright line if the objective is achromatic or over-corrected, whilst an under-corrected objective inverts the order of the colours or lessens their intensity. No special proof is therefore needed that this influence of chromatic aberration must affect the other interference lines also.

If the reflecting surface is spherically curved, the angle ϕ, which the reflected rays make with the reflecting plane—or, what is equivalent, with the corresponding tangential plane—has a greater value for every successive interference line. In Fig. 133, $A A$ is the reflecting plane corresponding to the point a, and $B B$ the one corresponding to the point b; ϕ and ϕ' are the respective angles of inclination of the incident rays. Hence, the distances of the dark lines become less, according to their proximity to the reflecting surface. This is observed in air-

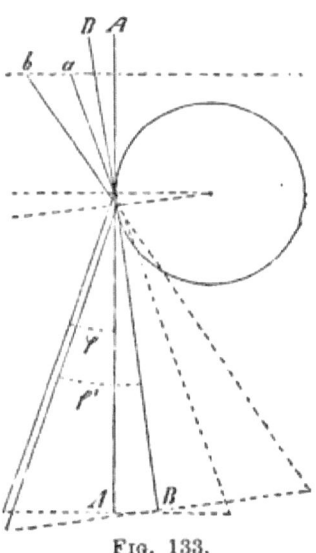

FIG. 133.

bubbles, mercury-globules, and other reflecting bodies—and with special distinctness if, instead of the usual circular diaphragm, a slit-shaped one is moved slowly backward and forward at a suitable distance from the object, to discover the most favourable position in relation to the reflecting surfaces.

In like manner the lines which are seen here and there on the walls in sections of tissue (*e.g.*, cork) may be very well shown. The increase or decrease of their distances, according to the inclination of the incident rays, proves that they belong to the above-described interference phenomena.

2.—Interference of Refracted with Reflected Light.

If $p\,q\,r\,s$ (Fig. 134) is the section of a refracting body—say, of a crystal—the parallel rays of light x and y, incident from below,

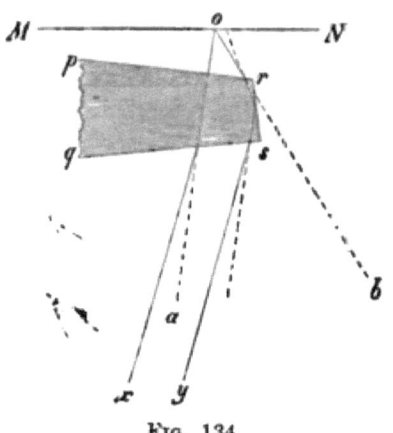

Fig. 134.

are refracted at the surfaces $q\,s$ and $p\,r$, and the latter (y) is reflected also by the oblique marginal surface. A point o in the plane of adjustment $M\,N$ is therefore equally illuminated from two sources of light a and b, whose position is determined by the laws of refraction and reflexion. There is the possibility, therefore, of interference, and it would be a matter of no difficulty to calculate, for any given case, the distance of the interference lines. The working out of such a calculation has no special value in practice, since the necessary data can very seldom be determined by direct observation, and must therefore be arbitrarily fixed. It is nevertheless evident that the interference lines seen from above are projected on the surface $p\,r$, and their distances from each other are the smaller the greater the distance of the source of light (diaphragm) from the plane of the reflecting surface. The fine lines, which in hollow cylinders take the place of the outer and inner bright line as soon as the rays illuminating the margin or the lumen contribute to the illumination,[1] have probably the same origin as the interference phenomena in crystals here mentioned.

[1] Flœgel has made the noteworthy attempt ("Archiv für mikr. Anat.," Bd. ix. p. 507) to refer the glowing yellow-brown colour of valves of *Pleurosigma angulatum* to the interference phenomena described under 1 and 2. The reflexion of the light would consequently take place at the walls of the small cavities (chambers), which, according to Flœgel, cause the familiar markings. The author himself considers his attempt as a mere supposition, but justly insists that the phenomenon is an optical one.

3.—Interference of Refracted and Direct Light.

The deflexion of incident rays towards a *directly* illuminated point p of the plane of adjustment (Fig. 135) may be produced as well by refraction as by reflexion, and since the refracted ray always travels over a longer path than the direct one, interference must then take place. It is not always easy to distinguish whether the refracted or the reflected rays are those which interfere with the direct ones, since the inclination of the reflecting or refracting surfaces, the refractive ratio, &c., in fact the necessary data for the determination of the path of the rays, in many cases cannot be ascertained.

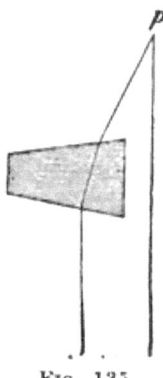

Fig. 135.

4.—Interference Colours of Thin Plates.

If a thin layer of air is enclosed between two refracting media, for instance, between a plate of glass and a convex lens of very shallow curvature, interference colours are produced which are usually termed Newton's rings. The interference takes place between the rays reflected at the upper and those reflected at the lower bounding surface of the layer of air. Since the latter have travelled over a longer path, they are on emergence generally displaced, as compared with the former, by a larger or smaller fraction of a wave-length. If the displacement, assuming the light to be homogeneous, amounts to half a wave-length, or an odd number thereof, the two systems of rays annul each other, and the corresponding part appears dark; in every other case a greater or less degree of brightness is produced, which reaches its maximum with a difference of phase of one entire wave-length.

The same phenomena are also produced by any thin layer of a liquid or solid substance, provided the incident light is reflected at both the bounding surfaces as just described. For estimating these effects of interference the fact is important, that by the reflexion half a wave-length is lost if the second medium is denser than the first, whilst the phase remains the same if the contrary is the case.

In addition to this difference of phase which depends upon the inequality of the paths, we have under all circumstances to take into account another half wave-length for the reflexion at the denser medium. If we suppose the given layer to be wedge-shaped, and illuminated by rays incident at right angles, it will appear dark at all parts where its thickness is equal to an even number of quarter wave-lengths, because then the difference in the paths of the two systems of rays is expressed by an even multiple of half wave-lengths, *i.e.* by whole wave-lengths, so that the difference of phase caused by reflexion decides the matter. The intensity of the light reaches its maximum in all places where the difference of path is equal to an odd multiple of quarter wave-lengths. If, therefore, the wave-length $= \lambda$, and the thickness of the effective layer $= d$, we get

Bright lines if $d = \frac{\lambda}{4},\ 3 \cdot \frac{\lambda}{4},\ 5 \cdot \frac{\lambda}{4},\ 7 \cdot \frac{\lambda}{4} \ldots$

Dark lines if $d = 0,\ 2 \cdot \frac{\lambda}{4},\ 4 \cdot \frac{\lambda}{4},\ 6 \cdot \frac{\lambda}{4} \ldots$

Since the wave-lengths are in inverse ratio to the indices of refraction, we can determine d for any substance whose refractive power is known. It is only necessary to divide the known thickness of the corresponding layer of air by the refractive index of the substance—the quotient is the desired thickness. A layer of water with the refractive index $\frac{4}{3}$ causes, therefore, exactly the same effect as a layer of air one-third thicker.

When white light is used, the maxima and minima of intensity for rays of different refrangibility do not of course coincide. Instead of the gradations from bright to dark, colours appear in consequence of this, just as in the above-mentioned phenomena of interference. They are the so-called Newton's colours, which, as is well known, follow one another in a definite order according to the increasing thickness of the layer. To each single colour there corresponds a definite thickness of the layer of air, from which can be calculated, in the same way as before, the thickness of any substance of known refractive power.

According to the researches of Flœgel on *Pleurosigma*, we are met by peculiar contradictions in these determinations if we compare them with direct micrometric measurements. The calculated thickness of the effective layer corresponds, as seems to result from the data of Flœgel, to exactly half of the real value, so that we are tempted to assume an error of half a wave-length in the determination of the difference of phase. Indeed, Valentin[1] gives, as the thickness of the layer of air for the first dark ring, one entire wave-length, that is, exactly double the value in the above formula; and similarly for the other parts and also for Newton's colours. But the figures given by Valentin are erroneous for reflexion in layers of air, glass, or liquids,[2] and there is at present no reason for assuming any other proportion with organic membranes.[3] We should rather suspect that the observations of Flœgel on *Pleurosigma* were influenced by the reflexion of light on the walls of the chambers. It appears to us *à priori* to be questionable whether a membrane of such peculiar structure will act as a homogeneous substance.

IV.

OBLIQUE ILLUMINATION.

WE have already stated that a cone of light incident obliquely may act more favourably with regard to the aberration of the objective than an axial one—that is, on the supposition that just those inclinations are represented for which the instrument is most perfectly aplanatic. We will entirely disregard here *this* consequence of oblique illumination, and assume we have a perfectly aplanatic Microscope, in which the oblique position of the incident cone of light influences the intensity of light, but not the distinctness of the image. There still remains to be discussed the question, how oblique, as opposed to axial, illumination acts upon the distribution of light in the microscopic image if the object is a substance composed of delicate layers or a membrane with slight elevations or depressions. Since both these cases may, with reference to their total effect, be reduced to the one—that corresponding portions of surface of the field of

---

[1] Valentin: "Untersuch. d. Pflanzen- u. Thiergewebe im polar. Licht" (1861), p. 120.

[2] Cf. Wilde, in Poggendorff's "Annalen," Bd. lxxx. p. 407; Bd. lxxxiii. p. 18.

[3] Cf. moreover Quincke: "Ueber die Aenderung der Phase bei der Reflexion der Lichtwellen." Poggendorff's "Annalen," Bd. cxlii. (1871), p. 192.

view act alternately as convex and concave lenses—our first task is to examine the resulting image from this more general point of view.

Let $a\,b$ (Fig. 136) be a surface-element which gives real images, and let us assume the illuminating mirror to be so placed that all incident rays are deflected to the same side of the vertical. Let the minimum of this deviation be δ, and the maximum δ'; the angle of aperture of the objective may be ω as before. On these suppositions the bounding lines of the umbra and penumbra are determined by the corresponding values of the angle of deviation ρ. Their dependence upon the inclination of the incident rays of light is more clearly exhibited if their distances from the centre of the refracting surface-elements are expressed directly by the focal length (f) of the latter. If F and F^* are the two focal planes, $s\,q$ and $s'\,q'$ incident rays, with the inclination δ', which are so refracted that they still just reach the objective, then the triangles $o\,p\,q$ and $o\,p\,q'$ give the trigonometrical relations

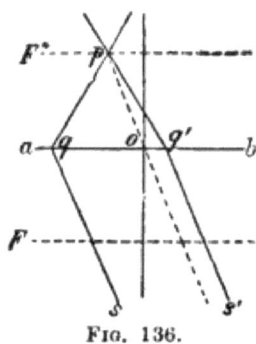

Fig. 136.

$$o\,p : o\,q = \cos\frac{\omega}{2} : \sin\left(\frac{\omega}{2} + \delta'\right);$$

consequently $o\,q = \dfrac{\sin\left(\frac{\omega}{2} + \delta'\right)}{\cos\frac{\omega}{2}} \cdot o\,p$, or, since $o\,p = \dfrac{f}{\cos\delta'}$,

$$o\,q = f\,\frac{\sin\left(\frac{\omega}{2} + \delta'\right)}{\cos\frac{\omega}{2}\cos\delta'} = f\left(\tan\frac{\omega}{2} + \tan\delta'\right);$$

and similarly we get

$$o\,q' = f\left(\tan\frac{\omega}{2} - \tan\delta'\right).$$

The bounding line of the umbra moves forward, therefore, on one side towards the optical centre, and coincides with it as soon as $\delta' = \dfrac{\omega}{2}$; it moves further from it on the other side, and soon attains a distance of $2f$, with high powers, if, for instance, $\omega = 80°$ and δ' is somewhat greater than 45°. The same reasoning holds good also for the penumbra; the formulæ remain the same, only δ is substituted for δ'. Hence it follows that oblique illumination with surface-elements, whose breadth amounts to less than $2f$, entirely obliterates the shadow on one side, while it intensifies it on the other. Where, therefore, many such minute convex lenses are joined to one another, the microscopic image will show the same number of bright points which alternate with dark ones; this is also the case with axial illumination. But while the latter may render only the penumbra visible, the oblique light will produce the much more distinct umbra, and will increase thereby the contrasts between brightness and darkness.

If, on the same suppositions, the minute convex lenses are replaced by concave ones, or by surface-elements which cause no deviation, the latter will remain bright, while the concave lenses will exhibit light and shade in the opposite order. The illuminated parts of adjacent surface-elements will therefore unite; each two elements will form together a bright and a dark part. The number of light and dark points in the microscopic image is therefore reduced to half, and it may happen that striæ become visible in consequence of this, which are not perceptible by axial illumination.

Although it is true these considerations do not apply to fine structural details of objects, since, as we have before shown, they are reproduced by interference images, yet it is important to understand the influence of oblique illumination upon the microscopic image, more especially with reference to the refracted rays. We will therefore explain, as a typical example, the effect which the vertical edge of a membrane, or of a cover-glass, produces with oblique illumination. If $q\,r$ (Fig. 137) is the vertical edge, and B the diaphragm which regulates the inclination

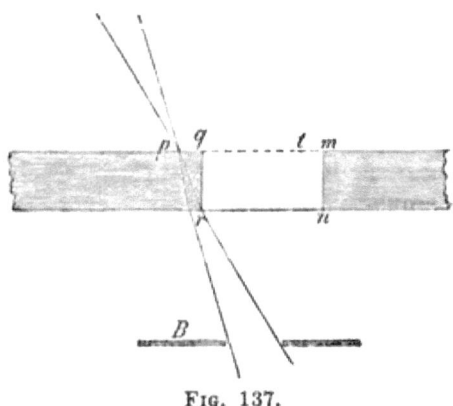

Fig. 137.

of the incident cone of light (placed laterally to the edge), then it is evident from the figure that if the Microscope is adjusted to the surface of the cover-glass, the point p will still appear fully illuminated. Further to the right the rays, produced backwards, are first partly, and then entirely, reflected towards the left from the vertical edge, or refracted so that they do not reach the diaphragm. The edge of the cover-glass is, therefore, in the shadow. Similarly, it may be shown that if the left edge m n of the cover-glass is observed with the same illumination, it will appear bright, and the adjoining portion m t of the plane of adjustment dark.

If, therefore, we take an object with projecting ridges (Fig. 138 A), and adjust the focus to the surface of the elevations, an image will be formed like that represented in Fig. 138 B. The right-hand sides of the ridges, as well as of the depressions between them, will fall in the shadow, and the latter becomes the broader the more oblique the illumination. The same effect

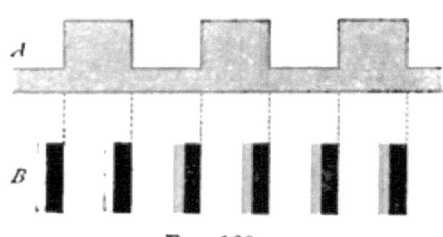

Fig. 138.

must necessarily be produced by alternate solid and aqueous layers with vertical bounding surfaces, if the Microscope is focused on the upper surface.

The importance of oblique illumination for the dioptric image lies therefore principally in the fact that it increases the contrasts between light and shadow, and in addition makes the shadow lines appear broader, and hence more palpable. In particular cases, moreover, if the illumination is sufficiently oblique, their number may be only half as great as by axial illumination, by which, of course, the power of perceiving them is raised in a still higher degree.

The angle of the incident cone of light, or, more strictly, of its axis, must be specially determined for each particular case by experiment. Its value is dependent not only upon the nature of the object and the aperture ω of the objectives, but also upon the difference between the extreme values of δ. Since the power of discrimination of alternately positive and negative impressions upon the sight is greatest when either the one or the other has a decided preponderance, that combination is always the most favourable by which the shadow lines and the bright lines are of nearly equal breadth, and the former, moreover, reach their greatest depth. Where both these conditions are not attainable simultaneously, a greater breadth of shadow can be produced only at the cost of its intensity, and a greater intensity only at the cost of its breadth. The eye must then decide what medium point is the most serviceable for observation.

As regards oblique illumination by the marginal rays of an axial cone of light of tolerably large aperture, such as may be obtained by the application of an annular diaphragm, we have not had the opportunity of becoming practically acquainted with its advantages in any of our researches. As far, however, as experiments made with test-objects admit of our judging, we should not value them very highly. Theoretically considered, the effect of this kind of illumination, as far as regards the dioptric part of the question, can only be to decrease the intensity of light in the image in the same ratio as that in which the transverse section of the effective cone of light diminishes through shutting off the central rays. The bounding lines of the umbra and penumbra retain their position under all circumstances, as they are in every case dependent only upon the inclination of the extreme marginal rays—*i.e.*, upon the maximum values of δ and ω. The intensity of light may also be reduced by directing the mirror to a less intense (yet sufficiently extensive) source of light, or by using a surface of white paper as a reflector, or by projecting the incident rays through a semi-transparent substance (tissue paper, ground glass, &c.).

An essentially favourable action of this method of illumination would be intelligible if the objective were aplanatic for marginal rays within certain limits of inclination only, but not for central rays, or if the peripheral interference pencils acted more favourably alone than in combination with the more central ones. In the former case an annular diaphragm would be advantageous not merely for particular objects, but for all without exception.

V.

THE PHENOMENA OF MOTION.

The observation of the Phenomena of Motion under the Microscope has led to many false views as to the nature of these movements. If, for instance, *Swarm-spores* are seen to traverse the field of view in one second, it might be thought that they race through the water at the speed of an arrow, whereas they in reality traverse in that time only a third part of a millimetre, which is somewhat more than a metre in an hour. It must not, therefore, be forgotten that the rapidity of motion of microscopic objects is only an apparent one, and that its accurate estimation is only possible by taking as our standard the actual ratio between time and space. If we wish, for the sake of exact comparison, to estimate the magnitude of the moving bodies, we may always do so; the ascertainment of the real rapidity remains, however, with each successive motion the principal matter.

If a screw-shaped spiral object, of slight thickness, revolves on its axis in the focal plane, at the same time moving forward, it presents the deceptive appearance of a serpentine motion. Thus it is that the horizontal projections of an object of this kind, corresponding to the successive moments of time, appear exactly as if the movement were a true serpentine one. As an example of an appearance of this nature we may mention the alleged serpentine motion of *Spirilium* and *Vibrio*.

Similar illusions are also produced by *Swarm-spores* and *Spermatozoa*; they appear to describe serpentine lines, while in reality they move in a spiral. It was formerly thought that a number of different appearances of motion must be distinguished, whereas modern observers have recognized most of them as consisting of a forward movement combined with rotation, where the revolution takes place sometimes round a central, and sometimes round an excentric, axis.[1] To this category belong, for instance, the supposed oscillations of the *Oscillariæ*, whose changes of level, when thus in motion, were formerly unnoticed.

In addition to these characteristics of a spiral motion it must,

---

[1] Cf. on this point Nægeli: "Beiträge," ii. p. 88.

of course, be ascertained whether it is right- or left-handed. To distinguish this in spherical or cylindrical bodies which revolve round a central axis is by no means easy, and in many cases, if the object is very small and the contents homogeneous, it is quite impossible. The slight variations from cylindrical or spherical form, as they occur in each cell, are therefore just sufficient to admit of our perceiving whether any rotation does take place. The discovery of the *direction* of the rotation is only possible when fixed points, whose position to the axis of the spiral is known, can be followed in their motion round the axis. The same holds good also, *mutatis mutandis*, of spirally-wound threads, spiral vessels, &c.; we must be able to distinguish clearly which are the sides of the windings turned towards or turned away from us.

If the course of the windings is very irregular, as in Fig. 139, a little practice and care are needed to distinguish a spiral line, as such, in small objects. The microscopic image might easily lead us to the conclusion that we were examining a cylindrical body composed of bells or funnels inserted one in another. The spirally-thickened threads, for instance, as they originate from the epidermis-cells of many seeds, were thus interpreted, although here and there, by the side of the irregular spirals, quite regular ones are also observed.

Fig. 139.

Moreover, it must not be forgotten that in the microscopic image a spiral line always appears wound in the same manner as when seen with the naked eye, whilst in a mirror (the inversion being only a half one) a right-handed screw is obviously represented as left-handed, and conversely. If, therefore, the microscopic image is observed in a mirror, as in drawing with the Sömmerring mirror, or if the image-forming pencils are anywhere turned aside by a single reflexion, a similar inversion takes place from right-handed to left-handed, and this inversion is again cancelled by a second reflexion, as in Oberhæuser's camera lucida, and in many multocular Microscopes. All this is, of course, well known, and to the practised observer self-evident; nevertheless, many microscopists still appear to be entirely in the dark about matters of this kind.

VI.

DIFFERENCES OF LEVEL.

Since only those objects can be distinctly seen through the Microscope which lie exactly in the plane of adjustment, it is possible to ascertain the distance of two object-points in the direction of the optic axis, by measuring the difference of level of the corresponding planes of adjustment with the help of a second Microscope placed horizontally, or by a micrometer screw constructed for this purpose. In this latter method there must be taken into account a source of error which influences to a very considerable extent the results of the measurement in cases where the objects are immersed in a solid or liquid medium—for instance, in water. If we determine the tube-displacement which is necessary in order to bring into the plane of adjustment the two given object-points alternately, the amount of the displacement will always be considerably less than the actual difference of level. We get for instance, as focal length of a spherical air-bubble in water ·98—1·04 (the radius taken as unity), whilst, in reality, it amounts for the central rays to about 1·3.

This error is due to the fact that the passage of the pencils from water into air (the cover-glass not here coming into account) produces with the microscope the same effect as with the naked eye—that is, a body situated in water is apparently raised, and, taken absolutely, the more so the deeper it lies. The virtual image of an air-bubble is therefore raised to a greater degree than its centre, the focal length being consequently shortened; and for the same reason in general the vertical distance of the two points is lessened. The amount of this diminution depends, of course, upon the angle of incidence of the effective rays, and is for small values (as long as the arcs are in the same ratio as the sines) expressed by $1 - \frac{1}{n}$, where n denotes the refractive index. Since, then, the incident cones of light are refracted in the objective as if they consisted of rays of definite mean inclination, and as this inclination does not exceed 12°—18° in the higher powers we have examined, the above expression is approximately accurate in most cases that occur, and a trigonometrical determination[1] of the error is consequently superfluous. In water, with the refractive index $\frac{4}{3}$, each difference of level is therefore diminished by a quarter of its real magnitude—*i.e.*, reduced to $\frac{3}{4}$. (Accurate calculation gives with an angle of incidence of 18° instead of $\frac{3}{4}$ the somewhat smaller value ·7395.)

There is, moreover, a simple method of eliminating this source of error : we replace the layer of air between the objective and the cover-glass by water ; the measurements will then agree exactly with the true values.

[1] If D is the real and d the apparent difference of level, a the angle of incidence, and a' the (larger) angle of refraction, we get

$$d = \frac{\cos a' \sin a}{\sin a' \cos a} \cdot D ,$$

and

$$D - d = \frac{\sin (a' - a)}{\cos a \sin a'} \cdot D .$$

PART V.

THE SIMPLE MICROSCOPE AND THE LANTERN MICROSCOPE.

THE compound dioptric Microscope now takes unquestionably the first place amongst the optical instruments employed for the observation of small bodies. It has become the special instrument of the naturalist, the chief weapon of the microscopist. The sphere of action for its former rival, the *Simple Microscope*, grows yearly less and less, and the *Lantern Microscope* (called, according to the kind of illumination adopted, Solar, Gas, or Photo-electric Microscope) has been of little or no importance to science hitherto. The reader will under these circumstances consider that we are justified in devoting only a relatively short supplementary discussion to these instruments.

I.

THE SIMPLE MICROSCOPE.

A. General Principles.

EVERY lens or combination of lenses, which is applied in such a manner that the eye directly observes the virtual image formed by it, may be regarded as in principle a simple Microscope. In practice, however, only the higher magnifying-lenses, or systems of lenses, which for more convenient use are attached to a stand, are usually included under this expression, while the low-power lenses which can be held in the hand are termed magnifying-glasses. It is obvious that this distinction must be somewhat arbitrary, since the magnifying-glass is often provided with a stand enabling it to be adjusted in the desired position relatively to the object; and conversely, the higher powers of the simple Microscope are occasionally held in the hand like the magnifying-glass, or attached to an ordinary holder. The customary appellation is founded, therefore, rather on the general form of the whole apparatus, than on its optical construction.

In theory it is immaterial whether the amplification is high or low, the formation of a sharp image is in all cases combined with the condition, that both kinds of aberration shall be as far as possible eliminated for the given object-distance, which is, of course, always somewhat less than the focal length. The manufacture of a good magnifying-glass or of a simple Microscope is therefore accompanied by precisely the same difficulties as those of an ordinary objective of about the same focal length. The fact that in the former case virtual images instead of real ones come into account, renders the task neither easier nor more difficult, although it must be taken into consideration. Our discussion of the combination of flint- and crown-glass lenses to form aplanatic systems, may therefore be applied here also, and the testing of the optical power and magnifying power may be effected as with the compound Microscope. It remains for us, therefore, to indicate specially those points which are essential to understand the path of the rays under the given conditions, and which differ from those we have previously considered.

1.—Aperture of the Effective Cones of Light.

The aperture of the cones of rays incident from the object-points is dependent, in a given system, upon the aperture of the pupil of the observing eye and upon the size of the diaphragms. If $a\ b$ (Fig. 140) is the object, F the anterior, and F^* the posterior focal plane, $E\ E^*$ the pair of principal planes, and P the pupil; then the diameter of the pupil is for all finite distances of the virtual image $a'\ b'$ evidently somewhat larger than the diameter of the surfaces in which the optically effective cones of light cut the principal planes. With a number of refracting surfaces each incident cone of light consequently meets the first of them (N^o) in a small circle, and each following one in a somewhat larger circle, until at length its diameter approximately equals that of the pupil. The position of this circle corresponds with the object-point; its distance from the axis increases and decreases with the distance of the latter.

If the eye is adjusted for infinite distance, the section of the

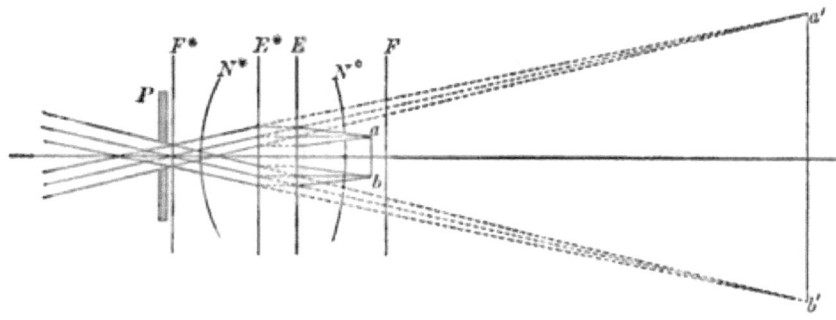

Fig. 140.

incident cones of light consequently being equal to the principal planes of the pupil, then the angle of aperture ω is obviously determined for the middle of the field of view by the formula

$$\tan \frac{\omega}{2} = r \cdot \frac{d}{2f},$$

where d denotes the diameter of the pupil, and f the focal length. Smaller angles of aperture are, therefore, approximately in inverse ratio to the focal length. If d is taken at 3 mm., calculation gives, for example, the following values:—

Focal Length in Millimetres	2	3	4	5	6	10	12
Angle of Aperture ...	73° 44′	53° 8′	41° 4′	33° 24′	28° 4′	17° 4′	14° 14′

With these values coincide approximately also those which we should obtain under similar assumptions for a finite distance of vision of 100—250 mm.

If there is situated beyond the last surface of the system a diaphragm, the aperture of which is less than the pupil, the size of this aperture evidently determines that of the incident cones of light. This also holds good in a certain sense for any other diaphragms inserted between the refracting surfaces; but in this case, as is readily seen from the construction, the peripheral pencils are cut off wholly or in part, while the middle ones pass through unaffected. Such diaphragms therefore determine the size of the field of view, and diminish more especially the cones of light which reach the retina from the peripheral points.

2.—BRIGHTNESS.

If the optically effective cones of light wholly occupy the aperture of the pupil, as is usually the case for the central object-points, they evidently possess, after the last refraction by which their point of convergence is moved to the distance of distinct vision, the same angle of aperture as those pencils which are received by the naked eye. The brightness of the virtual image is in this case (disregarding the loss caused by reflexion or absorption) nearly equal to unity—*i.e.*, the objects are seen through the simple Microscope nearly as bright as with the naked eye. The resulting brightness, as with the compound Microscope, is more accurately expressed by the formula

$$v = \left(\frac{\omega}{\rho m}\right)^2,$$

where v denotes the brightness, ω the angle of aperture of the system, ρ the angle of aperture of the naked eye for a particular length of sight, and m the coefficient of linear amplification for the same distance of vision. If $\rho = 1°$, as is, for instance, the case with a sight of 172 mm., and the aperture of the pupil is 3 mm., the above expression may be simplified into $\left(\frac{\omega}{m}\right)^2$, if unity is taken as the standard. Since the coefficients of amplification, calculating the distance of vision from the posterior focal point, are in the same ratio as the tangents of the half-angles of aperture (as both are in inverse ratio to the focal length), it follows that the brightness is the less the higher the amplification. A few examples in which the corresponding values of ω and m are placed side by side may perhaps elucidate this decrease.

Focal length	...	12	10	6	5	4	3	2
ω	...	14·2	17·1	28·1	33·4	41·1	53·1	73·7
m	...	14·3	17·6	28·6	34·4	43	57·3	86

3.—Curvature of the Field of View.

The so-called "curvature" of the field of view cannot be explained by the arching of the image-surface, but is due to the deviation of the pencils reaching the eye—that is, of their axes. On this point our previous discussion of the *flatness* of the field of view, and the action of the eye-lens, will hold good in general. The special application of those considerations to the cylindrical magnifying-glass, the doublet, &c., is facilitated inasmuch as the path of the incident and emergent pencils is almost exactly the same as in the Ramsden eye-piece. The point of convergence of the emergent pencils is determined by the position of the observing eye; it stands in the same relation to the point of intersection of the incident rays, produced backwards, as an object to its image, and may, therefore, always be chosen so that the latter, as in the eye-piece, is at a distance of about 200 mm. from the refracting surfaces. The coincidence is therefore complete.

For the elimination of the distortion of the image the refracting surfaces must be so combined that they form a system as far as possible aplanatic for the given focal lengths—*i.e.*, for the distances of the points of intersection. The arching of the image-surface is not dependent upon this; it is due to the fact that the object-points are not at equal distances from the refracting surface-elements.

4.—The Magnifying Power.

The magnifying power is given by the well-known formula $m = \frac{f - p^*}{f}$, or since p^* is here negative, $m = \frac{f + p^*}{f}$. Since p^* must be reckoned from the posterior principal plane, it is evident that with a given distance of vision the amplification is the lower the further the eye is from this principal plane, or indeed from the system of lenses. It is therefore the general rule to place the eye as near as possible to the last refracting surface.

5.—The Extent of the Field of View.

The extent of the field of view in a given system of lenses is dependent upon the aperture of the lenses and the position of the observing eye. The greater the distance of the latter, the smaller is the field of view. If again N^0 (Fig. 141) is the first, and

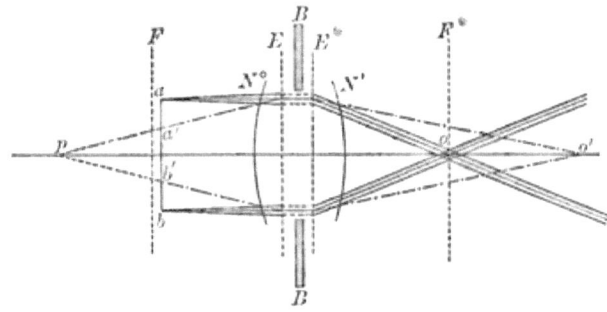

Fig. 141.

N' the last refracting surface of the system, whose principal and focal planes are denoted in the usual way, and if its aperture is determined by the diaphragms or the settings $B\,B$, the eye situated at o views the object $a\,b$ in its whole extent, because all pencils incident parallel with the axis are refracted towards o. But if the eye is placed at o', then only those pencils can reach it which diverge so that their point of convergence p is in the same relation to o' as an object to its image. The field of view becomes therefore necessarily less than the aperture of the system; the outermost points still perceptible are a' and b'. A withdrawal of the eye to a still greater distance involves a stronger divergence of the incident pencils, and therefore a further diminution of the field, and *vice versâ*. It is obvious that the field of view is dependent upon the aperture of the lenses, and would be reduced by smaller diaphragms.

B. The Optical Arrangement.

We will now pass to the description of the more popular forms which, in recent times, have been given to the magnifying-glass and the simple Microscope. A description of the devices of the past century lies beyond our sphere; we must refer our readers to the copious work of Harting in regard to this point and for the history of the Microscope.

Brewster was the first to hit upon the happy idea of making a magnifying-glass by grinding a tolerably deep groove in a glass sphere (Fig. 142), which gives images approximately free from aberration. Coddington facilitated their production by giving to the groove an angular form (Fig. 143), making the two halves of

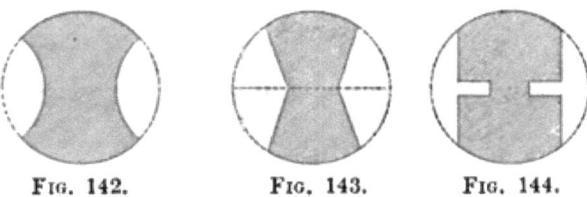

FIG. 142.　　FIG. 143.　　FIG. 144.

the sphere separately out of suitable pieces of plano-convex lenses, which were afterwards cemented together at their plane surfaces. The shape of the groove is of no consequence, and it is obvious that the construction represented in Fig. 144 answers the purpose in view equally well. These Coddington lenses (also called Coneopsids or Bird's-eye lenses), as manufactured by Lerebours, of Paris, met with general approval; according to Mohl, they give an amplification of upwards of 20 linear, and he commends them highly. They are, however, defective in two ways, which are important in many researches: the field of view is very limited, and the distance of the focus from the lower surface is only about $\frac{1}{4}$ of the diameter of the sphere. When, therefore, it is required to observe an object of rather large dimensions, which is the most frequent case in practice, the modern aplanatic combinations of lenses are preferable.

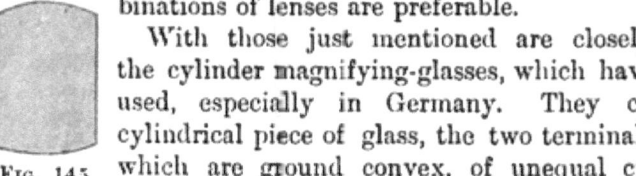

FIG. 145.

With those just mentioned are closely connected the cylinder magnifying-glasses, which have long been used, especially in Germany. They consist of a cylindrical piece of glass, the two terminal surfaces of which are ground convex, of unequal curvatures to diminish the aberration (Fig. 145). The shallower curvature is usually directed to the object. The cylinder lenses produce somewhat less perfect images than the Coddington, and are accompanied by the similar disadvantage of a short focal distance; nevertheless, they must always be regarded among the best and least costly magnifying-glasses. If the stronger curvature is turned to the object a greater object-distance is obtained, but the aberrations are then notably augmented, and the images consequently less distinct.

In recent times, since about the year 1830, systems composed of two or three plano-convex lenses (the so-called doublets and triplets) have come more into favour. It is true that earlier optical physicists, especially Euler and John Herschel, had already occupied themselves in developing such combinations of lenses, but with the impossibility of satisfying the results of calculation in the grinding of the lenses, the theoretically accurate proposals were of little practical importance. Wollaston was the first to discover a means of increasing the sharpness of the image rather by the relative distance of the lenses, determined by experiment, than by their particular form. His doublets consist of two plano-convex lenses with the plane surfaces turned downwards, their focal lengths in the ratio of about 3 : 1, and placed one above the other in such a manner that the distance of the plane surfaces is about 1·5 of the shorter focal length (Fig. 146). The success of these doublets, and the attention which they aroused, soon led to further improvements.

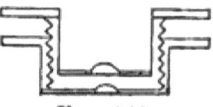

Fig. 146.

Pritchard found the most favourable distance of the two lenses to be equal to the difference of their focal lengths, but that these focal lengths might vary within certain limits (*e.g.*, between the ratios 1 : 3 and 1 : 6). He constructed doublets, which Mohl characterizes as remarkable for the clearness and sharpness of the image, up to an amplification of about 200 linear, and of which the low-power ones are especially well adapted for dissections, &c. The highest powers, which magnify somewhat more than 300 linear, exhibit distinctly the sixth band of Nobert's (30-band) test-plate.

Chevalier combined two plano-convex lenses of equal focal length but unequal size (Fig. 147), so that the larger was nearest to the object. Between the two was placed a diaphragm d with the aperture o. A considerable intensity of light was thus obtained, and at the same time a

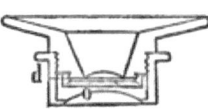

Fig. 147.

greater distance of the object from the first refracting surface. In order to further increase this distance with lower powers, such as are employed in dissections upon the object-stage, he applied above the doublet an achromatic concave lens, which not only answers the purpose in view, but also somewhat increases the amplification; a contrivance which was employed later on by Brücke also for his achromatic magnifying-glass[1] (Fig. 148).

Fig. 148.

The doublets of Chevalier, according to Harting's observations, are about equal to those of Pritchard. With an amplification of 48 linear, the lines of the first band were resolved on Nobert's (30-band) test-plate, and with 317 linear those of the sixth band.

Both Chevalier and Pritchard also made triplets, the construction of which is based essentially on the same principle. They consist of three plano-convex lenses of different focal lengths, and if the necessary care is bestowed upon their manufacture they give images somewhat more distinct than the doublets.

It is unquestionable that Wollaston's idea of directing special attention to the relative distances of the lenses, was a very happy one for practical Optics, and led immediately to important improvements. The doublet reached a degree of perfection which is still sufficient, even at the present time, in many practical cases. But the best combinations of this kind always suffer to a very considerable extent from faulty correction of spherical, and especially of chromatic, aberration,[2] which is, of course, all the more palpable the higher the amplification. As regards optical power they are

---

[1] "Sitzungsberichte der K. K. Akad. zu Wien," 1851, Bd. vi. p. 554.

[2] It is in principle impossible to eliminate chromatic aberration by a combination of two or more plano-convex lenses; by judicious selection of the relative distances the parallelism of the red and violet pencils, or of their axes, is at most effected, as we have already explained when describing the eye-pieces. An intersection of the optically effective pencils of rays between the refracting surfaces does not take place. In this respect the explanation given by Harting ("Mikroskop," p. 113, and Fig. 8 of the annexed table) is entirely erroneous, because the path of the rays assumed by him for the observing eye is impossible. How can the emerging pencils of light reach the retina through a pupil of 2 to 3 mm. in diameter, if they diverge as in the figure referred to?

therefore decidedly inferior to the modern achromatic systems. Their practical applicability is now, however, limited to the magnifying-glass and the dissecting Microscope, and they are of considerable use up to an amplification of 60—100 linear; but it seems probable that even for these purposes they will ultimately be superseded by aplanatic systems, which are already much employed for amplifications as low as 5—10 linear.

For a comparison of the performances of the more recent single lenses, doublets, &c., we give below a few of our observations which were made in the same way as in the testing of the compound Microscope. The focal lengths and object-distances (for the cylinder lenses those corresponding to both positions) are given in millimetres.

Optician.	Description.	Focal Length of the Anterior Lens.	Focal Length of the Posterior Lens.	Combined Focal Length.	Object-distance.	Amplification with Distance of Vision of 25 c.m.	Diameter in Mic. of the Meshes still just recognizable.
Zeiss	30	12·2	17	7·3	5·3	33	3·85
,,	60	5·2	8·5	3·78	2	64	2·3
,,	120	2·53	4·25	1·72	—	140	1·3
,,	Triplet	0·92	2·3 ; 3·7	0·73	—	330	0·7
,,	Single lens	40	38	21	11	12	17
Bénèche	No. 1	23·25	23·25	12·5	8·5	19	6·7
,,	No. 2	8·09	23·25	6·68	3·8	36	3·5
,,	No. 3	5·22	11·75	3·93	2·5	62	2·3
—	Cylinder lens	—	—	18·5	{ 8·5 / 10 }	13·5	16
—	,,	—	—	14	{ 6 / 8 }	18	12
—	Single lens	—	—	44	39	4·5	32

From this table it is evident the doublets of Zeiss and Bénèche very nearly equal the performance of a compound Microscope with equal amplification. The triplet of Zeiss is about equivalent to Hartnack's No. 7 objectives and Bénèche's No. 9—the latter tested with an eye-piece amplification of 4 linear; it shows also the transverse markings of *Pleurosigma attenuatum* by axial light with at least equal distinctness. But this, however, is true only when the object is exactly in the middle of the field of view; in every other position the image is valueless. Moreover, the vision is soon fatigued by observing through a diaphragm of only ·7 mm. in diameter, which Zeiss applies above his triplets and highest-power doublets.

C. The Mechanical Arrangement.

Of the mountings of lenses and systems of lenses, as well as of the stands, we shall treat very briefly, since every well-known optical establishment is able to satisfy on these points all demands which can reasonably be made. The observer must choose for himself among the various forms and arrangements supplied, that which best meets his requirements. If any very special end is in view, he will probably devise a suitable apparatus.

The lens-carriers described in micrographic works all agree in this point—that they are furnished with a movable arm, often jointed, which is attached to a box or to a vertical pillar so that its free end carrying the lens can be adjusted as required.

The stand of a simple Microscope admits of some variety of design. The principal requirements which the optician should bear in mind are the following :—(1) A firm—not too small—object-stage, to which a spring-clip or other suitable contrivance for holding the object-slides can be applied; (2) a convenient support for the hand whilst making dissections upon the stage; (3) an arrangement of the rapid focusing adjustment by which the lens is moved towards the object—the latter remaining stationary; (4) an illuminating apparatus providing cones of light that wholly occupy the aperture of the system of lenses or the pupil of the eye.

[Here, again, it has been thought preferable to omit the authors' descriptions of the simple Microscopes[1] for the reasons assigned on p. 122.]

[1] The following simple Microscopes are described in the German work :—

Opticians.	Microscopes.
Nachet & Son (Paris) ..	Lens-carrier, recommended by Mohl, "Mikrographie," p. 35 (Fig. 149).
C. Zeiss (Jena)	Small dissecting (Fig. 150).
,, ,, ...	Large ,, (Fig. 151).
Bénèche (Berlin) ..	Two, similar to Zeiss's, the fine adjustment like Nobert's.
Schieck (Berlin) ..	One, similar.
S. Plœssl & Co. (Vienna)	One, focusing by the stage.
Nachet & Son (Paris) ..	Dissecting (Fig. 152).
,, ,, ..	Binocular dissecting.
Quekett ..	Pocket and dissecting (Fig. 153).
Pritchard ..	Dissecting (Fig. 154).

II.

THE LANTERN MICROSCOPE.

We need not describe in detail the *Lantern Microscope*. The only importance it has for science consists in the fact that the magnified image of an object can thus be exhibited to several spectators at once, which is advantageous in public lectures. Its applicability is, however, limited to objects which can be seen distinctly with the compound Microscope and low powers; fine structural details that require more powerful objectives are not reproduced distinctly enough in the image to enable them to be demonstrated to the uninitiated. This is especially true when the illumination is effected by electric light, or by Drummond's lime-light, both of which are much inferior to sun-light in intensity. Moreover, the employment of this instrument is accompanied by conditions which materially diminish its practical value. We consider demonstrations are more satisfactorily made with the Microscope proper than by the indistinct images of a photo-electric or gas Microscope. A relatively favourable impression has been made upon us hitherto by the results obtained with low-power objectives only—particularly the enlargements of photographs upon glass.

The objectives of the lantern Microscope are constructed precisely like those of the compound Microscope, except that the lenses are not usually cemented together with Canada balsam, because this method of union might be injured by the heat produced by the illumination. Mohl states that he has employed ordinary objectives for the solar Microscope without injuring them, and that Plœssl has found no inconvenience arise from cementing the lenses together.

The mechanical arrangement of the illuminating apparatus varies, of course, according to the source of light employed. We think it unnecessary to enter into the details of construction, as the devices in use at the present time are described in most of the text-books of Physics, and are scarcely within our province.

PART VI.

TECHNICAL MICROSCOPY.

I.

THE USE OF THE MICROSCOPE.

It may be assumed, as a matter of course, that the various information regarding technical methods and contrivances which is presupposed in microscopical observation, may always be learnt most easily by practical instruction, and that individual skill, which the beginner gradually acquires, is always the main point. A short introduction, however, may not be superfluous, since it will afford some compensation to those who have not had sufficient of this practical training, or who may even lack the training altogether, and may enable them to avoid many errors of judgment. Everything relating to special researches in the departments of Anatomy, &c., and to the use of reagents, &c., we shall postpone to separate chapters under these headings, and here discuss only a few of the more general questions.

1.—Illumination.

For every microscopic investigation a favourable illumination is very important. As the illumination is partly dependent upon the *situation of the laboratory*, this should be so chosen that a large portion of the sky may be employed as the source of light, without direct sun-light. The most convenient is, of course, a room with north light and free prospect, or a room with windows on two sides, in which either side can be used according to the position of the sun. Reflected light from walls, &c., may also be of value under certain circumstances, yet its most favourable employment is limited to certain periods of the day. Hence, in towns, the upper stories of houses which allow of our utilizing the light near the horizon are preferable, and ground-floor dwellings in narrow streets should be avoided.

The most favourable *day-light illumination* is, in our opinion, uniform grey cloud; the pure blue sky also gives an agreeable light, but is not intense enough to develope the fullest power of the highest objectives. In difficult investigations, therefore, a bright white or yellowish white wall illuminated by the sun is more serviceable. Quickly-passing clouds are unfavourable, because the eye is soon fatigued by the continual change of brightness and colour. Direct sun-light should be avoided, or subdued by blinds.

As regards *placing the Microscope*, it is generally stated that a more favourable effect of light is obtained at a distance of six to nine feet from the window for critical work than at the usual distance of about three feet.

According to what we have already stated on illumination, the distance from the window is of importance only when the aperture of the incident cone of light is limited thereby, which is not the case on the application of small diaphragms such as are ordinarily used for high amplifications. In using oblique light, we must under all circumstances be careful that the light reflected from the table and stand is excluded, which may be effected by a vertical diaphragm specially contrived, or by lateral displacement of the usual diaphragm. In many experiments the light incident from above upon the object should be excluded by holding the hand or a screen before it. In short, the illumination should in all difficult cases be so arranged that no other light reaches the field of view than the incident cone of definite aperture and inclination.

What *aperture* or *inclination of the incident cone of light* may be most advantageous, must of course be found by trial in every case. The mirror, or if necessary the whole instrument, is moved in various directions, the observer at the same time looking through the eye-piece, and the effect of the different diaphragms is noticed; or they are moved up and down in the direction of the axis. These experiments are continued until the most favourable kind of illumination is found. As a general rule, striations, systems of lines, &c., are seen most distinctly when the light is incident as obliquely as possible; the relative position of mirror and object must always be arranged so that the direction of the incident rays is at right angles to the striæ.

As in most cases the source of light is limited, or unequally luminous, the *concave mirror* gives (for reasons previously stated) a more intense light than the *plane mirror*—i.e., it involves, with equal aperture of the incident cone of light, a greater brightness of the field of view. For high amplifications the concave mirror affords, therefore, decided advantages; for the lower amplifications its superiority might easily be exaggerated, because a field of view too glaringly bright soon fatigues the eyes. If the mirror is sufficiently large and can be adjusted near the object, the application of illuminating lenses or condensers is in most cases superfluous. Condensers are useful when the aperture of the incident cone of light is required to be enlarged.

If we use *reflected light* for the illumination, the Microscope should be placed as near as possible to the window, to obtain a moderately extensive source of light. With amplifications exceeding 100 to 120 linear, it is advisable to employ an illuminating lens, or the *Lieberkühn* mirror. In certain cases, also, sunlight may be utilized by inclining the Microscope so that the rays may strike the object directly. We doubt the accuracy of the statement of Schacht, that the three series of lines of *Pleurosigma angulatum* may with equal amplification be more distinctly seen in this manner than with ordinary transmitted light. We do not recommend this method of illumination for transparent objects.

Illumination by *lamp-light*, when subdued by ground-glass or cobalt-blue glass, is useful for many experiments, though it always strains the eyes more than day-light—the latter is also far more agreeable. Hence, for the preservation of the sight, prolonged microscopical observations should be limited to the hours of daylight. But in England and generally in Northern Europe, where the cloudy day-light is only of short duration, artificial illumination must be utilized. Under these circumstances the focal length of the concave mirror or of the illuminating lens must be duly considered, so that the convergence of the incident rays may reach its maximum; this matter hardly needs further explanation after what has been stated. In the various opticians' catalogues, microscope-lamps provided with a large condensing lens are mentioned. The same result may be attained with an ordinary moderator or petroleum lamp by placing before the flame a globular vase filled with a weak solution of ammonio-oxide of copper. In the opinion of competent judges, the light thus obtained is agreeable to the eye and at least equal to average cloud-light.[1]

2.—The Selection of the Magnifying Power.

The preparation to be investigated is first viewed with a moderate amplification, because a better survey of the material present is thus obtained, and the best part of the sections, or the most favourable objects, may be conveniently selected for more detailed observation. If the lowest-power objective at disposal magnifies too much, a further diminution of the image may be obtained by shortening the body-tube, or by removing the anterior lenses of the objective. The sharpness of the image suffers materially in the latter case, unless the objective is specially constructed.

In proceeding to higher amplifications, we must remember that the optical power is chiefly dependent upon the focal length of the objectives—not upon the power of the eye-pieces. The employment of deep eye-pieces—*i.e.*, such as magnify 10 linear or upwards—is, as a rule, advantageous only in cases where the determination of form-relations or the comparison of very small distances is required. New details in the image will be less perceptible the more imperfect the objective and the greater the length of the body-tube; in most cases the highest-power objectives exhibit decidedly less of the image than the preceding lower ones.

The most suitable amplifications for particular kinds of work will soon be discovered by those who use the Microscope for scientific purposes. It is manifest that immersion systems and similar high powers will be employed only in special cases, where very difficult peculiarities of structure, &c., have to be determined.

3.—Employment of Cover-glasses.

Most microscopic preparations are examined in water or other fluid medium, either to lessen the deviation of the rays of light, or to keep the object in a settled condition—soft, distended, &c.

[1] *Vide* Strasburger: "Ueber Zellbildung und Zelltheilung," p. 33.

But as these liquids, if the objects are not immersed in them, exhibit the well-known phenomena of adhesion, and therefore are very uneven on the upper surface, it is for optical reasons necessary, with the higher amplifications, to employ cover-glasses which render the upper surface flat. By their application a too rapid evaporation of the liquid is avoided, as also the immersion of the lenses. With lower amplifications the preparation may remain uncovered, and it is frequently advisable not to cover it, especially when a further dissection is contemplated.

If the preparations are of such a nature that it may be desirable to invert them and view them on the other side, they should be placed between two rather large pieces of cover-glass, which may then be turned about at will. This process is often of great advantage in the study of superposed cell-layers or other objects the structure of which is to be investigated.

The thickness of the cover-glass is immaterial with low powers; the choice must therefore be made rather in accordance with other considerations—such, for instance, as the pressure which is to be exercised upon the preparation. High powers, on the other hand, are mostly constructed for thin cover-glass, and correction-adjustment becomes necessary as soon as the thickness exceeds $\frac{1}{2}$ mm. With very powerful objectives only the thinnest cover-glass can be used, in consequence of the short focal distance.

When it is necessary to protect the preparation from all pressure, however slight, a suitable body (a hair, a quill, a strip of paper or tin-foil, &c.) is placed under the cover-glass, and of course the object-distance of the Microscope must then be taken into account. The beginner frequently discovers that he cannot focus on the object in consequence of the thickness of the layer of liquid over it; this may happen even to the practised microscopist in using high powers. This difficulty may be avoided by applying the cover-glass by one end and then pushing it further on until the object arrives at a sufficiently thin part of the wedge-shaped layer of water.

It is obvious that the cover-glasses must be as clean as possible, and must not therefore be touched by the fingers at the optically effective part of the surface.

4.—Preservation of the Instrument.

In order to preserve the Microscope in perfect condition, the lenses of the objective and of the eye-piece should be carefully cleaned after being used. The front surface of the objective should always be examined after use, for if solidifying liquids, &c., are left upon it they are very likely to cause injury. The greatest care is necessary in the employment of chemical reagents during observation, especially those which act upon the lead of the flint-glass. Chemical processes which develope sulphuretted hydrogen, chlorine, vapours of volatile acids and similar gases, ought never to be used in the room devoted to the Microscope. Chloride of calcium and nitric acid, fuming hydrochloric acid, &c., require cover-glasses as large as possible to protect the objective even at a normal temperature, and must not be placed in a heated state under the Microscope.

If, notwithstanding all care, a lens has been touched by the reagent, it must be washed with distilled water, and then cleaned in the usual way with a fine linen cloth softened by repeated washing, or with soft leather. Some microscopists use a fresh-cut surface of dry elder-pith; the small particles which adhere to the lens may be removed with a clean camel's-hair brush—certainly an efficient though somewhat complicated process.

If the Microscope is used daily it may be covered with a glass shade or a cloth to protect it from dust. Many observers are not accustomed to give much attention to these matters; they simply put the objective in its box, and cover up the eye-piece, leaving the instrument uncovered. In our experience, Microscopes which have been treated thus for years, even with the objective screwed on, have not been injured. It is, however, always better to err on the side of caution.

To prevent the formation of moisture on the eye-glass during observation, either the eye-piece may be slightly warmed or the whole instrument placed in a warm room before use.

5.—Care of the Eyes.

Those who have moderately good eyes will soon be able to devote several hours daily to the Microscope without injury to the sight; it is, however, advisable not to work at the instrument during the first hours of the morning, nor immediately after taking food. If fatigue or irritation of the eyes is experienced, work should be stopped for a few days.

It is immaterial whether the inactive eye be closed or open during observation. The observer should accustom himself to use both eyes alternately; they thus experience the same changes: hence, on the supposition, of course, that they were originally equal in strength, they will coincide both in power and distance of vision through life; otherwise the active eye usually becomes somewhat more short-sighted. The exclusive and continued use of one eye is said, moreover, to produce a peculiar irritation of the other. This is so far correct, that the eye which is usually inactive receives a far brighter light-impression on looking through the Microscope, and is more readily fatigued if we observe with it for any length of time—which obtains also with the healthy eye of the beginner. As regards the smarting which long-protracted work causes in the active eye, it is doubtful whether the alternate use of both eyes does not produce just as great inconvenience as the exclusive use of one eye. Many microscopists, however, have habitually observed with the same eye during several years without inconvenience.

6.—THE WORK-TABLE.

The convenience of a well-arranged work-table, as to height, size, &c., should not be under-estimated. It should be sufficiently large and massive, and furnished with one or more drawers for the various appliances, drawing prisms, glass vessels, &c. For dissections a slope let into the table is to be recommended, because most bodies are seen more clearly upon a black ground. We must pass over all other arrangements which may perhaps be serviceable in this or in that case, but which may in general be dispensed with.

II.

PREPARATION AND TREATMENT OF SPECIMENS.

There are only very few bodies which can be spread out at once upon the object-slide without further preparation, or be isolated by mere pressure on the cover-glass; the majority require a preliminary dissection into small portions, or must be examined in delicate sections, the preparation of which requires special skill. The proper treatment of objects, therefore, is always one of the principal points in microscopic work. In order to render a given object as accessible as possible for observation, and to become acquainted with all the processes requisite for this purpose, it is of the first importance to become expert in the manipulation of the preparing instruments. Verbal discussion of this point must naturally be confined to a short description of the means usually employed; matters supplemental must be reserved for practical instruction.

The *dissection of the object*, as well as the separate preparation of certain portions of it, is generally effected with a needle—in more difficult cases under a lens or a simple Microscope.

An ordinary needle may be used for this purpose if it is firmly fixed in a wooden handle. If better instruments are required, needles of different shapes and sizes may be obtained, some, for instance, with hook-shaped points, and others terminating in small blades. Needles for difficult dissections must have fine points, and when they become blunted should be sharpened; but those which are used merely for pressure upon the cover-glass or for raising it when the objects are being moved, and also for coarser dissections, should have blunt points, and should also be somewhat stronger and rigid. The handle, especially for fine needles, should permit their being changed at will and tightly fixed by means of a metal collar.

A skilful manipulation of the preparing needles is very important with many objects. The trouble caused by their preparation is in many cases more than counterbalanced by the accuracy of the observation thus attained. Beginners usually fail by operating with excessive quantities, and by not continuing the operation long enough.

If cellular or granular elements in a preparation are imbedded between fibrous ones, which latter are to be examined, the former may sometimes be removed by the brush—a method which is frequently employed in animal preparations. The specimen is placed in a suitable medium on a glass slide, and is lightly teased with a brush, held vertically (Fig. 155), until the liquid becomes disturbed and the tissue clears. This process is repeated several times if necessary, and the preparation is washed and turned about, fresh liquid being added until at length no further turbidity is seen.

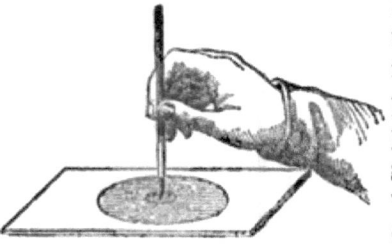

FIG. 155.

For the *preparation of sections* razors are generally employed; all other cutting instruments, such as double-edged knives, scalpels, &c., are superfluous for botanical purposes, and in our experience their value, even for animal substances, is very secondary. Those who can use the razor skilfully will find it sufficient—at any rate in the majority of cases. The blade should be ground flat or hollow, according to circumstances. For rather large sections through woody formations, a flat knife, which does not bend when pressed upon the nail, should be used, but for soft vegetable tissues hollow razors with pliable edges.

Before the actual cutting, the blade and the object should be moistened, because dry-cut sections are usually less satisfactory. Dry gelatinous substances, such as the thallus of many of the lichens, should be treated as exceptions, as they do not offer the necessary resistance when moist or distended; in this case a mere touch with the moist lip will suffice. In cutting, the knife must not be pressed through like a wedge at right angles to the longitudinal axis, but must rather be drawn through diagonally, in order that the sectional surface may be as smooth and clean as possible.

If the preparations are so thin or so small that their manipulation and the proper application of the knife are rendered difficult, they must be supported whilst sections are being made. With thin thread-shaped or lamella-like substances (leaves, &c.), several may be placed one upon the other and then pressed between the fingers. In most cases, however, it is advisable to squeeze them between cork or elder-pith, or, if the substance is of a suitable nature, it may be left to solidify in liquid gum. Thin roots, single leaves, the tips of stems, and similar objects with sappy tissue, may be cut between elder-pith, which is generally preferable to cork; the pressure must, of course, be regulated not to injure the object. Thicker objects may require a suitable groove to be cut in the pith. Longitudinal sections through delicate thread-shaped objects are generally most successful if the objects are bent round the edge of a flat piece of cork, one end being firmly pressed down and the other (which may be fixed on with a very little gum) carefully cut through—commencing at the edge. We have thus obtained thin median sections through the soft ends of roots, after numerous unsuccessful attempts between cork and elder-pith. The process of placing the preparation in gum can only be employed with those objects which can be dried without injury, and which re-assume their original shape on being moistened; with these, however, it is preferable to every other process. Pollen-grains with thick extine, starch-grains, the outer walls of epidermal cells, the thallus of lichens, the tissues of many mosses and algæ, &c., may thus be cut into the finest lamellæ without difficulty.

For decayed wood and similar tissues with loosely united parts, injection with melted stearine, which is afterwards removed on the slide with ether or benzine, is said to be serviceable under certain circumstances. Schacht states that he has obtained delicate sections of very decayed wood from grave-mounds by this treatment.

Finally, objects which consist of only a single row of cells or layer of cells—for, instance, filaments of algæ and leaves of moss—are usually placed upon the thumb-nail with a drop of water, and cut by a swinging motion of the razor—a process which may also be employed for several positions of the cells with moderately thick objects.

In order to convey to the slide the sections lying in the liquid on the blade of the knife or the thumb-nail, many observers employ a fine brush, to the moist point of which the single objects readily cling, whence they can just as readily be liberated again in water. The razor may also be laid flat upon the glass plate, the edge just touching the liquid; the separate sections may then be pushed into the liquid with the needle. In these matters much depends upon habit and practice. We never use the brush for this purpose.

If the objects are suspended in a liquid, the *pipette* (Fig. 156) is frequently used for collecting them and placing them on the slide. If the upper end of the glass tube is closed with the forefinger, the lower end, drawn to a fine point, may be immersed in a liquid without the latter entering. If then the finger is suddenly removed, a little of the liquid, together with the minute bodies suspended in it, will rush into the tube, and may be held there by reclosing the upper end. The pipette is especially valuable for taking up single specimens from various layers of liquid.

The *microtomes* so much recommended in recent times— devices which are said to facilitate the production of thin sections of hard tissues—may be usefully employed in many cases, especially where a greater extent and more uniform thickness of the sections is required; for instance, for the cabinet of microscopic specimens. For scientific purposes, however, we do not esteem them highly, since all we require is to cut small portions of an object, or to find suitable places for examining the sections. We agree in general with Mohl's opinion, that mechanical contrivances which are to aid the artificially strengthened eye by the artificially strengthened hand are of little scientific value, and that simple devices are quite sufficient in skilful hands, and are generally more serviceable than complicated ones. We therefore omit a description of the microtome—the more readily as the manufacturers usually furnish instructions with every instrument.

Fig. 156.

Scissors are much used with animal preparations, but rarely with vegetable ones; for the latter, ordinary small scissors are quite sufficient. Similarly, the chisel, double-edged chisel, saw, plane, &c., can hardly be reckoned among the instruments of the microscopist, neither are the pocket-knife, hand-vice, pliers, punch, &c., to be so regarded. The mode of employing these instruments is in most cases self-evident: thus, the saw is used with hard woods, bones, teeth, and similar bodies, for cutting them, or making thin sections or plates, which are afterwards ground flat; the plane is used with horny bodies for obtaining rather large sections;

the two-edged knife for cutting thin lamellæ of hard objects of considerable thickness, and so on. When the case arises that one or other instrument can be used with advantage, the most suitable shape and size are easily selected, and choice—if choice there be—may be made accordingly.

If the prepared sections are opaque in consequence of their containing air, we must endeavour to remove it. It may often suffice to leave the section some considerable time in the liquid, provided this is capable of absorbing more air; boiled water, for instance, is a perfectly safe medium, as also alcohol, according to Schacht. In other cases a slight heating of the preparation over a spirit-lamp answers the purpose. The most effective and convenient device, however, is the air-pump, a few strokes of which are sufficient to render the objects placed under the receiver quite free from air. If a large air-pump is too expensive, a smaller one in the form of a syringe can easily be made. Into a tube, of about one inch in diameter, closed at the bottom, is fitted a piston provided with a valve, which allows the air to escape on the downward stroke, but checks its entrance on the upward stroke. With this contrivance the object under investigation must be cut in small pieces and placed in the tube with a little water before its preparation, and must afterwards be fished out again, which is somewhat troublesome; nevertheless, a syringe of this kind is often serviceable in the absence of a better one.

If the opacity is due to enclosed particles, or incrustations on the membrane, recourse may be had to chemical reagents for their removal. We will give further details upon this point later on.

The *spreading out of preparations* upon the slide is usually effected by a slight pressure with the needle upon the cover-glass. This simple process affords the advantage of alternately increasing and diminishing the pressure at will, according to circumstances, and is otherwise serviceable for spreading out the object favourably for observation. If, further, a certain constant pressure is required to be maintained for some length of time, either for conveniently examining a large surface, or for making a drawing with a camera lucida, a mechanical contrivance, which admits of pressure by means of screws, is useful. Such contrivances—the so-called *compressors*—are now supplied by most opticians at a moderate cost.

For the observation of living cells in normal vegetation, or under particular conditions (*e.g.*, excluding the air, in a drop of liquid surrounded by carbonic acid, hydrogen, or mercury), a moist chamber, such as Geissler, of Berlin, prepares (Fig. 157 *c*, culture-cell), is most convenient. It has a central space for observation, which is connected at two opposite points with inlet and outlet tubes. The walls of the chamber in the centre are of the thickness of a cover-glass, and are flattened so closely together as to form a capillary space. When filled with a liquid, which is allowed to flow out again, a drop will always remain in the middle capillary space, which may be examined in its whole extent, even with high-power objectives. In this shallow chamber, therefore, a fungus-spore, yeast-cell, &c., may be placed with the proper nourishing fluid. Gases or liquids may be introduced during the process of development by means of an aspirator, which is connected with the outlet tube of the chamber.

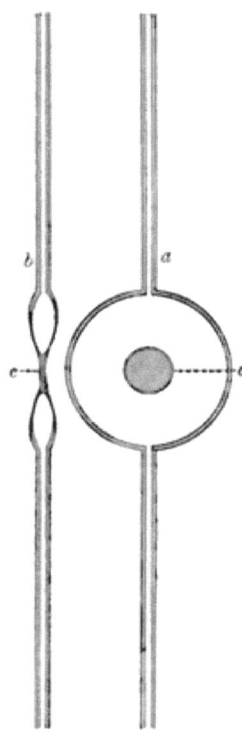

Fig. 157.

For the study of the processes of division in the lower algæ (spiro-gyra, &c.), which usually take place at night, the method applied by Strasburger[1] is to be recommended, by which the division can be delayed till the following day. The glasses in which the algæ are confined are placed in a temperature of less than 5° C. By this means the processes of division are interrupted, but they commence again on the following morning in the normal temperature of the room even with the brightest illumination.

III.

THE PRESERVATION OF MICROSCOPIC SPECIMENS.

THE collection of microscopic objects, especially those which we obtain by a careful preparation, or by a fortunate chance, has in recent times—owing to the improvement of our methods of

[1] Strasburger: "Zellbildung und Zelltheilung," p. 34.

preservation—reached an extent and importance, which make it the duty of every microscopist to become acquainted with the processes of making such preparations. Since it is in the first place necessary to select the proper medium for the preservation of the objects, the methods which are employed may be described with reference to these media. We specially note the following:—

(1.) *Air as a medium for dry substances*—for instance, scales of butterflies' wings, diatoms, crystals, and sections of various hard substances. After we have ascertained that these objects produce a satisfactory image in air, they are covered with a thin cover-glass, which is cemented upon the slide—preferably with thick fluid gum or Canada balsam. The whole slide is then covered with paper, which has a hole cut out over the middle of the cover-glass and projects somewhat beyond the edges of the slide.

(2.) *Canada balsam, for hard formations of every kind*—diatoms, fossil woods, polished bones and teeth, &c. Among the resinous substances which have been proposed as media for microscopic objects, none is so important or so generally employed as Canada balsam. For vegetable objects its use, however, is limited, since it renders the majority of cell-membranes too transparent to produce a distinct image; on the other hand, this quality considerably enhances its importance for animal preparations. The object to be mounted is thoroughly dried by warming or by being placed over sulphuric acid or chloride of calcium, and then soaked in oil of turpentine. A drop of liquid Canada balsam is put upon the slide; the preparation is then spread out on it and covered by a second drop. To remove the air-bubbles, the slide may be slightly heated (no boiling should occur), and this is continued, if necessary, after the application of the thin cover-glass, by placing it, for instance, near a stove; a more rapid hardening of the balsam thus takes place.

As, however, the balsam always remains somewhat viscous, it is advisable after the lapse of a few days to coat the edges of the cover-glass with a rapid-drying varnish which forms a firm non-adhesive ring. Solid Canada balsam may be liquefied with chloroform or oil of turpentine. Its solidification in air may be considerably retarded if kept in wide-necked glass bottles with ground stoppers. Any surplus of the balsam which oozes out at the edges of the cover-glass may be scraped off after it has hardened, and the glass then cleaned with oil of turpentine.

Instead of the usual thick balsam, a weak solution in ether or chloroform is, under certain circumstances, advantageously employed. Used cold, it may be spread on the slide with a brush; the preparation is laid on, more liquid is applied, and it is then finally covered.

(3.) *Glycerine, for most vegetable tissues.*—This is specially recommended for preparations where the preservation of starch-grains and chlorophyll is of importance. In order to limit the contraction of the primordial utricle as much as possible, the objects are first placed in a very weak solution with distilled water, which is either allowed to evaporate in the air or is gradually replaced by more concentrated solutions. The property of glycerine of absorbing water from the air up to a certain degree of concentration renders hermetic sealing superfluous; where, however, it is to remain constantly mixed with a rather large quantity of water, evaporation must be prevented by careful application of the varnish mentioned later on. There is no need for haste in this process; several preparations may be collected before we proceed to cement them.

Instead of pure glycerine others employ a mixture of glycerine and gum arabic or gelatine, because subsequent change of place of the objects, through the solidifying of this liquid, is prevented. This mixture is obtained by heating one part of gelatine with two parts of water, and then adding an equal volume, or even slightly more, of glycerine.

Farrants recommends a still more complex mixture, consisting of equal parts of glycerine, gum arabic, and a saturated solution of arsenious acid; it is used like Canada balsam. Cementing is not here necessary, as the outer layers soon become hard and prevent further evaporation.

(4.) *Chloride of calcium, for most vegetable tissues*, is employed in a similar manner to glycerine; if sufficiently concentrated, like glycerine, it does not need hermetic sealing. By carefully increasing the strength of the solution, the walls of the cells, even in young tissues, may generally be preserved; the colouring-matters, however, become more or less altered, the starch-grains swell, and the primordial utricle contracts. Nevertheless, chloride of calcium solution is in many cases to be preferred to glycerine. To prevent the formation of basic chloride of calcium, which has a tendency to crystallize on the preparation, the solution may be acidified with hydrochloric acid. The strength of the calcium solution must be regulated according to the nature of the tissues; woody cells bear complete saturation, whilst a weaker solution is preferable for young tissues.

(5.) *Sugar-water, for preparations which become too much altered by glycerine or chloride of calcium.*—The sugar element checks the otherwise invariable decomposition and the accompanying disturbance of the liquid. In order, also, to prevent any fungous growth, the solution may be mixed with a little corrosive sublimate. This, of course, always requires hermetic sealing.

In addition to these preservative liquids, numerous others have been proposed—chiefly adapted for animal preparations. A few of those most used may be mentioned here.

The mixture recommended by Pacini—a modification of the so-called Goadby's fluid—consists of

> Corrosive sublimate, 1 part,
> Pure chloride of sodium, 2 parts,
> Glycerine (25° Beaumé), 13 parts,
> Distilled water, 113 parts.

This mixture should be allowed to stand two months, and then be diluted with three times its weight of distilled water and filtered through blotting paper. For the preservation of blood-corpuscles, nerves, ganglia, cancer-cells, as well as delicate proteinous tissues, this is said to be excellent.

Similar mixtures are also employed in different proportions in the Pathological Institute of Berlin; according to Cornil[1] they are composed as follows:—

No. 1.	No. 2.	No. 3.	No. 4.
Corr. sublimate 1	Corr. sublimate 1	Corr. sublimate 1	Corr. sublimate 1
Chloride of sodium 2	Chloride of sodium 1	Chloride of sodium 1	Water 300
Water 100	Water 200	Water 300	

No. 5.	No. 6.	No. 7.	No. 8.
Corr. sublimate 1	Corr. sublimate 1	Corr. sublimate 1	Corr. sublimate 1
Acetic acid 1	Acetic acid 3	Acetic acid 5	Phosphoric acid 1
Water 300	Water 300	Water 300	Water 30

No. 1 is intended for vascular tissues of warm-blooded animals; No. 2, for those of cold-blooded; No. 3, for pus-corpuscles and

---

[1] Quoted by H. Frey: "Das Mikroskop," 5th ed. p. 127.

analogous formations; No. 4, for blood-cells; No. 5, for epithelial cells, connective-tissue, and pus-cells, if the nuclei are to be shown also; No. 6, for connective-tissue, muscles, and nerves; No. 7, for glands; and, finally, No. 8, for cartilage-tissue.

Beale mentions a fluid prepared according to the following recipe:—To 6 ounces of methylated alcohol and 3 drams of creosote, chalk-dust is added until the mixture forms a soft paste. This paste is diluted by careful pounding in a mortar with 64 ounces of water, which is gradually poured in, then mixed with a few small lumps of camphor and allowed to stand for about a fortnight. Finally, it is filtered, and preserved in a well-stoppered bottle. A similar mixture was used by Thwaites for the preservation of *Desmidiaceæ*.

We have enumerated the various preservative fluids employed, and will now describe the method of hermetic sealing which is necessary for most of them. The rules to be observed have reference, in the first place, to the base upon which the cover-glass is put to avoid pressure, and, secondly, to the permanent cementing of the edge of the cover-glass.

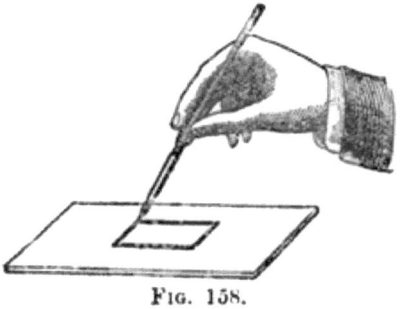

Fig. 158.

If the preparations are not too thick, it will suffice to paint a ridge of cement or varnish upon the slide by means of a brush (Fig. 158), and of such a shape and size that its outer margin projects somewhat when the cover-glass is applied. When the ridge of cement has partially dried, the preparation is placed in the enclosed space, and the latter is filled with the preservative fluid selected. If we have added too much of the fluid, the excess can be easily removed with a brush. It is convenient to put the cover-glass on first at one end only (Fig. 159), and then to lower it gradually to a horizontal position, so that it adheres to the frame of varnish by its edges.

Fig. 159.

If then a little fluid escapes, it may be removed with blotting paper; the edge of the cover-glass is then painted over with a somewhat thick varnish of the same kind. Instead of an enclosed ridge for the base, Schacht recommends two parallel lines of varnish upon which the cover-glass rests. The formation of air-bubbles is thus more easily avoided, but, on the other hand, the difficulty of securing perfect hermetic sealing is increased.

Square cover-glasses are more difficult to cement than circular ones. The latter are most easily cemented by the use of Frey's improved revolving-stage, shown in Fig. 160, the action of which

FIG. 160.

is briefly as follows:—The circular brass plate has a horse-shoe spring-clip to grip the slides. Concentric circles of different sizes are cut on the brass plate, corresponding with the cover-glasses, and indicate where the brush is to be applied perpendicularly to make the rings. Further, the brass plate is rotated rapidly by applying the finger to the milled edge beneath. A fine brush is dipped in the cement, and applied very lightly at first, increasing the pressure as the rotation slackens. A little practice will soon enable the operator to form regular rings. For circles the cement should be more liquid than that used for square cover-glasses.

If the preparations are so thick that the ridges of cement, though laid on several times, are not satisfactory, cells or troughs of gutta-percha, india-rubber, or glass, may be cemented on to the slide. The gutta-percha or india-rubber cells may be made from the thin sheets of commerce by cutting out square pieces, somewhat narrower than the object-plate, and boring them through with a punch (Fig. 161).

Glass cells of circular form are obtained by sawing off rings of suitable thickness from glass tubing, and grinding the surfaces flat; they require, of course, circular cover-glasses. We prefer small rectangular plates of glass with circular or oval aperture, like the gutta-percha cells, as also those which are composed of thin strips of plate-glass. The latter can be easily prepared by the student himself, by cutting two different kinds of strips, 18-20, and 12-15 mm. in length, and cementing them together to form a square framework. A single strip of glass might be bent at right angles over a blow-pipe and the ends fused together, though a practised hand is required for this process.

The final sealing is a most essential operation. Many cements and varnishes are apt to develope numerous fine cracks through which the air enters; the durability of the preparation depends chiefly upon the proper choice of the method of sealing. For fixing the glass cells, *marine glue* is best; it is composed of equal parts of shellac and india-rubber dissolved in benzine. A solution of gutta-percha and shellac in oil of turpentine is recommended

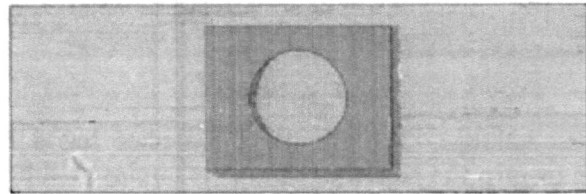

FIG. 161.

by Harting. These preparations are heated to the melting point and then spread with a brush on the edges of the cells or strips of glass, which are then applied and pressed into position under heat. For cementing the cover-glasses, however, the asphalt-varnish of commerce—a solution of asphalt in linseed-oil and turpentine—is usually employed; this has not proved entirely satisfactory, from its liability to crack and flake in the course of a few months. The *asphalt-varnish* of Bourgogne, of Paris, is, according to Frey ("Das Mikroskop," 5th ed. p. 133), much better; it dries comparatively rapidly, and is durable. The same author formerly recommended a white cement, invented by Ziegler, a painter of Frankfort, which may be diluted with oil of turpentine as required by slight heat, and which is spread on like asphalt-varnish with the brush. The cement is said to become hard very slowly, and to be adhesive even after some months; nevertheless, it forms a secure sealing. *Gold-size*, as used by gilders for leaf-gilding, is also recommended for temporary mounting; Harting states that he has employed it exclusively for a long time. Frey, on the other hand, recommends that the final sealing should be made with the cement of Ziegler, above mentioned.

Finally, for labelling the preparations and the whole arrangement of the collection, the shape and size of the slide itself are not quite without importance. In this respect the convenient and pretty design, which Bourgogne, of Paris, has chosen for his well-known preparations (Fig. 162), might be adopted as a

FIG. 162.

model. It is similar to that which has been introduced in English collections—slides of 72 mm. in length by 24 mm. in breadth (3 in. by 1 in.). The model proposed by the "Exchange Society" of Giessen (48 mm. by 28 mm.) is, in our opinion, unsightly and less convenient.

For the description of the preparation two labels are usually employed: one upon the left side of the slide with the name of the object, the other upon the right with the number of the collection, or similar notes. If it is of importance to find readily a particular portion in the preparation again, a space must of course be left for registering it. For this purpose, so-called *finders* have been constructed, yet Hoffman's suggestion still remains the most simple and serviceable, viz., to cut two crosses on each side of the opening in the object-stage, and draw two equal crosses upon the slide exactly over them, after the object has been adjusted in the middle of the field of view.

Microscopic preparations should be preserved in wood or cardboard boxes, the sides of which are provided with grooves in which the slides are inserted separately. To prevent the preparations from gravitating out of position on the slides, these boxes should be placed so that the slides lie horizontally. The slides may also be arranged to lie flat in shallow drawers or trays, which can either be drawn out, as in a cabinet, or lifted out one after another; they are thus very convenient for inspection.

IV.

THE MEASUREMENT OF MICROSCOPIC OBJECTS.

WE need not expatiate on the importance of the determination of the magnitudes of the objects in microscopical investigations; nor is it necessary to describe in detail the different micrometric methods which have been proposed from time to time. We confine ourselves to measurements with the glass micrometer and screw-micrometer, and to the determination of angular magnitudes by the goniometer.

The *glass micrometer*, as generally employed at the present time, consists of a glass plate applied within the eye-piece mounting, or arranged to slide into it; sufficiently fine divisions are ruled upon it with a diamond. The larger spaces are denoted by longer division-lines, as on an ordinary scale (Fig. 163); the divisions into square meshes formerly made appear to have gone out of date. The eye-piece micrometers of the German and French opticians are generally constructed on the metric scale, and in most cases are divided into spaces of one-tenth millimetre.

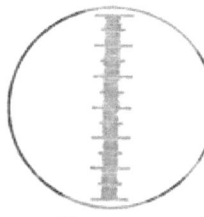

Fig. 163.

It is, of course, evident that the divisions of the eye-piece micrometer can, in the first place, serve for measuring the magnitude of the real image only. In regard to the object itself, their value becomes the less the higher the objective amplification, and must therefore be specially determined for the individual objectives. If, for instance, the magnifying powers of two objectives, in combination with the eye-piece, are taken respectively as 20 and 50, a length of 1 mm. seen on the eye-piece micrometer corresponds to a length of 5 mic. and 2 mic. in the object.

To determine the *relative* value of micrometer-divisions, it is best to use a stage-micrometer—a small glass slip, upon which the millimetre is divided into 100-250 parts or more, so that the distance of the division-lines = 10 to 4 mic., or even less. If we place one of these stage-micrometers under the Microscope, and adjust the eye-piece micrometer so that its divisions can easily be compared with those of the object, then the distance of the divisions required can obviously be determined as is usually done with an ordinary unverified scale by comparing it with a standard one. If, for instance, 10 divisions in the eye-piece micrometer coincide with a space of 25 mic. on the stage-micrometer, we get a value of 2·5 mic. for one division.

The errors which are incidental to this method of determination, and the measurements based upon it, are of course proportional to the accuracy of the two micrometer-divisions; they are not, however, of much importance in view of the perfection of modern dividing machines. In the eye-piece micrometer, where the equality only of the intervals—not their absolute magnitude—comes into account, they may be regarded as almost infinitely small, as may be easily seen if the micrometer-divisions are slid over the real image of any object,—the latter always covers the same number of division-lines. On the other hand, however, as shown by Harting, slight differences occur in the sense that the intervals do not correspond exactly to the values specified by the optician, but appear somewhat too large or too small. In the glass micrometers of Oberhæuser the difference is said to amount to ·041, in those of Plœssl to ·009 of the asserted value, so that in order to be exact the results of the measurements must be diminished by these fractions. To what extent these statements hold good for micrometer-divisions of recent date we need not inquire; we recollect, however, having repeatedly measured one and the same object, employing Plœssl's and Oberhæuser's graduations, without meeting with essential differences. The differences which occur in the micrometers of the principal makers appear to us of very minor consideration in the determination of the size of microscopic objects (disregarding quite special cases).

In many investigations where measurements are necessary, it is indeed only a matter of relative accuracy—*i.e.*, of the comparison of the results of the same observer. He is supposed to know what importance the possible error might attain, and of course will try to avoid those sources of error which would impair the accuracy of his conclusions. Generally speaking, he is satisfied by a few rapidly made measurements—in more difficult cases, by a greater number of careful measurements and calculation of the arithmetical mean; the testing of the measuring apparatus itself will be requisite in exceedingly rare instances only. The errors which are due to the instrument are almost always infinitely small in comparison with those which the observer commits by inaccurate adjustment. In questions where these errors of adjustment might be of importance, no microscopist would be willing to compare the measurements of another with his own; he would, at most, take into account only the figures of the same series of observations. It might with confidence be affirmed that if measurements of different observers admit of comparison in any investigation, this comparison still exists if the respective quantities are taken 5-10 per cent. greater or less.

When observers state that they have refined the accuracy of direct measurement up to 1 mic. by continued repetition of the process, such assertions are to be regarded as deceptive, because the nature of the adjustment is dependent upon the subjective perception within much wider limits. Suppose two experienced observers to make repeated measurements of one and the same object with the same Microscope; let the object be sharply outlined and the measurements accurate to 1 mic. according to the usual method of calculation: except by sheer chance, the comparison of the mean values found will yield differences which amount, with objects of 10 mic. diameter and upwards, to from five to ten times the asserted error. This accuracy is therefore not necessarily real.

The mounting and arrangement of the eye-piece micrometer admit of much variety. In small instruments it is generally fixed immovably in the plane of the diaphragm; hence it is not equally clearly seen by eyes of different focal distance, moreover, the adjustment of a division-line to a certain point necessitates the displacement of the object. In this respect the micrometers adapted to slide in the eye-piece mounting are somewhat more convenient; the same eye-piece is thus available for ordinary observations by removing the micrometer. [The authors here mention several micrometers constructed by Continental opticians which possess more or less efficient means of adjustment; but here again the matter is dealt with somewhat incompletely from an English point of view, and has therefore been omitted.]

The determination of the size of microscopic objects which only partly fill the interspaces of the micrometer-scale, can only be approximately effected. Thread-shaped objects should be adjusted to lie across the division-lines at right angles; the proportion of the interspaces to the thickness of the thread can thus be easily determined within a slight error. Spherical objects should, where possible, be arranged consecutively in a series; the length of the whole series is then measured and divided by the number of the objects. Similarly, the mean thickness of layers in membranes, starch-grains, &c., is estimated; and also, where practical, the sectional magnitude of the elementary organs in tissues.

The *screw-micrometer* is now almost entirely superseded by the more convenient and much less costly glass micrometer. The arrangement of the instrument is based upon the principle much applied in machinery (*e.g.*, dividing engine, &c.), that the movement of a screw is proportional to the angle at which any point of it is turned. The object is placed upon a sliding plate, which is moved backward and forward, by means of a micrometer-screw having a very fine thread (*e.g.*, 5 revolutions to 1 mm.), until the opposite edges of the image coincide successively with a thread extended in the eye-piece. The distance travelled from one point to the other can thus be read off on the graduated screw-head, since the value of one division can easily be calculated from the known value of an entire revolution, and may therefore be regarded as given.

Another form of this device, well known in England, is the *eye-piece screw-micrometer*. In this, two parallel threads are seen stretched vertically across the field of view of the eye-piece, of which the one can be made to traverse the field by the micrometer-screw, whilst the other remains stationary. If the edges of the objective-image are now brought to coincide with these threads its diameter can clearly be much more accurately determined than is possible with the help of a micrometer-scale upon glass. The measurement is, however, in this case also only an indirect one; the relative value of the screw-thread must be specially determined for every objective and for every tube-length, just as with the glass micrometer.

Upon the principle of Ramsden, of measuring the objective-image instead of the object, is based also the screw-micrometer of H. von Mohl, recently proposed ("Archiv für mikr. Anat.," Bd. i., 1865). Mohl finds the usual kind, where this principle is employed, liable to error, because the image of the object suffers a greater or less distortion through the eye-piece, in consequence of which the amplification appears higher or lower according to the distance from the centre; whence, of course, as Harting has already pointed out, the relative value of the micrometer-scale undergoes an alteration. In order to avoid this defect of the eye-lens, Mohl does not measure the size of the objective-image under the fixed eye-piece by moving the spider-web, but moves the eye-piece with the crossed-threads right across the image by means of the micrometer-screw, so that on each adjustment only its centre is optically effective.

It appears to us that the asserted defect of the eye-lens, which is said to be eliminated by Mohl's suggestion, does not in reality exist. For it is perfectly immaterial whether the eye-lens magnifies uniformly or not, provided the image which it gives possesses the distinctness necessary for focusing. The plane of the objective-image coincides of course with the plane of the micrometer-scale, and the space between the two margins, through which the micrometer-screw has to pass, is a quantity entirely independent of the distortion of the image by the eye-lens; both image and scale suffer the same changes through the eye-lens. If the margins of the objective-image are sharp and clearly defined so that the threads in the eye-piece can be adjusted exactly to them, then, even if the eye-lens were to magnify twice as strongly at the edge as at the middle, the result of the measurement would remain unaffected. The determinations of size made by Harting,[1] according to which ten divisions of a glass micrometer, measured separately in succession and in the centre of the field of view, corresponded in section to 75·8 degrees of the index, while the total extent of the ten divisions, measured directly, amounted to 73·5 degrees (instead of 75·8), prove only that the objective-image measured was not magnified uniformly—but less towards the edge.[2] This is a defect which is met with to a greater or less degree in all Microscopes, and the injurious influence of which can only be eliminated, in cases where extreme accuracy is required, by determining the relative value of the micrometer-divisions or

---

[1] "Das Mikroskop," 2nd ed. ii. p. 240.

[2] We cannot understand how Harting concludes that these differences are to be explained by *higher* magnifications of the marginal parts. On the causes of lower magnification towards the margin, *vide* p. 64 *et seq.*

screw-threads by experimenting with an object which occupies in the field of view the same space as that to be measured. If, for instance, it is required to determine accurately the size of an object whose image covers five divisions in the eye-piece micrometer, we select for the determination of the micrometer-divisions or screw-threads a distance upon the stage-micrometer which corresponds to the same number of divisions in the eye-piece. Hence it also follows that the object to be measured must always be brought into the middle of the field of view.

The eye-piece screw-micrometer of Mohl cannot, therefore, afford any other advantages than those which are in all cases combined with the greater distinctness of the image in the middle of the field of view; and it is not to be supposed that these advantages are capable of demonstration in objects of at most 20–30 mic. in diameter, for which the instrument is intended.

That very accurate measurements can be made with well-constructed screw-micrometers is beyond doubt. According to Harting the probable error of a single measurement amounts to about $\frac{1}{8}$ mic. only in one of Plœssl's instruments, and in the middle of a series of measurements would be from $\frac{1}{13}$ to $\frac{1}{10}$ mic. only. In this case it is, of course, assumed that the observer is accurate in the adjustments. If interference lines are mistaken by the observer for the outlines of the object, or if he is confused by optical effects of any other kind, he will obtain erroneous results even with the most accurate instrument, especially if the object to be measured is very small. Perfect vision in the observer, as already stated, is in all cases the essential condition.

Finally, as regards the means of denoting micrometric dimensions, we think that Harting's proposal to use the micromillimetre ($= \cdot 001$ mm.) as the standard of unity deserves general acceptance. It would be advantageous to decide once for all upon a standard of unity smaller than most of the objects which have to be measured, because we are accustomed in other matters also to measure both time and space by such unities. As the French scale is undoubtedly the most generally employed in science, and the micromillimetre forms, moreover, a convenient unit even for the smallest objects which have to be measured, we do not think there is anything valid to be said against Harting's suggestion. Larger dimensions are always expressed in millimetres, and still larger in metres, so that the determining number never exceeds three figures. If it be preferred to express the actual values in millimetres, a decimal fraction is the most convenient in practice, for example, ·0048 mm. (= 4·8 mic.), since everyone knows that the units of the micromillimetres are to be found in the third place of decimals. Vulgar fractions, such as $\frac{1}{657}$ or $\frac{1}{250}$ mm., give, in our opinion, just as vague an idea of the actual size of an object as the expression $\frac{1}{657}$ or $\frac{1}{180}$ of a year does of the corresponding time. They are also unquestionably less convenient than integers or decimal fractions in calculations. Vulgar fractions, whatever may be the numerator, such as $\frac{7}{105}$, $\frac{3}{230}$, &c., are wholly objectionable, since the relation of the numerator to the denominator has first to be determined by division before we can form an approximate idea of the measured dimensions.

For the comparison of the different units of measurement which are usually employed for statements of the dimensions of microscopic objects, in addition to millimetres and micromillimetres, the following table of reductions will be of service:—

Paris line	… … =	2·2558	millimetres
„ inch	… … =	27·07	„
English duodecimal-line	=	2·1166	„
„ inch	… … =	25·3997	„
Rhenish line	… … =	2·1802	„
„ inch	… … =	26·1622	„
Vienna line	… … =	2·1952	„
„ inch	… … =	26·3419	„

For the opposite ratio we thus get:—

One millimetre	=	·4433	of the	Paris line
„ „	=	·0369	„	„ inch
„ „	=	·4724	„	English line
„ „	=	·0394	„	„ inch
„ „	=	·4587	„	Rhenish line
„ „	=	·0384	„	„ inch
„ „	=	·4555	„	Vienna line
„ „	=	·0379	„	„ inch

The determination of angular magnitudes is best effected with the *goniometer*, of which a very convenient arrangement is that of C. Schmidt (Fig. 164). It consists of a graduated circular disc attached to the eye-piece; to this disc a vernier is applied, with which crossed-threads extended in the eye-piece revolve round the point of intersection. In practice the apex of the angle to be measured falls upon this point of intersection, while the two threads are brought alternately to cover the legs of the angle. The angle of the rotation is read off by the vernier. Other opticians apply the graduation to the rotating stage, so that in making measurements the crossed-threads in the eye-piece remain stationary while the stage is rotated with the object. This ar-

Fig. 164.

rangement has the disadvantage that the object changes its place in the field of view in consequence of defective centering of the stage. To remedy this defect the stage should be centered very accurately by adjusting screws acting in two directions. We therefore prefer the eye-piece goniometer.

In default of a goniometer, a drawing may be made of the angle with the camera lucida; the two legs are then produced by means of a rule, and the angle measured with an ordinary protractor.

Whatever method of measuring we adopt, there always remains a source of error to be counteracted, which is not dependent upon the accuracy of the measuring instrument. Though we should entirely disregard the numerous irregularities which influence the distinctness of the lines and disturb the parallelism of edges which ought to be parallel, or if we suppose the legs of the angle required adjusted with all desirable accuracy, yet it cannot be affirmed with certainty whether the plane of the angle coincides exactly with that of the field of view or is inclined to it. It is evident that an elevation or depression of the apex makes the angle appear too large, whilst turning it round one leg makes it too small; hence the results of measurement may differ considerably according to position, and unequal angles may even appear equal. This disadvantage cannot be wholly eliminated in crystallographical researches; it may be checked to some extent by repeating the measurements of equivalent surfaces. In many cases, also, it is possible to control these inclinations so that we know certainly whether the actual measurement of an angle differs in

a + or − direction from the real size. If we are examining tabular crystals which are to be regarded as truncated clinorhombical columns (Fig. 165), the equal breadth of the column-surfaces (shaded in the figure) is a proof that the terminal face is not inclined either to the right or left. In this position, therefore, the angles a and b cannot, at any rate, appear smaller than they really are, since an inclination in the direction of a to b would in all cases magnify them. Similar data are also furnished by many other crystal-forms, and the observer must discover them in every given case and take note of them. Of what value are goniometers which exhibit the arcs of the revolutions accurately up to one minute if the errors arising from the object amount to 1–2 degrees? No one would expect that such errors in small octahedrons, rhombohedrons, &c., could be easily avoided. Repeated experience has shown us that a skilful observer is required to effect the accurate determination up to $\frac{1}{3}$ degree of an angle in crystals such as are commonly found in vegetable cells, or as they arise after the addition of reagents to the fluid under observation. The goniometer might in such circumstances be regarded as perfect.

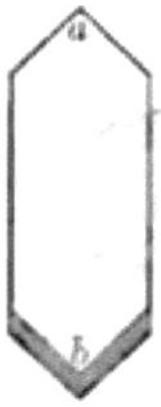

Fig. 165.

V.

THE DRAWING OF MICROSCOPIC OBJECTS.

THOSE who are much occupied with microscopical investigations ought to be able to make drawings of what they have observed agreeing in all essential points with the mental impressions they have received. We say with the impressions received, not with the object itself; for a truth to nature surpassing this relative one is not imaginable. We cannot reasonably demand that the hand of the draughtsman should figure correctly what the eye of the observer has seen incorrectly. The acquirements which are usually expected of the draughtsman—correct conception, understanding the nature of the object, truth to nature without subjectivity, &c.—must therefore be strictly demanded of the observer; the draughtsman's part is merely to call up in the mind of another the impression which the observer has appropiated to himself, whether it be correct or incorrect. Absolute truth to nature must always remain for the microscopist an ideal, which he indeed strives to approach but can only approximately attain. His drawings are true to nature, in the most favourable case, in those points alone for the delineation of which they were intended. The author himself, if he refers to them again after the lapse of some years, will not expect to find more in them.

As the accurate reproduction of the outlines in complicated drawings is a difficult matter even for the practised hand, numerous aids have been invented by means of which it is possible to project the microscopic image upon the plane of the paper. The better

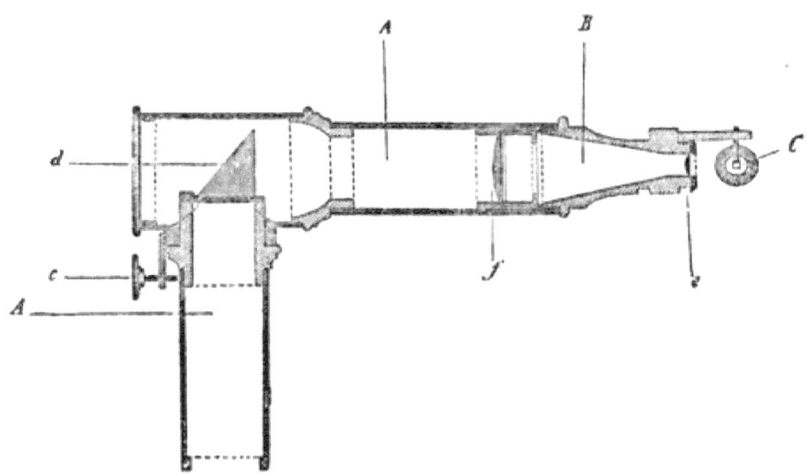

Fig. 166.

known of these appliances may be brought under the following categories :—

(1.) Devices which so reflect the pencils of light issuing from the eye-piece that the image is projected upon a vertical or more or less inclined surface. The last reflecting surface is hence provided with a small aperture, or is of very small extent, so that in either case the eye sees through a portion of the pupil the point of the pencil with which the outlines of the image are being traced. The pencil is therefore seen directly, whilst the microscopic image is seen by reflexion. But since a single reflexion involves, as we have already stated, a half-inversion of the image, those appliances are the most advantageous in which two reflexions take place in the same plane, *e.g.*, prisms acting by two internal total reflexions.

To this category belong Sœmmerring's mirror, the older drawing-prism of Nachet, and the equilateral prism—all acting by a single reflexion; and further also the camera lucida of Wollaston, Milne-Edwards, Amici, and Oberhæuser—all acting by two reflexions. The Oberhæuser camera lucida is provided with a right-angled tube $A \, A$ (Fig. 166); the drawing paper, serving as the plane of projection, may therefore be placed flat upon the work-table, a special stand not being required. The vertical arm of the apparatus is inserted in the body-tube after the eye-piece has been removed, and the prism C is so adjusted by turning the tube B that its plane of reflexion coincides with that of d. In all the other devices the plane of projection must be so situated that the rays emerging in the direction of the axis of the Microscope meet it perpendicularly after the last reflexion, if produced backwards; consequently a special drawing stand must be employed, otherwise the projected image appears distorted.

(2.) Devices which project the rays of light from the point of the pencil by two reflexions, so that they appear to come from points in the field of view. Here the microscopic image is seen directly through an aperture in the last reflecting surface or near to it; on the other hand, the point of the pencil is seen by reflexion. This principle may of course be applied with various modifications similar to those of the preceding devices. Among the contrivances of this nature we may mention the drawing apparatus of Gerling, the camera lucida of Nachet, Chevalier, Nobert, &c., and Hagenow's Dicatopter (Fig. 167). The last is fitted with a horizontal tube, similar to the Oberhæuser camera, so that the microscopic image undergoes a

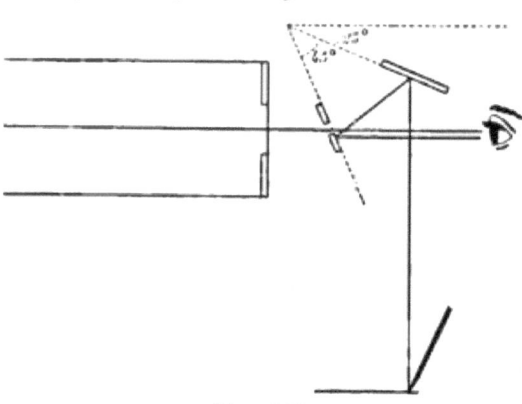

FIG. 167.

half-inversion. In practice, the rule also holds good, that the plane of projection in the reflected image must coincide with, or at least be parallel to, the field of view.

It cannot, therefore, be once for all decided which of these contrivances is the most suitable for convenient manipulation, for, obviously, familiarity and practice may lead different observers to various opinions upon this point. So much, however, will be evident to all, that a half-inversion, such as a single reflexion involves (whereby, for instance, a spiral winding to the right is changed into one winding to the left), must be looked upon as a real disadvantage which not only adds to the difficulty of making the drawing, but which may also, under certain circumstances, give rise to many errors. The reflected image obtained in this manner should therefore be again inverted, which may conveniently be done by tracing it on the reverse side of the paper against a window.

The method recommended by Harting, to draw the real image produced by the objective on a screen at a distance of 400–500 mm., and then make a tracing of it with tracing-paper, as also Stilling's method[1] (which would not be applicable for high amplifications), involve such troublesome minutiæ that they are not likely to find many supporters.

Since the amplification of the projected image is dependent upon the distance of the plane of projection, it must be determined directly for the usual distance and also for different distances (cf. p. 179). This should be done for each objective; the amplifications should be tabulated and preserved in the microscope-case, together with the table of the relative values of the micrometer-divisions.

The scale of the amplification should be given with each figure thus: (60), (350), and so on, or in the form of a fraction, as $\frac{60}{1}$, $\frac{350}{1}$, though the addition of the denominator is superfluous,

---

[1] Harting ("Das Mikroskop," 2nd ed. ii. p. 295) describes Stilling's method as follows:—"A small piece of tracing-paper is attached with gum upon the cover-glass, under which is the microscopic preparation. This is now brought under the Microscope, and the outlines of the magnified object are drawn upon the tracing-paper with a fine point. A network of fine lines is afterwards drawn upon it, dividing the whole into small squares; a similar network of larger squares is made upon a piece of paper, and the drawing on the tracing-paper is copied."

since the scale is based on the assumption that the diameter of the object is the unit. When the image is smaller than the object the scale must of course take the form of a fraction.

The drawing obtained by means of the camera lucida must be carefully compared with the microscopic image, and corrected before the details are filled in. It may easily happen that important dimensions—for instance, the thickness of the walls in cellular tissues—or that certain peculiarities of form may appear somewhat inaccurate in the drawing, and that the aid of the free hand is absolutely requisite for perfect reproduction of the outlines. The drawing must therefore be completed with the free hand; this part of the task is the most difficult, for experience and skill are required on the part of the draughtsman. In difficult cases the unpractised hand cannot produce even an approximately true image; it is hence strongly recommended that all intending microscopists should learn from the very commencement to delineate accurately what they see. Books give but slight assistance in this matter; a teacher or constant practice will be of far more service.

The character of the drawings should be variously modified according to the purpose in view. In many cases a mere outline drawing showing the morphological or anatomical proportions is quite sufficient; in others, it is necessary to add a purely conventional shading of certain parts of the drawing, to which special attention is directed. Similarly, with stratified or fibrous substances the course of the strata or fibres may be represented by straight lines; it must then be remembered that the lines do not represent the thickness or fineness of the strata, but merely indicate their presence and illustrate their course.

The task is very different if the drawing is to be true to nature in the smallest details visible—that is to say, as far as the distribution of light and shade may reasonably be represented. In this case we have choice of two methods of procedure. The nearest approach to nature would of course be made by exact reproduction of the gradations of light as they really appear in the microscopic image; the field of view itself would thus receive a tint corresponding to its brightness. The denser layers of a membrane would therefore be represented as lightest, the more open layers somewhat darker, and the shadows at all points of transition approaching a deep black. In this way a skilled hand can so far heighten the realistic execution of the drawing, that it may almost be mistaken for the microscopic image.

The opposite process of delineation—by which the brightest spots are regarded as the darkest, and *vice versâ*—would be of equal value for scientific purposes. But, just as in chartography, the process resolves itself to this— that the drawing must be executed according to a certain system, whatever may be decided upon. The section of a cylindrical cell, the membrane of which consists of two dense outer layers and one less dense (red) layer, may consequently be just as well represented by Fig. 168 *b* as by Fig. 168 *a*. Generally speaking, a white background is more serviceable where the multiplication of the diagram is contemplated, and is attended with other practical advantages also.

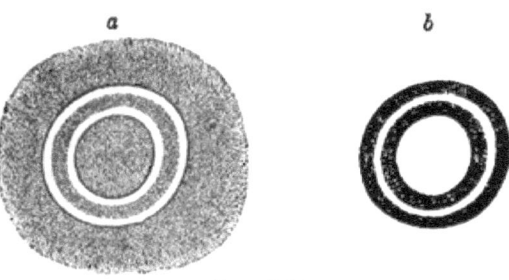

Fig. 168.

If, in accordance with these explanations, we now examine the anatomical figures given in the more recent works on the subject, we shall at once perceive that neither the one principle nor the other has been strictly adhered to. Most draughtsmen (we do not exclude even ourselves) commence work, consciously or unconsciously, in an intentionally eclectic manner; for instance, they represent thick cell-walls by light tints, while they represent threadlike concretions, and the starch-grains contained in the lumen, &c., by dark tints. Everyone who will carefully examine finished anatomical drawings with reference to the method of representation will almost always meet with these incongruities. As an example we will just mention the elegant plates of Schacht's " Lehrbuch der Anatomie und Physiologie." It contains many figures which at first sight might seem to be very perfect drawings, but which on closer examination turn out to be mere outlines, in which the most arbitrary shading serves to direct attention to certain parts of them. The concentric circles, by which, for instance, the stratification of bast-cells is indicated (Plate V., 3), correspond neither to the looser nor to the denser layers; for if we were to assume that this were the case, then the relative thicknesses of the two layers would be wholly inaccurate. This is also true of other stratified membranes (III., 18, 26–28), as well as in the illustration of starch-grains. As regards cell-masses, sometimes the double-outlined membranes are figured as light, while those of the surrounding cells are shaded, and so on; indeed, all the microscopic figures of Schacht that we have seen are represented diagrammatically. It is almost superfluous to remark that we by no means wish to disparage them on this account, for diagrammatic representation has just as much claim to general approval as any other method. We merely wish to establish a fact which appears to be but little known.

The individual discretion which the observer may use in the combination of different focal adjustments must be added to the arbitrariness which he usually permits himself to exercise in delineating microscopic objects. Sections through parenchymatous tissues, for instance, are not seldom represented as though the single cells were situated exactly at the same height, and had been cut right through the middle. In spiral vessels, annular vessels, &c., the surface view is frequently combined with the central sectional view; occasionally, too, we find the opposite sides combined together in one view. Other draughtsmen are accustomed mentally to enlarge microscopic objects to any extent, and then to draw them as they would appear to the naked eye, with reflected or transmitted light. Fig. 169, for instance, is a spiral thread, depicted upon this principle. The greatest freedom is here taken by the draughtsman, for he supposes cell-membranes, which are at the same time transparent and opaque,—transparent, inasmuch as the projecting concretions (annular and spiral threads, &c.) are figured as if visible; opaque with regard to the course of the spiral and the shading of the surrounding sheath (at the points where they cross only the front side is drawn). The spiral itself is shaded partly on the one side and partly on the other, though no really definite effect of illumination is assumed in either case. Briefly, in the delineation of microscopic objects, disregarding their mere form, which each reproduces according to his ability, there exists a degree of arbitrariness which no words can express.

Fig. 169.

The contradictions affecting the usual methods of representing objects are by no means easily explained. On the one hand, plane microscopic vision with transmitted light and large aperture of the cone of rays reaching the eye; on the other hand, the endeavour to depict the objects as they appear in nature, and yet combined with different sectional views; and again, the habit of keeping oneself at the distance of distinct vision with the naked eye or with transmitted light; and finally, the difficulty or rather impossibility of producing the conditions of microscopic vision for the naked eye,—we are unable to suggest how all these opposing conditions can be reconciled in a satisfactory manner. The natural reproduction of the microscopic image in a photographically true, or even in a rationally diagrammatic manner, is and remains conceivable only for a definite focal adjustment; with solid figures it is an impossibility.

For some years *photomicrographic representations* of objects, with drawings of them in juxtaposition, have attained a considerable importance. Preparations, the production of which involves great difficulties, may in this way be rendered permanent to a certain extent, and thus serve for subsequent investigations from new points of view, whilst the draughtsman accurately reproduces only those special points which engage his attention. This advantage of the photomicrographic process has, however, been felt rather by the zoologist than the botanist. Moreover, since the manipulation of the photomicrographic apparatus presupposes a knowledge of the photographic processes—on which subject special treatises have in recent years appeared—we must refer those of our readers who would learn more of this branch of technical microscopy to the special treatises.[1]

---

[1] The most complete work upon this subject is "Die Photographie als Hülfsmittel mikroskopischer Forschung," translated from the French work of Dr. A. Moitessier, with numerous additions by Dr. B. Benecke (Brunswick, 1868). Cf. also Gerlach's book with the same title (Leipzig, 1863).

PART VII.

THE PHENOMENA OF POLARISATION.

I.

ARRANGEMENT OF THE POLARISING MICROSCOPE.

In the examination of microscopic objects in polarised light a suitable arrangement of the *polarising Microscope* is of the first importance. We shall assume that our readers are aware that this instrument must possess two polarising contrivances—a so-called *polariser* beneath the object, and an *analyser* above it. We have to explain how these pieces of apparatus may be most serviceably contrived, and how they act under given circumstances; and we shall inquire what additional apparatus, &c., is desirable for convenience and accuracy in experimenting. Before we consider these points, however, it should be observed that in the following explanations we confine ourselves to actual Microscopes—that is to say, instruments which can be used for ordinary investigations after the polarising apparatus has been removed. Polarising apparatus, such as the so-called **Nœrrenberg's** polarising Microscope, shall be treated of separately, so far as microscopic objects are concerned therewith.

1.—The Polariser.

If the polariser is required for the higher amplifications, it must provide a cone of light whose peripheral rays deviate at least 8 to 10 degrees from the vertical; for since only half of the incident light is transmitted, a greater convergence is necessary for the illumination of the field of view. Such a cone of light may be produced in two ways: either it is provided directly by the polariser, since this apparatus is capable of polarising rays of different inclination equally completely; or the emerging polarised rays being approximately parallel, are subsequently made to converge by a convex lens applied between the object and the polariser. The latter arrangement is found in the so-called *polariscopes* as manufactured by Dove, Nœrrenberg, and others; it is in all cases applicable where there is room for inserting a condenser above the reflecting or refracting contrivances, and hence, as was previously said of illumination in general, does not require any further explanation. Hartnack, Merz, and others have in recent times applied a condensing lens also to the polarising Microscope. Polarisers are in general use which furnish completely polarised light with sufficient convergence without condensers.

The best known polarising contrivance of this kind is *Nicol's prism*. We may assume its construction to be known to the student, as it is described in every text-book of Physics; we will merely add by way of illustration that in Fig. 170 ab and ef represent the two end faces which are artificially cut to an angle of 68° to the edge ad; while in bc and de we still see indicated the faces of an obtuse rhombohedron. The dividing face be, where the two halves are cemented together, forms with the end faces angles of 89° 17', which may therefore be considered right angles. All these faces are vertical to the principal section of the crystal coinciding with the plane of the paper.

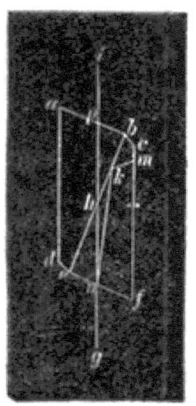

Fig. 170.

The optical action of Nicol's prism is as follows:—The ray gn, incident from below, divides into two rays, of which the *ordinary* ray nk undergoes total reflexion at the layer of Canada balsam, and is absorbed at the blackened lateral face of the prism, while, on the other hand, the less refracted *extraordinary* ray passes through and reaches the field of view in the direction pi. Both rays are polarised, the effective extraordinary ray perpendicularly to the plane of the principal section; its plane of polarisation is called the *polarising plane of the Nicol*. In the cross-view of the prism it is

x

determined by the longer diagonal, while the shorter one is parallel to the principal section.

The greatest possible inclinations which the incident rays can attain towards the different sides are evidently determined, under the given relations, by the corresponding limiting angle of the total reflexion at the layer of balsam. Let us first trace the direction of the rays deviating in the plane of the paper towards the right. Assuming the refractive index of calc-spar to be 1·6583, and that of Canada balsam (according to Brewster) 1·549, then for the ordinary ray the sine of the limiting angle $= \frac{1\cdot6583}{1\cdot549}$, precisely what 69° 4′ gives for the angle itself. Hence, if kf (Fig. 171) is an incident ray deviating to the right, and fh the ordinarily refracted one, the latter, if it is a limiting ray, forms the above-mentioned angle of 69° 4′ with the normal $h\,g$. If we now draw through f a line perpendicular to dc and produce it to g, then δ is the angle of incidence, γ the angle of refraction, and z the inclination of the incident limiting ray to the axis. But in the rectangle $dfhg$ we stated above that $a = 89°\ 17'$; therefore, since the angles at h and f are right angles, $g = 90°\ 43'$. Whence we obtain

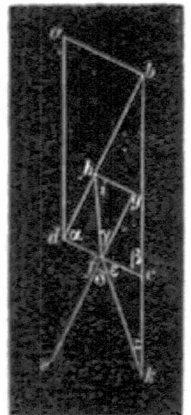

Fig. 171.

$$\gamma = 180° - (69°\ 4' + 90°\ 43') = 20°\ 13';$$

and since
$$\sin \delta = \cos \epsilon = 1\cdot6583 \sin \gamma,$$
$$\epsilon = 55°\ 2'.$$

Further, since $\beta = \epsilon + z$, that is $z = \beta - \epsilon$, we get as maximum inclination
$$z = 12°\ 58'.$$

If again the refractive index of the Canada balsam is assumed to be 1·528, according to Wollaston, we obtain for the limiting angle of the total reflexion 67° 7′. Hence $\gamma = 22°\ 10'$, and $z = 16°\ 44'$.

The calculation for the rays inclined towards the other side is not so simple. Here we have the transmitted extraordinary rays, whose limiting angle determines the greatest deviation from the perpendicular; for the ordinary rays meet the layer of balsam more obliquely the greater the inclination, and therefore always undergo total reflexion. The refractive index of the extraordinary rays is, however, a varying factor which, according to Rudberg, amounts at its minimum to 1·48635, and gradually increases with the deviation to the left until the direction of the principal axis of the crystal is reached. Since the latter in the given case forms an angle of 63° 44′ with the perpendicular, each value for the different deviations is determined by the curve of an ellipse whose major axis coincides with the principal axis of the crystal, and is in the ratio to the minor axis as 1·6585 : 1·48635. Hence we obtain as coefficient of refraction 1·5636 for a ray deviating in the crystal 15° to the left; this value exactly suffices to cause total reflexion at the layer of balsam which the ray meets at an angle of 82° 17′ (the refractive index of the balsam assumed to be 1·549). If the refraction at the lower face of the Nicol is also taken into account, we get about 11° as the maximum inclination for the incident rays. In reality the inclination might be found from two to four degrees less, as the refractive index of Canada balsam is given as somewhat lower by Wollaston and Young.

From the above investigations the whole aperture of the incident cone of light, in the direction from right to left, therefore amounts to 12° 58′ + 11° for the one, and 16° 44′ + 7° for the other extreme—that is, in both cases, to nearly 24°. This magnitude agrees tolerably accurately with that determined experimentally by Bénèche with a Nicol prism. In the direction vertical to the plane of the paper the angles of inclination are somewhat larger; they are here determined solely by the dimensions of the prism.

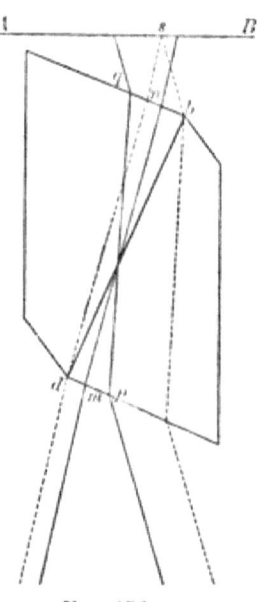

Fig. 172.

This is not, however, equivalent to saying that the whole of such a cone of light reaches the field of view of the Microscope. The path of the rays, accurately drawn in Fig. 172, would indicate that this is only possible under the most favourable circumstances. The two lines $m\ n$ and $p\ q$ represent two limiting rays, which pass through the prism in the directions determined above, and suffer the deviation figured at the end faces. If we now draw through the terminal points b and d, of the layer of Canada balsam, two lines parallel to these limiting rays and produce them upwards to their point of intersection s, it is evident that if s lies in the field of view $A B$, this is the only point which is illuminated by the full cone of light. A portion only of this cone of light reaches the other points, and this diminishes in inverse ratio to the distance of s. It is also clear that if the plane of adjustment of the Microscope is situated at a somewhat higher level, as is usually the case, the aperture of the cone must be the more reduced for the middle of the field of view the greater its distance from $A B$.

The practical result involved in our explanation is that Nicol's prism should be brought as near as possible to the plane of adjustment, since its withdrawal to a distance from that plane reduces the cone of light. But as the optician has also to bear in mind that sufficient space must be left for the insertion of the selenite plates, he cannot exceed a certain limit of proximity. The optically effective portion of the cone of light is consequently reduced to about $29°$ aperture. A polariser of Bénèche gave at the level of the stage in the middle of the field of view $21° 30'$. Where an aperture of this size is not sufficient, the convergence of the rays must be increased by a condensing lens applied above the Nicol.

If the polariser is required also to illuminate sufficiently the marginal points of the field of view, the diameter of the end face must not be too small. Since the length of a Nicol prism is almost three times this diameter, such a prism is necessarily costly.[1] Foucault endeavoured to overcome the difficulty by replacing the Canada balsam by a stratum of air. The limiting angle of the total reflexion is thereby reduced to $37° 5'$ for the ordinary ray, and the refractive indices are the most favourable when a (in Fig. 173) $= 51°$, and $\beta = 70° 52'$, which latter

---

[1] Hasert (in Poggendorff's "Annalen," Bd. cxiii. p. 188) states that he has devised an improved and less expensive construction of the Nicol prism; instead of Canada balsam, he selects a medium whose refractive index is equal to that of the extraordinary ray in the calc-spar. He affirms that the sectional angle is thereby reduced from $90°$ to $81°$, which, of course, amounts, with equal bases, to a corresponding shortening of the prism. He also states that the edge of the polarised zone appears neither blue nor red, but shows only a faint secondary tint, just like a well-corrected objective—which points we are unable to follow.

magnitude corresponds to the natural inclination. If we assume $cd = 1$, we obtain by trigonometrical calculation $bc = \cdot915$. Hence the Foucault prism is relatively less than one-third the length of Nicol's. It may therefore be constructed of smaller crystals of calc-spar, and hence it is much less expensive. On the other hand, it has the disadvantage of a very small aperture, for the greatest possible deviation from the vertical amounts to only about 4° against 12° in that of Nicol. A prism of this kind could hardly be used on the Microscope without a condenser.

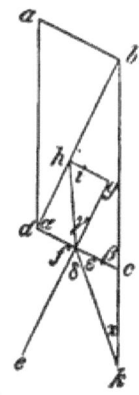

FIG. 173.

A few years ago Hartnack and Prazmowski[1] invented a construction, by which not only is the prism shortened, but at the same time the angle of aperture is increased. The following are the essential points of this construction:—(1.) The sectional face which separates the two halves is vertical to the axis of the crystal. (2.) The faces of incidence and emergence are ground at right angles to the axis of the Microscope. (3.) The two halves are united by linseed oil or copaiba, whose indices of refraction are taken as 1·485 and 1·507 respectively. (4.) The angle which the sectional plane—in which the two halves are in contact—makes with the incident and emergent surfaces is then calculated for linseed oil 73° 30′, for copaiba 76° 30′, and for the angle of aperture we get 35°. The length of the prism is hence reduced to 34, or at most 37 mm.

Besides prisms of calc-spar, *tourmaline* has been much employed in polarisation researches. Tourmaline, when cut in plates of about 1 mm. in thickness parallel to the axis, possesses the property of absorbing the ordinary image—transmitting the extraordinary refracted rays only. This property may be very readily employed with the tourmalines of red to red-brown or green colour, for the crystals are not only sufficiently transparent, but they may be easily obtained so that plates of 5 mm. or more in diameter can be slit off. But as these plates never give the pure polarisation colours—for the latter are modified by the colour of the tourmaline itself—they have not come into use as polarisers, to our knowledge. They are, however, occasionally employed as analysers.

[1] "L'Institut" (1866), pp. 28, 29.

The same drawback is experienced with *Herapathite crystals*, as recently recommended by Haidinger. Herapathite certainly polarises the light very perfectly even in plates of $\frac{1}{20}$ mm. in thickness; its green colour is, however, detrimental. Moreover, large plates are difficult to obtain, and are therefore expensive.

Under these circumstances the Nicol prism is preferable as a polarising apparatus. As ordinarily constructed its aperture is large enough to provide extraordinary rays for a sufficiently extensive field of view; whilst its utility has been considerably increased by the improvements of Hartnack and Prazmowski.

The Nicol may be mounted to drop into the stage opening from above, or it may slide in beneath after the manner of a cylindrical diaphragm. The diaphragm should be placed as near as possible above the terminal face of the prism, and, if necessary, should be replaced by a condenser to increase the convergence of the rays, or it may be attached to the condenser. Achromatic condensers are as superfluous here as with illumination by ordinary light; a plano-convex lens of sufficiently large diameter answers the same purpose.

As regards the position of the Nicol in relation to the source of light, Mohl[1] recommends that the end face of the rhomb be adjusted vertically to the axis of rotation of the mirror, in order to bring into play the light partially polarised by the reflexion at the surface of the mirror. In opposition to this, Valentin[2] states that the loss which the mirror produces by partial polarisation varies with the position of the sun, because the light from the sky (as Brewster has shown) is already partly polarised in a plane passing through the visible expanse of the sky, the sun and the eye, or in a plane which intersects the eye and the expanse of sky vertical to it. From this we see that no general rule can be laid down as to the most favourable position of the Nicol.

In our opinion Valentin's objection is without foundation. For the loss which the light proceeding from the sky suffers through the illuminating mirror, is wholly independent of that which the reflected light suffers by its passage through the prism. The former is determined for the observer by the position of the room he is working in, and the position of the mirror

---

[1] Poggendorff's "Annalen," Bd. cviii. (1859), pp. 181-183.
[2] "Die Untersuch. der Pflanzen- und Thiergewebe im polaris. Licht," p. 101.

at each moment dependent thereon. Whether the loss be great or small, the polarised portion of the reflected light is, under all circumstances, polarised in the plane of reflexion, and therefore passes through the prism undiminished, or is wholly absorbed by it, according as the plane of polarisation of the extraordinary ray coincides with the plane of reflexion of the mirror or is at right angles to it. The position of the prism consequently determines the polarised portion of the incident light, although half the non-polarised portion is evidently transmitted in any position.

Theoretically regarded, Mohl's rule is therefore based upon facts. As far as regards its practical value, we should not feel inclined to estimate it highly, since the differences of intensity in the light which are observed on turning the polariser, are always so very slight that they may practically be disregarded. The reflexion at the end face is of greater importance, inasmuch as the light incident from above and from the side is not completely cut off. In order to avoid this, it is well to turn the oblique terminal face away from the source of light.

2. The Analyser.

Of the various modifications of this instrument which may be suggested, we will first consider the different positions it may occupy with regard to the refracting surfaces of the Microscope. Theoretically it is, of course, quite immaterial at what point above the object the analysis takes place, provided it extends to all the rays that reach the eye. The analyser may therefore be applied above the eye-piece or immediately above the objective, or at any point in the body-tube, or between the object and the objective, provided its angle of aperture is sufficiently large. The question is only what advantages or disadvantages are combined with these different positions.

If the analysing Nicol is situated immediately *above the objective*, as in the Microscopes of Oberhæuser, then, according to Brücke,[1] it is only possible to produce a perfectly dark field of view, with rectangular crossing of the Nicols, by means of strong double-refraction. According to this statement we lose in intensity of

[1] "Denkschriften d. Wiener Akad." xv. (1858), p. 69.

light what we gain in perfection of polarisation. Moreover, it is only to be expected that the objective image must lose somewhat in sharpness through the interposition of so large a body, and, of course, to greater extent, the less perfectly its faces are polished. On the other hand, as stated by H. von Mohl,[1] we shall obtain a larger and more convenient field of view by this position than by any other.

Valentin[2] endeavours to explain the imperfect darkening of the field of view by the divergence of the rays after their passage through the objective. In the sentence we refer to, it is stated that when this divergence exceeds the angle of aperture of the Nicol, ordinary rays pass through together with the extraordinary ones, and consequently give a bright field of view with crossed Nicols. This is, of course, quite correct; but since the limiting angle of the total reflexion amounts to about 12°, or, according to Valentin, even 14°—18° for the axis of the Nicol, it would be entirely unjustifiable to assume so important a divergence. The image-forming cones of light, which proceed from the objective, have their base situated upon its second principal plane, and occupy a linear aperture of 3 to 5 mm., according to the magnification. Their maximum inclination is determined by the size of the diaphragm in the eye-piece setting, and in the Microscopes known to us does not exceed 4°; adding half the aperture of a single cone of light—say 1°—we find that the greatest possible inclination of the marginal rays to the axis of the Microscope is about 5°. Hence there can be no question of the transmission of the ordinary rays.

The supposition of a too great divergence of the emerging cones of light would not, moreover, suffice to account for the phenomenon in question, even for the case where it was perfectly well founded. For since each cone of light is directed towards a definite point in the image, the middle portion of the surface of the image, as far as it is projected by rays of less than 12° inclination, should appear completely dark with crossed Nicols. A faint light becoming more and more bright towards the edge would first be noticeable at the margin, and this marginal portion might easily be cut off by a corresponding diminution of the field of view.

The subsequent inferences which Valentin draws from this

[1] Poggendorff's "Annalen," Bd. cviii. (1859), p. 181.
[2] Ib. p. 97.

imaginary divergence of the cones of light need no contradiction after what has been stated above. It is evident that the magnifying power of the objective, as well as the size of the diaphragm of the illuminating apparatus, under the relations given in the Microscope, exercise practically no influence upon the inclination of the emerging rays, but act upon the brightness only of the field of view. The length of the body-tube, likewise, need not enter into the consideration, as the differences it produces are infinitesimal. It is also clear that the divergence of the optically effective rays between the objective and the eye-piece must in all cases be the same, for their path is in a straight line; and that, on the other hand, the section of the illuminated space gradually increases upwards, and at length nearly equals the diameter of the eye-piece. From this it follows, however, that it is quite immaterial at what part of the body-tube the analyser is inserted, on the hypothesis that the diaphragm is always of the same relative size—*i.e.*, is in the same relation to the above-mentioned section of the whole body of the rays. The application of the Nicol at the lower end of the eye-piece, as recommended by Harting, cannot possibly afford advantages which are not attained if it be placed lower in the tube; moreover, this position is attended with the practical inconvenience that the prism must be considerably larger to produce an equivalent effect.

We cannot recommend that the analyser be inserted between the refracting surfaces, because the prior survey of the field of view and the critical observation of particular objects are thus rendered more difficult. It may be important in investigations by polarised light, and it is always convenient to make a preliminary examination of the preparation under bright illumination and with the greatest possible distinctness of the image— that is, without analyser, and after it has been brought into the middle of the field of view to examine it with the analyser. We know no better method for satisfying these requirements than that already recommended by Talbot,[1] viz., to apply the upper Nicol *above the eye-piece*. With regard to the perfection of the polarisation we have not, however, found that an analyser applied in the body-tube transmits light with crossed position to a noticeable degree; the field of view was so completely darkened that we

---

[1] Poggendorff's "Annalen," Bd. xxxv. (1835), p. 330.

should not give the preference to any other position of the Nicol.

The position of the analysing Nicol above the eye-piece necessitates, moreover, in most Microscopes that it shall be of considerable size, so that the field of view may not be too contracted. Further, the position of the eye here enters into consideration, and the convergence of the cones of light connected with it. We will examine more closely the influence of these two factors.

Confining ourselves to a definite case. In Bénèche's No. 2 eye-piece, the diameter of the optically effective portion of the eye-lens is about 3·5 mm. The peripheral cones of light which correspond to the marginal points of the field of view are inclined at an angle of about 16° to the optic axis, and the eye-point is situated about 6 mm. above the eye-piece. If we demonstrate this relation by an accurate construction (Fig. 174, where the pencils of light are delineated by simple lines), and suppose a Nicol applied with its mounting B, inasmuch as the displacement of the rays parallel with themselves may be disregarded, the following conclusions immediately obtain:—

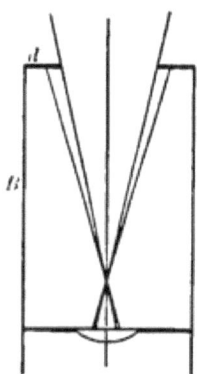

Fig. 174.

(1.) The peripheral pencils of rays must be cut off by the diaphragm at d up to an inclination of about 12° (the limiting angle of the total reflexion for the ordinary rays). The field of view is thus reduced.

(2.) As the diaphragm is situated considerably above the level of the eye-point, the eye surveys at one time only a small portion of the field of view, and must be moved over the diaphragm in order to perceive all parts of the field. This inconvenience is the greater, the longer the mounting of the Nicol in relation to the distance of the eye-point. We might, however, construct the eye-piece so that the eye-point would lie in the plane of the diaphragm, which arrangement would also involve a less inclination of the rays to the axis. The eye would then survey at one time the whole field of view of the Microscope from d, and it would be unnecessary to cut off the peripheral rays. The diameter of the Nicol would of course have to correspond to the effective part of the eye-lens.

If we are satisfied with a small field of view, such as is given by

Bénèche's No. 2 eye-piece, under the relations represented in Fig. 174, the Nicol may also be replaced by a simple prism of spar, by which the image given by the ordinary rays is displaced so far laterally that the eye situated in the axis sees only the image given by the extraordinary refracted rays. A prism of this kind is in many cases preferable to Nicol's, since it renders possible the comparison of the complementary ordinary image with the extraordinary one, and facilitates the true interpretation and designation of the polarisation colours. For more convenient observation of the ordinary image the mounting is arranged to admit of a slight lateral motion of the prism, sufficient to enable the observer to view the one or the other image in the direction of the axis of the Microscope. Obviously, too, it must satisfy the demands which are otherwise made of the analyser.

In Hartnack's analyser of recent construction (Fig. 175) the mounting is united to the eye-piece $b\ c$, and there is a graduated disk a, in which the tube with the lenses and the analysing prism can be rotated. The pointer d, which rotates with the prism, indicates the angular magnitude of the revolution. Leitz, and Seibert and Krafft, have also constructed this apparatus of Hartnack's, adding a vernier and crossed wires; and Merz, Wasserlein, and Vérick have adopted the graduated disk. Such contrivances are useful for certain observations (for instance, on circular polarisation); but in most cases with crossed Nicols and stationary plates of selenite, it is even more important to bring the object by means of a rotating plate to the different positions with regard to the planes of polarisation of the Nicols, and thus be enabled to determine the angles.

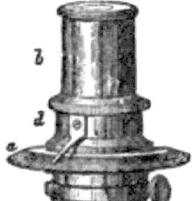

FIG. 175.

3. THE APPARATUS FOR ROTATING THE OBJECTS.

The observation of the changes which the polarisation colours undergo on rotating the objects is so important in researches with polarised light, that the usual devices for effecting the rotation are not adequate for the purpose. A complete polarising Microscope should be provided with means for securing a slow rotation round a perpendicular and a horizontal axis, and also for reading the angles.

Rotation round a perpendicular axis is most simply effected by means of Welcker's rotating-plate arranged for centering and graduated for determining the angles. Revolving stages of the ordinary (Continental) construction are less convenient, because the simultaneous turning of the body-tube changes also the position of the analyser, unless the latter is prevented from moving by the hand or by a special contrivance. Moreover, the rotation of the objective lenses may give rise to deceptions where slight differences of colour are the criteria, for if the object is moved a more or less evident change of colour frequently takes place, owing to the anisotropy of the glass. It is therefore advisable, under all circumstances, to rotate the object alone.

The determination of the azimuth, in which an object has been inserted with regard to the polarisation planes of the Nicols, is, moreover, quite impossible, even with the aid of a graduated scale, with the same degree of accuracy which is otherwise attained in angular measurements. A given diameter of the rotating-plate may certainly be brought tolerably accurately into the diagonal plane of the polariser, since the position of the latter is determined by the edge of the prism, and similarly no difficulty is connected with the adjustment of the analyser. The right-angled crossing of the Nicols is accurately determined to within one degree by observing the greatest possible darkness in the field of view. The result is quite different if the azimuth of a microscopic object, such as the edge of a crystal, is to be measured by the aid of the graduated scale upon the rotating-plate. It is here a question of producing parallelism of the lines, of which only one can be seen with the right eye, the other with the left, at one and the same time ; and in this circumstance lies a very considerable source of error for the novice. If, for instance, we look into the Microscope with the right eye, and place a pencil or a rule upon the rotating-plate in such a manner that looking with both eyes it lies parallel to, or exactly coincides with, the object situated in the field of view, and then look with the left eye into the Microscope, the supposed parallel directions make an angle of from 6° to 10° if viewed again with both eyes. It is clear that under such circumstances we can no longer rely upon our own vision. The cause of this striking phenomenon, which obtains also in stereoscopic vision, is of a physiological nature which we do not purpose discussing here;[1] we may, however, state that such an illusion—be it greater or less—is unavoidable, and the only point of importance to us is,—how under these conditions the *true* azimuth of a microscopic image may be most certainly determined.

In most cases we should be approximately correct if we regarded the line of bisection of the above-mentioned angle of 6° to 10° as the measure. With normal (non-squinting) eyes, which undergo approximately the same amount of turning on looking in the direction of the axis of the Microscope, the possible error is extremely slight. Where great accuracy is desired, this process is *not* sufficient, but should be verified by other means. The following methods may then be employed :—

(1.) If the eye-piece is provided with an aperture for the insertion of the micrometer, a small rule, a **strip of glass**, or other suitable object is slid in this aperture so as to cover the margin of the object under investigation. The projecting portion then yields the azimuth required, and it is easy to place an indicator upon the graduation parallel with it.

(2.) The micrometer-division, or the margin of a cover-glass placed upon the diaphragm, is adjusted in the field of view parallel to the direction in question; the object is then removed, and in its place a sufficiently long glass plate is laid, the margin of which, appearing in the field of view as a dark line, is brought into the same position determined by the micrometer. The production of this margin then gives the reading of the azimuth upon the graduated scale. If the analyser is fixed upon the body-tube, it should be applied in such a manner that its position remains unaltered on turning the eye-piece.

Rotation round a horizontal axis requires a special contrivance adapted to lie upon the stage or to be screwed upon it. The apparatus represented in Fig. 176 may be recommended.

To the brass plate $A B$, having a large central opening, is attached vertically a graduated semicircle, in the centre of curvature of which the rotating-pin with milled head D is applied. This pin carries on the other side of the semicircle two sprung

---

[1] The questions opened up here have recently been explained in several ways; cf., for instance, the papers on this subject by Helmholtz in the "Archiv für Ophthalmologie," 1863 and 1864, and the older treatises quoted therein.

brass plates, between which the object-slide is inserted so that an object lying thereon is nearly in the axis of rotation, and consequently suffers little or no lateral displacement on rotation.

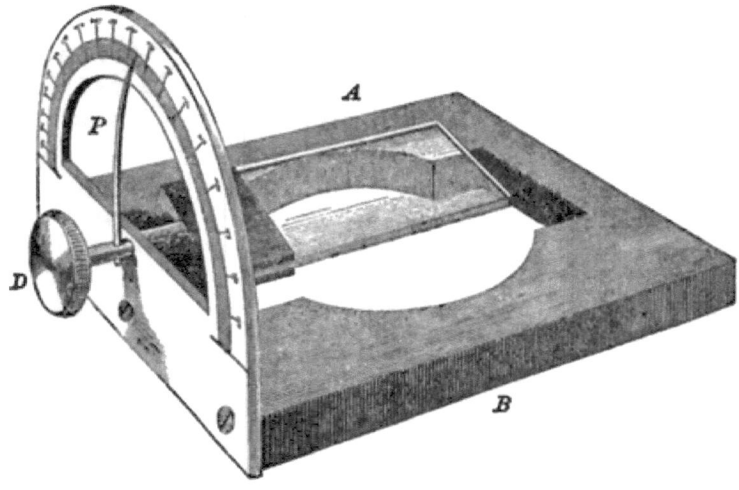

Fig. 176.

The indicator P, connected with the milled head D, moves along the graduated scale on rotation, and thus gives the reading of the

Fig. 177.

angle. It is advisable, for certain purposes, to arrange the apparatus so that the object can be turned under water or other liquid. This is the case, for instance, in the trough apparatus of

Ebner,[1] which in other respects is constructed on the same principle as ours.

Valentin's *object-stage with double rotation* (Fig. 177) does not answer satisfactorily the purposes required, as it affords no angular determinations; it may nevertheless be used in many cases. It is arranged to be clamped to the ordinary stage, and is provided with adjusting screws for centering. The disk $h\ i\ k\ l$ can be revolved in its own plane, and likewise round the horizontal axis g.

II.

THE ACTION OF ANISOTROPIC CRYSTALLOID BODIES, VIEWED SEPARATELY.

WE will now investigate the phenomena which anisotropic crystalloid bodies produce in polarised light, assuming that the reader is acquainted with the fundamental principles of double-refraction and polarisation as explained in the ordinary hand-books of Physics. We have considered it advisable firstly to develope the relations and laws, which are of special significance in the application of microscopical methods of observation, for crystalline media, in order to prepare the student for the more difficult investigation of organized substances. Many conceptions and expressions, which we should otherwise have to explain and determine later on, may be more easily and more directly deduced from the phenomena produced in crystals than would be the case in the more complicated structures of animal and vegetable tissue. They form, at the same time, the natural starting-point for the solution of our task.

1. THE ELLIPSOID OF ELASTICITY.

It is well known that the optical action of double-refracting crystals finds its explanation in the property which they possess of

---

[1] Cf. Ebner: "Untersuchungen über das Verhalten des Knochengewebes im polarisirten Lichte." ("Sitzungsberichte d. k. k. Akademie d. Wissenschaften in Wien," third division, 1874.)

propagating the waves of light in the different directions of space with unequal velocity. That is to say, this propagating power reaches its maximum in a certain direction, and its minimum in a direction at right angles; between these two directions the regular transitions lie. If we suppose that from a given point within the substance lines are drawn (Fig. 178), representing the relative conducting power in the corresponding directions, the terminal points of these lines will lie in a surface of ellipsoidal form, of which the given point is the centre. This is the so-called *surface of elasticity* or *ellipsoid of elasticity*.[1] The geometrical axes of this ellipsoid, which are termed *axes of elasticity*, coincide with the axes of the crystals when the latter are rectangular, but deviate more or less from them in oblique-angled systems. In monoclinic crystal-forms the deviation, however, occurs only in the plane which divides them into two symmetrical halves.

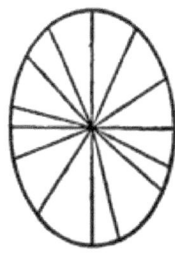

Fig. 178.

In crystals with one principal axis and equal sub-axes, to which the tetragonal, hexagonal, and rhomboidal forms belong, the conducting power is equal in all directions when perpendicular to the principal axis, and greatest or least when parallel to it; the ellipsoid of elasticity here represents a surface of revolution, whose axis coincides with the principal axis of the crystal. On the other hand, in crystals with three unequal axes, such as the rhombic, monoclinic, and triclinic systems, the geometrical axes of the surface of elasticity are also unequal,—hence the latter is an ellipsoid with a major, a mean, and a minor axis.

[1] Strictly speaking, the surface of elasticity, as Fresnel determined analytically, is not an ellipsoid, but a surface of the fourth order, whose equation, with regard to rectangular axes, is

$$(z^2 + x^2 + y^2)^2 = a^2 x^2 + b^2 y^2 + c^2 z^2.$$

The diametral sections through these surfaces are, however, approximately ellipses, and in two particular positions circles—as in the ellipsoid. The same also is true for the surface of elasticity of pressure, according to Neumann. In mathematical Optics the ellipsoid, being far easier to manipulate than the real surface of elasticity, is hence made the basis for further calculations.

For the following considerations a definite assumption as to the form of the surface of elasticity is not at all necessary. It is sufficient to know that its sectional surfaces are, in general, oval figures with two unequal axes, and in two special cases circles.

THE ELLIPSOID OF ELASTICITY.

In order to bring these properties into a definite connection with the inner structure, it is well to recall to memory the analogous properties of compressed or expanded glass. We know that in a parallelopiped of glass, which is compressed by means of a screw-vice, the molecules become condensed or approach each other in the direction of the active force. A sphere, which we may imagine within the glass before compression (Fig. 179 *A*), becomes

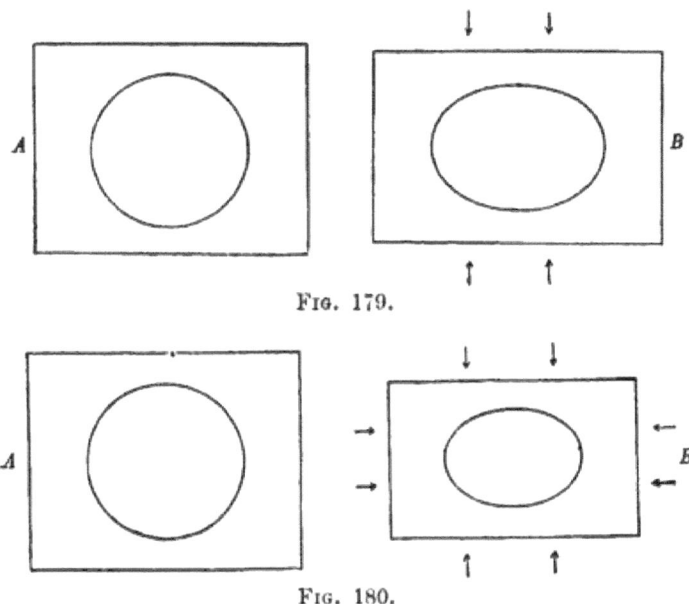

FIG. 179.

FIG. 180.

flattened in consequence of the pressure in the direction indicated in Fig. 179 *B*, and hence becomes an ellipsoid of revolution whose principal axis is parallel to that direction. If the compression takes place in two directions at right angles (Fig. 180 *B* compared with *A*), only that diameter of the sphere which is perpendicular to the plane of the paper retains its original form (strictly speaking, it is somewhat lengthened by the lateral pressure), whilst the two others are shortened by an equal or unequal fraction, according as the active forces are equal or unequal. Hence, equal forces transform the sphere into a lengthened ellipsoid of revolution; unequal ones into an ellipsoid with three unequal axes.

The changes of form which the imaged sphere suffers in consequence of compression may therefore be readily determined. For since in the compressed glass the attraction of the molecules increases with their mutual approximation, and the repulsion of the atmosphere always keeps it in equilibrium, it must be assumed that the density of the air undergoes a change through the compression in a manner similar to that of the substance. But with the density of the air the velocity of propagation of the light is also changed; and even if we leave unexplained how this alteration is effected, it is nevertheless apparent that the resulting ellipsoid exhibits the optical properties of the compressed glass in a manner analogous to the surface of elasticity of the crystals.

There is, moreover, no difficulty in demonstrating experimentally the agreement for all the different cases imaginable. If the glass is compressed or expanded in *one* direction only, it acts as a crystal with one principal axis, which lies in the direction of the active force; the surface of elasticity is an ellipsoid of revolution. If we then apply perpendicularly to the first a second force unequal to it, the ellipsoid of revolution is changed into another having three unequal axes; the glass now acts as a crystal without a principal axis. We may then state generally, that a parallelopiped of glass may always be so compressed that its optical action corresponds to a given positive or negative crystal with one or two axes.

The manner in which the atmospheric density acts upon the motion of light is, in general, explained by the above-determined relations. But a definite formulation of the precise correspondence has not yet been found. Those relations are not sufficient to construct the ellipsoid of elasticity in a given medium with known optical properties, as far as they are determinable by *observation*; this can be done only on assumptions which are derived from the *theory* of the propagation of light, and must therefore not be confused with facts such as direct observation furnishes. The undulatory theory in its present form assumes that the velocity of propagation of light is dependent merely upon the composition of the atmosphere in the direction of the vibrations, but not in the direction of the rays. The ellipsoid of elasticity is hence so constructed, that its diameters, which are parallel to the directions of vibration of any rays, correspond to the velocities (calculated from the refractive indices) with which those rays traverse the crystalline medium. It appears therefore flattened like the globe in positive uniaxial crystals, and oval in negative uniaxial crystals.

We have preferred to pass over these conceptions, and to connect our explanations with the ellipsoid of glass, without theorizing how it arises from the sphere by pressure or tension. We accordingly ascribe to the negative uniaxial crystals an ellipsoid of revolution flattened at the poles, since they act in layers, which are cut parallel to the optic axis, like a glass plate compressed in the direction of that axis. On the other hand, the positive uniaxial crystals act as an ellipsoid lengthened in the direction of the optic axis, since they behave optically like a glass plate expanded in a similar direction. The comparison of double-refracting media with compressed or expanded glass will enable us, in general, to determine accurately the form and position of the ellipsoid of elasticity.

We think we may conveniently retain the phrase "ellipsoid of elasticity," although we by no means imply thereby an algebraically formulated conception such as that where the motion of light is determined through the radii of our ellipsoid. The connection, which has significance for our purposes, does not require to be expressed either numerically or through formulæ; it may just as well be a reciprocal as a direct one; but it is obviously everywhere the same.

2. The Phenomena of Polarisation in Relation to the Ellipsoid of Elasticity.

We now proceed to the determination of the relations which exist between the ellipsoid of elasticity and the phenomena of double refraction and polarisation; we must premise, however, that we treat only of those cases which are of importance in microscopical observation. Facts which can be of interest to the physicist or mathematician alone are not discussed here.

Let us assume that the ellipsoid of elasticity, which we will regard as in the substance, has three unequal axes; let a be the major axis, b the mean, and c the minor. Then the sectional surfaces, which we assume to be in any direction through the centre, are in general ellipses whose excentricity is greatest when they lie in the plane of the major or the minor axis. In two cases, however—as we learn from analytical geometry—these ellipses become circles, and these *circular sections* of the ellipsoid always pass through the mean axis or are parallel to it. Their inclination to the axis a is determined by the ratio $a:b:c$, and we get

$$\tan \delta = \pm \frac{c}{a} \sqrt{\frac{a^2 - b^2}{b^2 - c^2}},$$

δ denoting the angle of inclination in question. The *normals* of the circular sections—*i.e.*, the lines which cut them perpendicularly—therefore lie in the plane of the major and the minor axes; they intersect the former at angles which are the complementary of δ to $90°$.

We will endeavour to demonstrate these relations by a diagram. Let $a\,a$ (Fig. 181) be the major, and $c\,c$ the minor axis; the mean

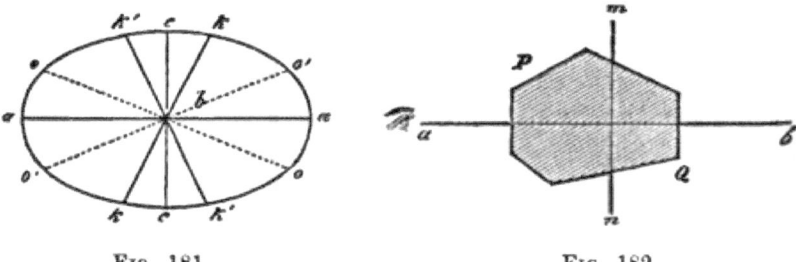

Fig. 181. Fig. 182.

axis, which is perpendicular to the surface of the paper, appears therefore at b as a point. Let the two circular sections be $k\,k$ and $k'\,k'$; they are also perpendicular to the surface of the paper, and are consequently seen as lines which form right angles with the normals $o\,o$ and $o'\,o'$.

The optical action of double-refracting substances is connected with these properties of the ellipsoid of elasticity, as follows:—If $P\,Q$ (Fig. 182) is a piece of double-refracting substance with any known ellipsoid of elasticity, its action upon the pencil of parallel rays $a\,b$ is dependent only upon the ratios of elasticity in a plane $m\,n$ perpendicular to the direction of the rays, and is consequently determined by the ellipse which represents the diametrical section of the ellipsoid parallel to that plane. If, therefore, we suppose any given object rotating under the Microscope, so that its ellipsoid of elasticity (Fig. 183) gradually assumes all possible inclinations to the rays of light incident from below, then the effect which it produces is determined for each position by the sectional surface $m\,b\,n\,b$, which is parallel to the plane of the field of view. In other words, a very thin section cut in any shape from the substance and laid upon the object-slide always acts as a section made in the same direction through the ellipsoid of elasticity. Such a section, therefore, to a certain extent replaces the ellipsoid; it gives a perfect image of the elasticities effective under the given relations, and hence may be denoted as an *effective ellipse of elasticity*.

The connection in point is thus reduced from space (three dimensional) to the plane, and may be readily demonstrated. If $p\,q\,r\,s$ (Fig. 184) is the effective ellipse of elasticity, and $p\,q$ the major, $r\,s$ the minor axis, the incident rays of light (perpendicular to the plane of the paper) are divided into two systems of waves which are polarised in the planes of the two axes. The vibrations of the one system therefore take place parallel to $p\,q$, those of the other parallel to $r\,s$. The relative magnitude of the axes conditions at the same time the velocities with which the two systems traverse the object; the greater $p\,q$ is in relation to $r\,s$, the greater is the difference between the two velocities of propagation, as also the difference of phase thereby conditioned with a given thickness of the object.

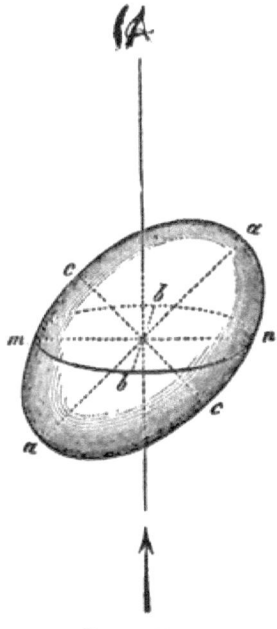

Fig. 183.

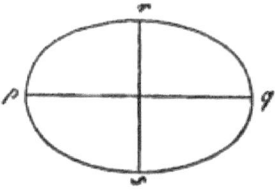

Fig. 184.

On turning the object round, the ratio between $p\,q$ and $r\,s$ is, of course, altered, and it is important to examine the process of these changes. We will first suppose that the axis of rotation coincides with the mean axis of the ellipsoid, and that the rotation itself takes place in our side view (Fig. 183) in the plane of the paper; $a\,a$ is then the major, $c\,c$ the minor axis of the ellipsoid; the mean axis $b\,b$ appears perspectively shortened. If we now consider the position in which $a\,a$ coincides with the plane of the field of view, the effective ellipse of elasticity passes through the axes $a\,a$ and $b\,b$ of the ellipsoid. It assumes, therefore, nearly the form represented in Fig. 185 A, since $b\,b$ is of course greater than $c\,c$ and less than $a\,a$. If we now turn the ellipsoid round the axis $b\,b$, and let it revolve like a rolling sphere from left to right, then another ellipse of elasticity evidently represents each position. In all ellipses which here become successively effective, however, the one axis has the constant length $b\,b$, since it is the axis of revolution, while the other gradually assumes all values between $a\,a$ and $c\,c$, and after half a revolution again becomes equal to $a\,a$. In the plane of the circular section ($k\,k$ and $k'\,k'$ in Fig. 181) the axes

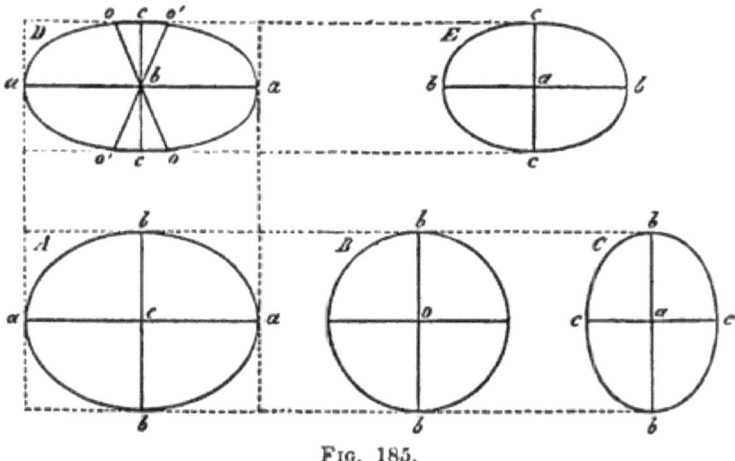

Fig. 185.

are of course equal in length (Fig. 185 B); then the transverse axis becomes less than $b\,b$, and decreases continuously till a rotation of $90°$ has been made, when it coincides with the axis $c\,c$ of the ellipsoid (Fig. 185 C). On continued rotation all possible forms of the sectional surface repeat themselves in opposite sequence, till $a\,a$ for the second time lies in the plane of the field of view. All possible positions of the ellipsoid are thus exhausted; for since the two vertical angles are equal, the return to the horizontal position is equivalent to the return to the starting point.

The positions of the ellipsoid, in which the circular sections lie horizontal, may be further considered. The elasticity of the ether is in these positions equally great in all directions throughout the field of view, precisely as in an isotropic substance.

Accordingly, the light incident from below undergoes ordinary refraction only, and polarisation does not take place. Hence the normals upon the circular sections correspond to the directions in which the light moves as in a single-refracting medium; these are the *optic axes*. The line which bisects their acute angle is termed the *middle line*; according as the latter coincides with the major or with the minor axis of our ellipsoid, the bodies are usually called *optically positive* or *optically negative*.

Let us now return to our starting-point (Fig. 185 *A*), and from thence trace the rotation round the axis. We will suppose that the ellipsoid is rolled upon the surface of the paper, so that the axis of revolution, after a quarter of a turn, coincides with $a\,a$ in Fig. 185 *D*. In this position $b\,b$ will evidently be perpendicular to the surface of the paper, and the minor axis $c\,c$ will become optically effective. The axes of the ellipse of elasticity are therefore $a\,a$ and $c\,c$. In our figure the optic axes are also represented ($o\,o$ and $o'\,o'$), since they lie in the plane of the drawing. On continued rotation the minor axis of the ellipse would, of course, again increase till after half a revolution it had a second time reached its maximum value $b\,b$.

If, finally, we turn the ellipsoid round the third axis $c\,c$, then $c\,c$ obviously forms the one axis of the effective ellipses of elasticity, which represent the different positions, while the other axis gradually assumes all values between $a\,a$ and $b\,b$. In our figure the ellipse which comes into play with a rotation of 90° is shown (Fig. 185 *E*).

We have still to trace the rotation round a line which does not coincide with either of the three axes. We will suppose that the ellipse has first been turned round the axis $b\,b$ till it has attained the inclination shown in Fig. 183 or any other inclination, and then begin the rotation round the line $m\,n$ from this position. After a rotation of 90°, we obtain an ellipse of elasticity whose major axis is $a\,a$, and whose minor axis is $c\,c$ (Fig. 186). This ellipse is, however, differently explained, according to the direction of the rotation; it is inclined to the right (*A*), if the upper vertex in Fig. 183

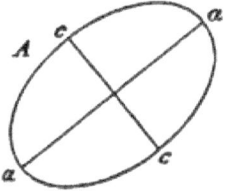

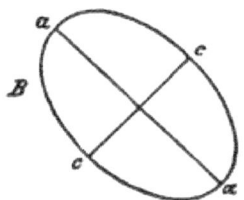

Fig. 186.

is allowed to sink below the plane of the paper; and to the left (B), if it is raised above the paper. The ellipses, which represent angles of rotation between 0 and 90°, go through the range of transitions in regard to form and position between the two terminal members. The angle 0° must evidently be represented by an ellipse, which may be situated thus ⊂⊃ or thus 0, according as $m\,n$ is longer or shorter than the mean axis. The transitions will therefore occur, in the first case, in such a way that the ellipse rises with its right vertex and thus becomes lengthened (Fig. 187 A, B, C), whilst in the second case a

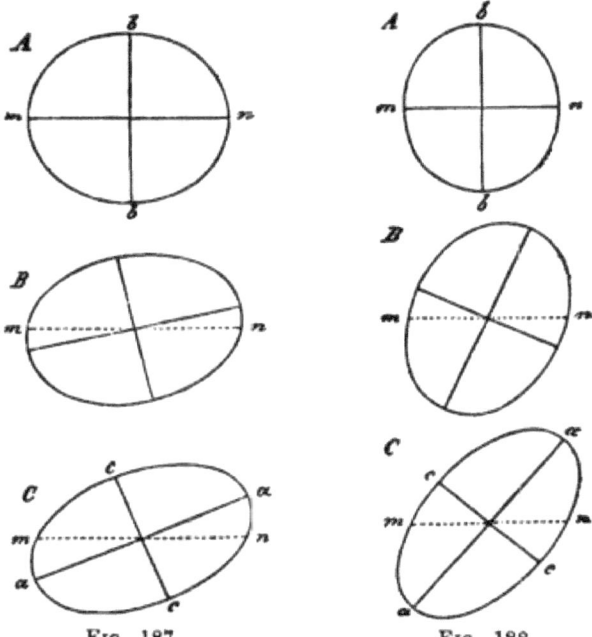

Fig. 187.   Fig. 188.

gradually increasing inclination to the right is combined with a similar extension (Fig. 188 A, B, C).

We can just as easily determine the sectional surfaces which come into play on rotation round any other given diameter in the plane of the field of view. It is only necessary to compare the known ellipse of the initial position with the equally known one after a revolution of 90°; the intermediary members then make in form and position the transitions.

The axial directions of the effective ellipses of elasticity remain altogether unchanged if the rotation takes place round a principal axis of the ellipsoid situated in the plane of the field of view; on the other hand, they suffer a continuous alteration if any other diameter or an inclined axis forms the axis of revolution. The approximate process of this change is determined through the known or easily constructed sectional surfaces, which become effective after a rotation of 90°.

There is no difficulty, therefore, from what has been stated above, in approximately determining the position of the axis of the ellipse of elasticity for any inclination of our ellipsoid, and in defining, as far as necessary, the changes with regard to position and excentricity which take place during revolution. The planes of polarisation of the two systems of waves are, however, also given with the axial position; and with the excentricity the difference of phase, which they attain in a medium of given thickness, also rises and falls. The difference of phase, as is well known, involves interference colour, and thus we arrive at the conclusion that *the optical action of a double-refracting medium, whose ellipsoid of elasticity is given, may be pre-determined for any given direction of the penetrating rays of light.*

We have intentionally combined the above explanations with the general case that the three axes of the ellipsoid of elasticity are unequal, since the special cases which may be imagined are contained in them. It may be not quite superfluous, in addition to what has been stated above, to briefly mention these special cases. If we suppose, first, that the difference between the mean and the minor axes of the ellipsoid gradually diminishes and at length becomes *nil*, then the two circular sections approach nearer and nearer to the plane of these axes, since the angles they make with this plane become more and more acute, and at length coincide with it. The two optic axes consequently move in an opposite direction; when they have reached the major axis of the ellipsoid they coincide and form a single axis. The refracting medium becomes, therefore, *optically uniaxial* and *positive*, since it was also positive in a biaxial state before the axial angle was 0, in accordance with the above-given definition. If, on the other hand, we suppose the mean axis to become gradually equal to the major axis, then the optic axes at last coincide with the minor axis of the ellipsoid, and the negatively biaxial medium becomes *negatively uniaxial*.

The surface of elasticity is consequently always an ellipsoid of revolution in uniaxial media; moreover, the axis of revolution coincides in optically negative media with the least diameter, and in optically positive media with the greatest. The sectional surfaces of such an ellipsoid are generally ellipses likewise; they become circles only when at right angles to the axis of revolution. The changes of position and excentricity, which the optically effective ellipse undergoes during the rotation of the ellipsoid, may be so easily traced—since the problem is considerably simplified owing to the equality of two axes—that a special exposition of them appears to us necessary.

3. Determination of the Axes of Elasticity.

The estimation of the above-developed dependence of the optical action on the position and form of the ellipsoid of elasticity is based upon the principle of reciprocity. If it is possible to predetermine from the directions and relative magnitudes of the three axes the process of the phenomena which we observe on rotation round any given perpendicular or horizontal axis, it must, conversely, be also possible to deduce the directions and relative magnitudes of the axes of elasticity from the known optical phenomena, and thus to construct the ellipsoid of elasticity, as it were, in the refracting medium. This is precisely the task which presents itself in researches in polarised light; with its solution the object, which is usually the question in point, is fully attained.[1]

We have therefore to discuss how the determination of the axes of elasticity is to be practically effected. The observer must take into consideration two unknown factors, which, however, we will regard separately, viz., the directions of the axes and their relative magnitudes. As regards the first point, the *axial directions*, it is hardly necessary for us to make any introductory remarks for those readers who are acquainted with the phenomena of polarisation as explained in the text-books of Physics, since it

---

[1] Further questions—for instance, whether an irregular distension alters the relation of the axes of elasticity, and how this takes place, whether the cause of double refraction lies in the individual micellæ (molecular groups) of organised bodies, or in their arrangement, &c.—are, at any rate, of secondary consideration, and take for granted the solution of this problem.

is only a question here of applying the fundamental laws. We therefore confine ourselves to a few remarks.

(1.) If the polariser and analyser are applied to the Microscope in such a manner that they cross at right angles, it is well known that the field of view is dark. A double-refracting object therefore appears dark also, if the axes of the ellipse of elasticity lie in the polarising planes of the Nicols; in every other position, however—as when a circular section of the ellipsoid is not parallel to the surface of the field of view—it appears more or less illuminated, most intensely when they deviate by $45°$ from those planes. We will hereafter denote this latter position as *diagonal*, and the former, in which the object appears dark, as *orthogonal*. It is evident that each of these positions determines the axial directions of the ellipse of elasticity. Since, however, the greatest darkness is always more certainly perceived than the greatest light, the orthogonal position is preferable in angular measurements.

(2.) In order to decide whether an object, which remains dark on rotation round a perpendicular axis, belongs to the single-refracting substances, or whether possibly a circular section of the ellipsoid of elasticity may be effective, it is only necessary to repeat the observation with different directions of the object, such as we get by revolution round horizontal axes. Double refraction would then be exhibited—that is, if it takes place, and of course always on the supposition that the effective layer is powerful enough to produce a visible effect.

(3.) If the object is double-refracting, and the axial direction of the effective ellipse of elasticity known, it may be further questioned whether perhaps the one of its axes or both at the same time are axes of the ellipsoid. In order to discover this, the object is brought to a diagonal position, and then turned, by means of the contrivance we have described (p. 326), round the axis to be tested towards the opposite directions successively. If the same changes take place here whether we turn it in one direction or the opposite—*i.e.*, if we observe in both cases the same change of colour with equal angles of revolution—the direction vertical to the axis of rotation is an axis of the ellipsoid. We arrive at this inference through the converse of the statement that two sectional surfaces which form equal angles with an axis of the ellipsoid are equivalent. In the orthogonal position the same rule holds good in all cases where revolution produces increased illumination; and if the field of view remains dark, then the axis of rotation itself is an axis of the ellipsoid.

(4.) If, on the other hand, the changes which are observed on revolution in one direction or the opposite are unequal, none of the axes of the ellipsoid lie in the plane of the field of view. The same test is then repeated at the other sectional surfaces, and observation is continued until a surface is found which satisfies the above-mentioned conditions.

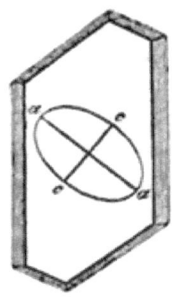

Fig. 189.

(5.) If one of the axes of the ellipsoid is determined, then the two others evidently lie in a sectional surface perpendicular to it. It is therefore only necessary to bring this sectional surface into action, and to determine by the known method its axes of elasticity, which are at the same time the axes of the ellipsoid.

We give examples explanatory of the above. A tabular crystal of selenite (Fig. 189), in which the truncation face of the acute edges of the prisms is the predominant one, appears dark in the polarising Microscope with crossed Nicols, if the directions $a\,a$ and $c\,c$ (the former making an angle of 50° with the side lines) are parallel with the polarising planes of the Nicols, and consequently brightest if they are brought into the diagonal position. These directions are consequently the axes of the ellipse of elasticity. On rotation round the axis $a\,a$, the changes are the same in either direction; similarly round the axis $c\,c$. Both are therefore simultaneously axes of the ellipsoid; hence the third axis is at right angles to the tabular surface. If we turn the crystal round the left or the right edge, so that the third axis is brought into the plane of the field of view, then this axis can be recognized in a similar manner; the direction in the field of view at right angles to it is not in this case an axis of the ellipsoid, as is at once manifest on turning the crystal round its transverse axis.[1]

Flat membranes of free vegetable cells act in a precisely similar way. The axes of the ellipses of elasticity which are effective in surface views, always show themselves on rotation to be axes of the

---

[1] For such observations only small needle-shaped crystals should be used. Larger plates, in vertical position, appear white, or give too high interference colours to enable one to observe the changes caused by the rotation round a horizontal axis.

ellipsoid. The differences of direction which occur, therefore, relate only to their position within one plane—*i.e.*, to the angles which the axes form with the longitudinal and latitudinal directions of the cell. Accordingly, in sections cut perpendicularly to the surface of the membrane, one axis of the effective ellipse of elasticity lies in the direction of the layers, the other is at right angles to it, and is at the same time an axis of the ellipsoid. This obtains likewise with the membranes of cellular tissues, so far as we know them; here, however, observation is rendered more difficult by the circumstance that the two lamellæ, of which the dividing walls consist, really represent two objects, whose axes of elasticity are possibly very differently situated.

The second point that comes into consideration in the determination of the ellipsoid of elasticity, viz., the *relative magnitude of the axes*, assumes, under certain circumstances, a more exact acquaintance with the phenomena of polarisation. It is of primary importance that the observer should be familiar with the scales of the interference colours, at least up to the third or fourth order, which appear on gradual increase of the difference of path between the ordinary and the extraordinary rays, and that he should be quite clear about the factors which produce the variation of the tints. We must refer those who are not conversant with the subject to the text-books of Physics, since a detailed exposition of these phenomena here would lead us too far from our subject. We consider it, however, within our province to present the more important points in their mutual connection, and to elucidate their relations to the ellipsoid of elasticity.

If $A\,B$ (Fig. 190) is a parallelopiped of glass compressed in the directions of the arrows—for example, one end of an object-slide—then the position of its effective ellipse of elasticity is represented in the figure. An object-slide of this sort appears, on rectangular crossing of the Nicols, most brightly illuminated when it forms an angle of 45° with the planes of polarisation, and is consequently situated in a diagonal position. It then produces in white light an interference colour, which, with given thickness, increases in pro-

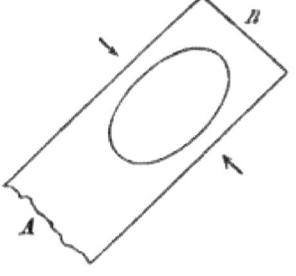

Fig. 190.

portion to the force of the lateral pressure—in proportion, consequently, to the excentricity of the ellipse. If the apparatus admits of a gradual increase of pressure, the resulting colours form a series, which coincides with that of Newton's rings. The same series would also be produced if we were to gradually increase the thickness of the refracting medium by superposing a greater number of equally strong but slightly less compressed plates of glass; for it is evident that the difference of path which the lowest of the plates would produce becomes greater by a certain amount through each succeeding one. The action of a wedge is to be similarly explained; the gradually increasing thickness produces the colours in their natural sequence.

Inasmuch as the crystalline media act in regard to these relations analogously to compressed glass, we arrive at the general law that *the interference colour rises and falls with the excentricity of the effective ellipse, and with the power of the refracting medium.*

Since the colours of the higher orders vary merely in red and green tints, which become more and more faint the higher they rise, and consequently graduate less distinctly from each other, the first three orders are of paramount importance for microscopical observation, and therefore call for a more detailed examination. For this reason we have enumerated in our table the undistinguishable shades of the lower colours with somewhat greater completeness than is usual in the text-books of Physics. In the second column we have given the complementary colours for each order, which correspond to Newton's rings in transmitted light.

First Order.		Second Order.	
Black	Bright white	Purple-red	Light green
Iron-grey	White	Violet	Greenish yellow
Grey-blue	Yellowish white	Indigo	Bright yellow
Light grey-blue	Brownish yellow	Blue	Orange
Light blue	Yellow-brown	Greenish blue	Orange-brown
Greenish white	Brown-red	Green	Light carmine
White	Reddish violet	Lighter green	Purple-red
Yellowish white	Violet	Yellowish green	Purple-violet
Yellow	Light indigo	Greenish yellow	Violet
Brown-yellow	Grey-blue	Pure yellow	Indigo
Brownish orange	Blue	Orange	Dark blue
Red-orange	Blue-green	Bright orange-red	Greenish blue
Red	Pale green	Dark red-violet	Green
Dark red	Yellowish green		

DETERMINATION OF THE AXES OF ELASTICITY.

Third Order.		Fourth Order.	
Violet	Greenish yellow	Light violet	Light greenish yellow
Blue	Yellowish orange	Bluish green	Light pink
Green	Red	Green	Light red
Yellow	Violet	Light greenish yellow	Lilac
Rose-red	Greenish blue	Light yellowish red	Light greenish blue
Red	Green	Light red	Light green

Fifth Order.		Sixth Order.	
Light blue	Light pink	Light blue	Light pink
Light green	Light red	Very light blue	Light red
Whitish	Whitish	Whitish	Whitish
Light red	Light green	Very light red	Light greenish

If we place two double-refracting plates of equal thickness and with equal effective ellipses of elasticity together, in such a manner that they intersect at right angles (Fig. 191), they act as a single-refracting medium. The undulations of light which take place in the lower plate parallel to the major axis coincide in the upper plate with the plane of the minor axis— and conversely. The ratio of the velocities with which the two systems of waves traverse the lower plate is consequently inverted on entering the upper one, and the difference of phase attained is therefore eliminated during the passage through.

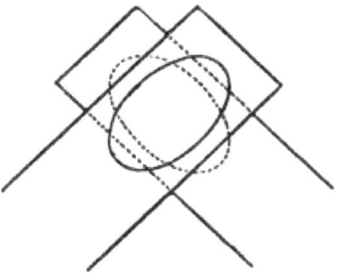

Fig. 191.

If the one plate were thicker than the other, by the magnitude d, the combined effect would clearly be the same as if a single plate of the thickness d were observed.

The same method of reasoning may be applied also to plates of any thickness and of any ellipse of elasticity. The effects will always be added together if the two ellipses are equally situated, and become partially or entirely cancelled if they intersect at right angles. Let us assume, for instance, that the interference colour of the one plate is red of the first order, and that of the other yellow of the first order (of course, on the supposition that the position is a diagonal one), then we obtain as the combined effect of both for the interval of acceleration a colour of the second order (yellow II.), and for the interval of retardation a colour of the first order (light blue I.), which in regard to the yellow is one grade lower than red.

We shall presently return to these acceleration and retardation colours, and tabulate them for a series of combinations as they commonly occur in practice; here it is only a question of estimating them for the determination of the relative magnitudes of the axes of elasticity, and for this purpose the facts we have mentioned amply suffice. It is obvious that the comparison of any given medium with a compressed glass plate, whose ellipse of elasticity is known, must present a simple means for correctly explaining the unknown ellipsoid of elasticity. We will denote its three axes, whose directions we assume to be known, by a, b, and c; then it is only necessary to combine the sectional surface passing through $a\,b$ with the glass plate: the position in which acceleration or retardation takes place then decides whether a or b is the greater. In like manner we determine in sectional surfaces, which are cut parallel to $b\,c$, the ratio of b to c, and in others cut through $a\,c$, that of a to c. Thus, as nearly as possible according to this method, our problem is solved,—we know what directions correspond to the least, the mean, and the greatest elasticity.

The examples already adduced will serve here for further elucidation. If the tabular crystal of selenite (Fig. 192) is so placed upon the compressed glass plate, that the direction $a\,a$ is parallel to the major axis of elasticity in the glass, we observe an increase of the interference colour—the effects of the two media are added together. On the other hand, if we make a rotation of 90° retardation sets in, and the interference tint is lowered. Hence of the two axes of the ellipsoid, which are parallel to the directions $a\,a$ and $c\,c$, the latter is the shorter. The discovery of the third perpendicular axis is combined with some difficulties. From the unsymmetrical position of the axes $a\,a$ and $c\,c$ it is evident that the crystal is biaxial, which excludes the possibility of the third axis being equal to one of those already determined; yet we have always to discover whether it is the major, the minor,

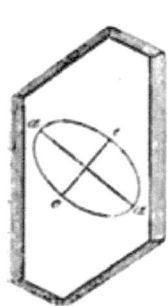

Fig. 192.

or the mean axis. But since sections cannot be made through the crystal in the direction of the principal planes, rotation upon the two horizontal axes is the only method of testing applicable here. The observer must bear in mind the following theoretical deductions:—(1.) If the third is the *major* axis, then by rotation upon $a\,a$ a position must be reached in which an optic axis coincides with the direction of the transmitted light. The interference colours must therefore descend rapidly, becoming black in the given position, and, on further rotation, again rise. (2.) If the third is the *minor* axis, the same changes must take place by rotation upon $c\,c$. (3.) If the third axis is the *mean* one, the colour cannot in any case descend to black. The effective ellipse of elasticity becomes less excentric by rotation upon $a\,a$ as well as upon $c\,c$, though without ever becoming a circle. But since the longer path which the rays of light describe in the inclined plate of crystal necessarily increases the difference, this influence can become preponderant under certain circumstances, and can produce a rise of the colours notwithstanding the slight excentricity of the ellipse.

We learn from observation that the colour of a plate of selenite rises by rotation upon $a\,a$, and falls by rotation upon $c\,c$. This latter circumstance proves that the third cannot in any case be the major axis; but whether it is the minor or the mean one can hardly be determined with certainty from its total action, since the inclination of the transmitted rays can only be increased to a certain limit, owing to the refraction at the surface, the direction of the optic axis therefore being possibly not reached. We should have to carry out the rotation in an approximately homogeneous medium, such as oil, and moreover investigate the rise and fall of the colour more exactly in order to arrive at certainty upon this point.

[1] If we rotate a uniaxial crystal of suitable shape upon the horizontally placed optic axis, the ellipse of elasticity remains unchanged, and the rising of the colour is due merely to the increased length of the path in consequence of the inclination. The comparison of this change in the colour with that of an unknown crystal must therefore show whether the difference of path in the latter increases in higher or lower ratio, from whence it is at the same time decided whether the excentricity of the ellipse of elasticity increases or decreases during rotation. In this manner we learn with tolerable certainty that the third axis in the selenite is the mean one. In general, however, the

With flat membranes, as indeed with all objects which can be cut in every direction, the problem is much simplified. If, with an object of this nature, two out of the three axes are known, their ratio to the third axis is directly determined by transverse sections cut parallel to the corresponding principal planes. If, for instance, a, b, c, d (Fig. 193) is the ellipse of elasticity of a flat piece of membrane, and if a transverse section parallel to $c\ d$ shows that the third axis is less than $c\ d$, then it is of course the minor axis of the ellipsoid; if it is greater than $c\ d$ the ratio to $a\ b$ decides whether it is the major or the mean axis.

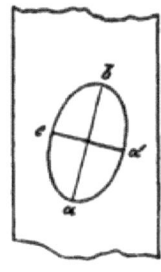

FIG. 193.

Together with the relative magnitude of the axes of elasticity the *plane of the optic axes*, as already stated, is given, since it always passes through the major and the minor axes of elasticity. A very thin plate cut parallel to the mean axis must therefore, on rotation upon this axis, come twice into such a position that the rays incident from below enter it in the direction of an optic axis; it must consequently act as a single-refracting medium, and must appear black with crossed Nicols. Hence it is only necessary to measure the inclinations of the plate corresponding to these positions, and also to take into account the deviation at the bounding surfaces of the plate, in order to determine approximately the angle which the optic axes make. Such determinations are not absolutely accurate, yet it is generally possible to distinguish whether the object is positive or negative.

In order to avoid erroneous interpretation of the interference colours, it is always advisable to compare the positions of acceleration and retardation accurately with one another. If the interpretation has been made correctly, retardation must yield a colour which stands above, as regards the higher of the two com-

results of observation appear too vague for a more detailed investigation to be justifiable upon these and similar theoretical inferences.

It is moreover sufficiently clear from the above that the deductions which Valentin ("Die Untersuchung der Pflanzen- und der Thiergewebe im polarisirten Licht," p. 144) makes from this rise and fall of colour in selenite plates are entirely incorrect. Thus, he states that the plane of the optic axes corresponds, "therefore," to the diagonal direction $+ 45°$ (which in our figure is denoted by $c\ c$), whilst in reality it coincides with the plane of the plate.

bined colours, to just the same extent as the latter do in regard to the acceleration colour. Thus, if we combine blue II. with yellow I., we get red in the one position, which a skilful observer will recognize as red II.; in the other position a yellow, which agrees with yellow I. In our scale of colours also red II. and yellow I. are actually about equidistant from blue II., and a repetition of the combination with known selenite plates affords a perfect proof that blue II. and yellow I. give the above-mentioned colours in the positions of acceleration and retardation.

Since this repetition of the combination always affords the best check, we have appended two tables, in which the colours of acceleration and retardation are collected together for a series of cases. For the first three orders their names are based upon their classification into six colours each, whose distances upon a polished crystal wedge are about equal. If we therefore denote the colours of the first series by 1—6, those of the second by 7—12, and so on, we can pre-determine any combination colour by addition and subtraction of the corresponding numbers, whereby rapid and accurate interpretation of the observed colour is considerably facilitated. The colours 6 (red I.) and 3 (white I.) give, for instance, in the position of acceleration $6 + 3 = 9$ (green II.); and in the position of retardation $6 - 3 = 3$ (white I.). We notice, moreover, that the colours of the selenite plates of commerce do not always agree exactly with the theoretical gradations, and that the same numbers of the same maker often produce a sensibly different tint.

The second table is of special practical value from the fact that a selenite plate, which gives red I. as a rule, most easily admits of the recognition of the rise and fall of the colour.

TABLE I.

Combined Colours.		Acceleration Colour.	Retardation Colour.
Grey I.	Grey I.	Light blue I.	Black
Light blue I.	Grey I.	White I.	Grey I.
—	Light blue I.	Yellow I.	Black
White I.	Grey I.	Yellow I.	Light blue I.
—	Light blue I.	Orange I.	Grey I.
—	White I.	Red I.	Black
Yellow I.	Grey I.	Orange I.	White I.
—	Light blue I.	Red I.	Light blue I.
—	White I.	Indigo II.	Grey I.
—	Yellow I.	Blue II.	Black

Combined Colours.		Acceleration Colour.	Retardation Colour.
Orange I.	Grey I.	Red I.	Yellow I.
—	Light blue I.	Indigo II.	White I.
—	White I.	Blue II.	Light blue I.
—	Yellow I.	Green II.	Grey I.
—	Orange I.	Yellow II.	Black
Red I.	Grey I.	Indigo II.	Orange I.
—	Light blue I.	Blue II.	Yellow I.
—	White I.	Green II.	White I.
—	Yellow I.	Yellow II.	Light blue I.
—	Orange I.	Orange II.	Grey I.
—	Red I.	Red II.	Black
Indigo II.	Grey I.	Blue II.	Red I.
—	Light blue I.	Green II.	Orange I.
—	White I.	Yellow II.	Yellow I.
—	Yellow I.	Orange II.	White I.
—	Orange I.	Red II.	Light blue I.
—	Red I.	Violet III.	Grey I.
—	Indigo II.	Blue III.	Black
Blue II.	Grey I.	Green II.	Indigo II.
—	Light blue I.	Yellow II.	Red I.
—	White I.	Orange II.	Orange I.
—	Yellow I.	Red II.	Yellow I.
—	Orange I.	Violet III.	White I.
—	Red I.	Blue III.	Light blue I.
—	Indigo II.	Green III.	Grey I.
—	Blue II.	Yellow III.	Black
Green II.	Grey I.	Yellow II.	Blue II.
—	Light blue I.	Orange II.	Indigo II.
—	White I.	Red II.	Red I.
—	Yellow I.	Violet III.	Orange I.
—	Orange I.	Blue III.	Yellow I.
—	Red I.	Green III.	White I.
—	Indigo II.	Yellow III.	Light blue I.
—	Blue II.	Pink III.	Grey I.
—	Green II.	Red III.	Black
Yellow II.	Grey I.	Orange II.	Green II.
—	Light blue I.	Red II.	Blue II.
—	White I.	Violet III.	Indigo II.
—	Yellow I.	Blue III.	Red I.
—	Orange I.	Green III.	Orange I.
—	Red I.	Yellow III.	Yellow I.
—	Indigo II.	Pink III.	White I.
—	Blue II.	Red III.	Light blue I.
—	Green II.	Bluish green IV.	Grey I.
—	Yellow II.	Green IV.	Black
Orange II.	Grey I.	Red II.	Yellow II.
—	Light blue I.	Indigo III.	Green II.
—	White I.	Blue III.	Blue II.
—	Yellow I.	Green III.	Indigo II.
—	Orange I.	Yellow III.	Red I.
—	Red I.	Orange III.	Orange I.
—	Indigo II.	Red III.	Yellow I.
—	Blue II.	Violet IV.	White I.
—	Green II.	Blue IV.	Light blue I.
—	Yellow II.	Green IV.	Grey I.
—	Orange II.	Green IV.	Black

Combined Colours.		Acceleration Colour.	Retardation Colour.
Red II.	Grey I.	Violet III.	Orange II.
—	Light blue I.	Blue III.	Yellow II.
—	White I.	Green III.	Green II.
—	Yellow I.	Yellow III.	Blue II.
—	Orange I.	Pink III.	Indigo II.
—	Red I.	Red III.	Red I.

TABLE II.

Object Without Selenite Plate.	Combined with Selenite Plate Red I.	
	Acceleration Colour.	Retardation Colour.
Grey I.	Indigo II.	Orange I.
Light blue I.	Blue II.	Yellow I.
White I.	Green II.	White I.
Yellow I.	Yellow II.	Light blue I.
Orange I.	Orange II.	Grey I.
Red I.	Red II.	Black
Indigo II.	Violet III.	Grey I.
Blue II.	Blue III.	Light blue I.
Green II.	Green III.	White I.
Yellow II.	Yellow III.	Yellow I.
Orange II.	Pink III.	Orange I.
Red II.	Red III.	Red I.
Violet III.	Light red-violet IV.	Indigo II.
Blue III.	Bluish green IV.	Blue II.
Green III.	Green IV.	Green II.
Yellow III.	Light greenish IV.	Yellow II.
Pink III.	Light pink IV.	Orange II.
Red III.	Light red IV.	Red II.
Light violet IV.	Light red IV.	Violet III.
Bluish green IV.	Light violet-red V.	Blue III.
Green IV.	Light blue V.	Green III.

III.

THE ACTION OF TWO SUPERPOSED CRYSTALLOID BODIES, WHOSE PLANES OF VIBRATION INTERSECT OBLIQUELY.

The considerations we have so far discussed may be immediately applied in all cases where the object under investigation represents a crystalloid medium; they are therefore sufficient for flat pieces of membrane and other organised structures which act as crystals. But practically it is not always possible to prepare the objects for examination in such a manner that the optically effective part satisfies that condition—

i.e., that it may be compared with a simple crystal. If, for instance, we are dealing with small horizontally situated cells or fibres consisting of concentric layers of molecules whose ellipses of elasticity are inclined towards the longitudinal direction of the cells, the portions of the walls situated in the median zone act singly as simple crystals, though we should be compelled to split up the fibres in order to be able to observe these effects separately; and such a division of the fibres is in many cases impracticable. The same holds good also for somewhat thickened membranes which form walls between adjoining cells, and which are therefore to be looked upon as composed of two lamellæ. For since the thickenings of these membranes (*e.g.*, spiral threads, obliquely-situated pores, &c.) are, as a rule, unsymmetrically placed, which indicates a corresponding situation of the series of molecules, it is also probable that the axes of elasticity of the two lamellæ do not lie in the same plane.

The observer has not unfrequently to deal with two inseparable superposed media, whose axes of elasticity intersect at various angles. He is thus presented with a new problem, which is apparently more complicated than the foregoing one; and the first question to be decided is whether it is really capable of solution by the same method. We will endeavour to answer this question experimentally.

Let AB and CD (Fig. 194) be the ellipses of elasticity of the two superposed bodies, and e the angle between their major axes (assumed to be variable). The focal adjustment should always be so chosen that the bi-sectional line mn of that angle is always at 45° to the crossed Nicols. On this assumption we get experimentally the following results for the different values of e:—If e is very small the effects of the two bodies increase almost as if they were observed in the position of acceleration, where $e = 0$; the resulting colour undergoes no considerable alteration up to an angular magnitude of several degrees. On further increase of e, however, a marked degradation in the interference colour soon follows, which expresses itself up to a mean value of 45° by a distinct change in the tint. From this point, however, the colour remains ap-

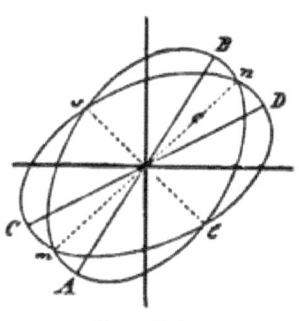

FIG. 194.

proximately the same, though the field of view becomes more and more dark, until at length with an angle of 90° complete darkness, or at any rate—when the two bodies are not exactly equal—the greatest darkness, is found. The passage from one tint to the other certainly takes place in a somewhat different manner to that in Newton's Table, generally as if the two colours were mixed in ever-varying proportions; in the majority of cases, however, the rise or fall cannot be mistaken.

If, for instance, we place two selenite plates, each of which yields light blue of the first order, together in the given way, we obtain for $e = 0$ yellow I., for $e = 22\frac{1}{2}°$ light yellow, for $e = 45°$ white; finally, for $e = 67\frac{1}{2}°$ bluish white. These colours evidently represent only those tints taken in reversed order to yellow I. A few additional examples are collected together in the following table :—

Two Selenite Plates.	$e = 0$.	$e = 22\frac{1}{2}°$.	$e = 45°$.	$e = 67\frac{1}{2}°$.
Grey I.	Light blue	Darker blue	Bluish	Dark blue
Light blue I.	Yellow	Light yellow	White	Bluish white
White I.	Violet	Light violet	White	Yellowish white
Yellow I.	Blue	Light green	Yellow-orange	Yellow
Orange I.	Yellow	Yellow	Darker yellow	Dark yellow
Red I.	Red	Red	Darker red	Dark red
Blue I.	Greenish yellow	Blue-green	Blue	Blue
Green II.	Red	White	Green	Darker green
Yellow II.	Green	Yellow	Darker yellow	Dark yellow

It is evident from this table that the colour we obtain in the position $e = 45°$ agrees tolerably throughout with that of the single plate, or very approximately. Inasmuch as the acceleration colour for the position $e = 0$ is to be regarded as given, we can predetermine the changes which the gradual increase of e produces in any given pair of plates. These changes always consist in a gradual but direct passage of the acceleration colour into the colour of the single plate.

It now remains for us to combine the action of a firmly bound pair of plates with that of a selenite plate, or in general of a third double-refracting body. We have seen that the pair of plates act, to a certain extent, like a single body whose ellipse of elasticity approaches the more nearly to the circle the greater the angle e. If we suppose that the axes of this ellipse are the lines joining the points $m\ n$ and $s\ t$ (Fig. 195), in which the ellipses corresponding to the two plates intersect, this supposition is so far true that the difference between the major axis $m\ n$ and the minor $s\ t$ diminishes continuously, in conformity with the interference colour, when e increases, and finally becomes nil when $e = 90°$; presumably, therefore, it represents also the combined action of a pair of plates and a selenite plate. The position in which the imaginary ellipse of elasticity is situated similarly to that of the selenite plate will here also produce an acceleration colour, and when the homologous axes intersect at right angles a retardation colour is produced. Thus in fact it actually behaves, as may be seen from the following table :—

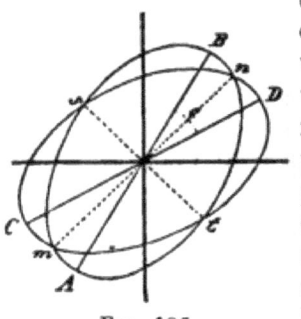

Fig. 195.

Two Crystal Plates.	Angle e.	With Selenito Plate Red I.	
		Acceleration Colour.	Retardation Colour.
Blackish grey I.	0	Indigo-violet II.	Orange I.
	$22\frac{1}{2}°$	Violet	Reddish orange
	45°	Reddish violet	Red-orange
	$67\frac{1}{2}°$	Red-violet	Dark red-orange
Grey I.	0	Blue-green II.	Light yellow I.
	$22\frac{1}{2}°$	Greenish blue	Orange-yellow
	45°	Blue	Orange
	$67\frac{1}{2}°$	Indigo	Reddish orange
Light blue	0	Yellow II.	Light blue I.
	$22\frac{1}{2}°$	Green-yellow	White
	45°	Light green	Yellowish white
	$67\frac{1}{2}°$	Light greenish blue	Light reddish orange
White I.	0	Red II.	Black
	$22\frac{1}{2}°$	Light red-orange	Greenish
	45°	Greenish yellow	Greenish white
	$67\frac{1}{2}°$	Greenish white	White
Yellow I.	0	Blue-green III.	White I.
	$22\frac{1}{2}°$	Green	White
	45°	Yellow	White
	$67\frac{1}{2}°$	Orange	Bluish lilac
Orange I.	0	Light yellow	Yellow I.
	$22\frac{1}{2}°$	Yellow-orange	Light orange
	45°	Orange	Whitish
	$67\frac{1}{2}°$	Red-orange	Lilac

The gradations of colour do not of course even here agree with those of Newton's Table, though the rise and fall on increase of e are, as a rule, distinctly expressed. The highest acceleration and the lowest retardation colours are evidently given by the position $e = 0$; if e becomes equal to $90°$, we get in all positions the red of the selenite plate. Where the change of colour is exceptionally indistinct, it may be rendered more perceptible by the application of another selenite plate, e.g., grey I.

If we turn the pair of plates, without altering the angle e, upon the selenite plate round a perpendicular axis—for instance, by gradually moving them from the position of acceleration to that of retardation—we observe that the changes of colour do not agree

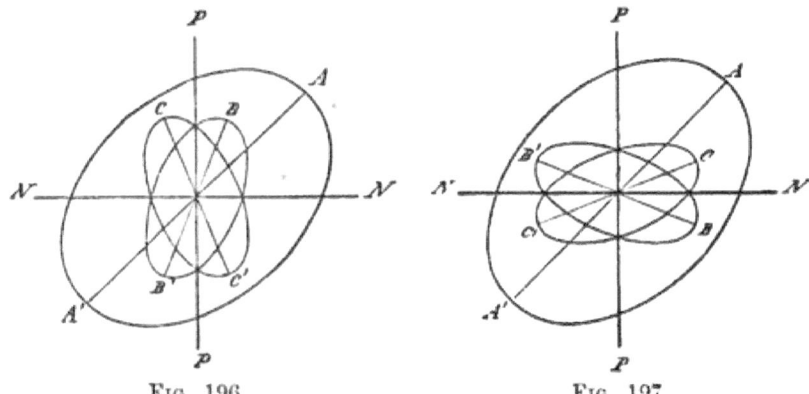

Fig. 196.   Fig. 197.

either with Newton's Table or with the colours of a single plate in the corresponding positions. While a single plate, by rotation upon a selenite plate, produces tints which may be obtained on the palette by mixture of the acceleration and the retardation colours with that of the selenite plate, the pair of plates produce a change of colour of complicated nature, which can be definitely formulated only for the more simple cases. The practical significance which attaches to this change nevertheless demands a detailed explanation.

The difference of colour in the two orthogonal positions, where the bi-sectional line of the angle e coincides with the polarising plane of the lower or upper Nicol, may be denoted as the chief peculiarity of the changes which are observed on rotation of the pair of plates upon a selenite plate. In fact, these positions cannot, under any circumstances, be exactly equivalent, as may easily be shown by demonstration of the respective positions of the three effective ellipses of elasticity. If, for instance, AA', BB', and CC' (Fig. 196) are the ellipses of the selenite plate and of the two superposed plates of crystal in their sequence from below upwards, and if, further, PP and NN are the polarising planes (at right angles to each other) of the two Nicols; then it is evident from the figure that in the position indicated the homologous planes of vibration of the ellipses follow one upon the other like the steps of a spiral staircase, in so far as the alphabetical order of the letters ABC or $A'B'C'$ at the same time represents the sequence of the corresponding axes in the direction of a spiral turning to the left. This is the orthogonal *consecutive position*. If, on the other hand, we take the pair of plates rotated to 90° (Fig. 197), CC' falls between BB' and AA', and the vertices ABC or $A'B'C'$ of the ellipses no longer appear as steps, but make a zig-zag one above another—a position which is denoted as the *alternative position*, in contradistinction to the other one.

The difference of the colours which these positions produce is then in many cases of such a nature that their comparison with the acceleration and retardation colours affords the requisite data

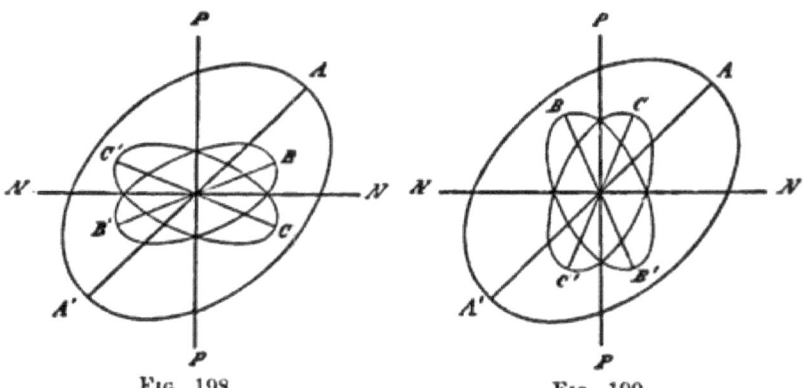

Fig. 198. Fig. 199.

for recognizing a given orthogonal position as a consecutive or alternative one. When this is the case, the solution to the question immediately follows,—how the effective ellipse of elasticity in the upper and in the lower plate is inclined. If, for instance, Fig. 198 is recognized as a consecutive position, and Fig. 199 as an alternative one, we know that the ellipse BB' belongs to the lower and CC' to the upper plate.

THE ACTION OF TWO SUPERPOSED CRYSTALLOID BODIES. 347

To facilitate the comparison of the colours, which is here the question at issue, the following synopsis may be of service; it consists of observations with different pairs of plates, which intersected at angles of $22\frac{1}{2}°$, $45°$, and $67\frac{1}{2}°$. In the first combination series the acceleration colours move in the second order—*i.e.*, they are colours which for $e = 0$ follow the second order of Newton's Table. The second combination series refers, on the other hand, to acceleration colours which belong to the third order. The selenite plate, upon which rotation was made, was red I. The colour of the consecutive position is denoted by C, that of the alternative by A.

FIRST COMBINATION SERIES.

Acceleration Colour.	Colours of the Orthogonal Positions.	Retardation Colour.
Red-violet	C. Dark red / A. Light red	Dark red-orange
Reddish violet	C. Violet-red / A. Orange-red	Red-orange
Violet	C. Dark red / A. Light red	Reddish orange
Indigo	C. Violet-indigo / A. Red-orange	Reddish orange
Blue	C. Indigo / A. Dark orange	Orange
Light green	C. Blue / A. Yellow-orange	Yellowish white
Green-yellow	C. Blue-indigo / A. Dark orange	White

SECOND COMBINATION SERIES.

Acceleration Colour.	Colours of the Orthogonal Positions.	Retardation Colour.
Green	C. Red-orange / A. Blue-indigo	White
Greenish yellow	C. Green / A. White	Greenish white
Yellow	C. Orange / A. Bluish white	White
Yellow-orange	C. Red-orange / A. Dark indigo	Light orange
Orange	C. Red-orange / A. Light lilac	Bluish lilac
Orange	C. Reddish orange / A. Violet	Whitish
Red-orange	C. Red-orange / A. Violet-red	Lilac
Light red-orange	C. Greenish / A. Light red	Greenish

In this synopsis only the first combination series—in which each plate separately gives less than white I., the combination for $e=0$ therefore giving not quite red I.—admits of the recognition of a certain conformity to law. In this case the colour of the orthogonal consecutive position, like the diagonal position of acceleration, always remains *above* the red of the selenite plate, the colours of the alternative and retardation positions, however, always *below* it; "dark red" denoting the red which forms the passage to violet II., "light red" that to red-orange.[1]

The second combination series is obtained by pairs of plates of which each one separately gives white I., or a higher colour. The colours of the different positions do not here admit of interpretation, since they lack a definite acceleration or retardation character.

Let us now reverse the problem, and ask ourselves what data the relations we have developed afford us for determining from known polarisation colours the unknown ellipses of elasticity of two superposed plates. First, it is clear that the colours of the maximum acceleration or retardation, which are observed on rotation upon a selenite plate, always correspond to the two positions which we have denoted by the diagonal position of acceleration and retardation, and that they consequently disclose to us the position of the acute angle e, made by the major axes of the ellipses of elasticity. The bi-sectional line of this angle is thence determined.

Secondly, observation decides whether the colours of the two orthogonal positions are sufficiently distinctly outlined that the consecutive or alternative position may be recognized as such. Where this is the case, the position of the lower and of the upper ellipse of elasticity is determined in the sense explained above; where it is not the case, the investigation can, as a rule, be repeated with smaller objects of the same kind which give rise to a considerably stronger colour. With most organised structures (membranes, fibres, prosenchymatous cells, &c.), which act as a pair of plates, we can thus succeed in producing a change of colour which unmistakably belongs to the first combination series of the above synopsis, and which therefore admits of an accurate interpretation.

[1] Cf. Nägeli: "Die Anwendung des Polarisationsmikroskops," Beiträge III. p. 98.

The possible positions of the ellipses of elasticity are now limited to the space which the acute angle leaves vacant. In order to determine this angle we may endeavour to observe separately the colours of the two lamellæ in sections cut perpendicularly to the surface, so as to draw further conclusions from their rise and fall on rotation in different directions. In the majority of cases, however, we shall arrive at the conviction that this part of the problem does not admit of solution.

IV.

THE ACTION OF CYLINDRICAL AND SPHERICAL OBJECTS COMPOSED OF CONCENTRICALLY GROUPED ANISOTROPIC ELEMENTS.

A FURTHER problem, frequently occurring in practice, is the determination of the axes of elasticity in spherical and cylindrical structures, whose component parts lie in radial series round the axis or the centre, as in starch-grains, cylindrical fibres and tubes, &c. The difficulties to be surmounted are similar in nature to those of the previous case; they consist in the inequality of the positions of the double-refracting elements to the transmitted light, and in the different hypotheses which may be based thereon. The process of investigation is therefore prescribed: we must endeavour to clearly grasp all the conceivable possibilities, and then by observation reduce them to the narrowest circle.

1.—CYLINDRICAL OBJECTS.

The term *cylinder*, or hollow cylinder, does not of course refer merely to the outward form, but rather to the inner structure—that is, to the arrangement of the double-refracting elements. We will assume that the elements of equal radius agree with reference to the position of their planes of vibration and to the effective elasticities; that all radii lying in a plane through the axis are equal; and that all the radii, upon a section at right angles to the axis, act as if the said radius were carried round in a circle.

On these suppositions a transverse section of the cylinder gives the simplest combination, since the effects of the superposed elements become added together as in a simple crystal; moreover, the positions of the ellipses of elasticity effective upon a radius agree with each other (Fig. 200). The combined effect is at once evident from these facts. Each diameter of the section acts as a crystal needle (or, in thicker sections, as a plate of crystal placed vertically), and the interference colours, which come into view successively on rotating the crystal needle round a perpendicular axis, appear side by side at the same time upon the sectional surface.

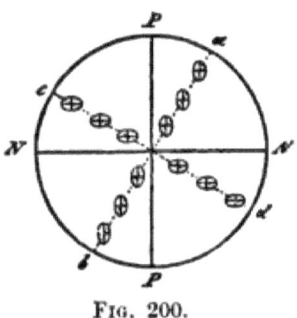

Fig. 200.

Two diametral zones $a\ b$ and $c\ d$, in which the axes of the ellipses of elasticity lie in the polarising planes $P\ P$ and $N\ N$ of the Nicols, consequently act as single-refracting media; without the selenite plate they appear black, and with it they appear illuminated by its normal interference colour. The middle lines of the intermediate quadrants of necessity, therefore, show the brightest colours, and, in combination with a selenite plate, pairs of acceleration or retardation colours, according as the homologous axes of the ellipses coincide or intersect at right angles. Upon a selenite plate, red I., whose ellipse of elasticity takes the position represented in Fig. 201, the quadrants of the cylinder denoted by A would therefore increase the colour, whilst those denoted by S would diminish it.

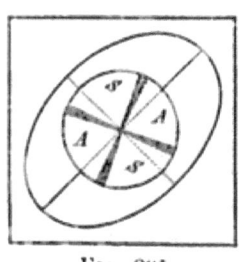

Fig. 201.

Hence, in transverse sections of cylinders and hollow cylinders, the axial position of the effective ellipses of elasticity is easily discovered. We will only add, that all hitherto known cases so far agree, that the one axis is situated radially, the other at a tangent, while the neutral zones which form the dark cross always lie in the polarising planes of the Nicols. The deviation (from this) shown in Figs. 200 and 201 is not therefore actually observed, but is merely assumed for the sake of generality.

The transverse section of the cylinder gives us, therefore, the means for deciding whether the axial directions of the ellipses of elasticity coincide with the radius and the tangent, or intersect obliquely, and in the former case informs us which is the greater—the tangential or the radial axis. But what have we gained by this? Granted that the two axes really are parallel to the radius and to the tangent, then it must further be asked whether perhaps one of them is at the same time the axis of the ellipsoid, or whether we are dealing with a diametral section? This question is not always easily answered. The above-mentioned test, by means of rotation upon the two axes, can indeed guide us only in cases where the section is not too thick in proportion to the diameter of the cylinder. With thicker pieces this process is not applicable, for when they are appreciably oblique they no longer act as sections. In order to obtain further data, we shall, therefore, in the majority of cases, have to resort to longitudinal sections, and, where we cannot obtain these, longitudinal views.

In the first place, as regards *longitudinal sections*, it is evident that a middle lamella $B\,B$ (Fig. 202), if it lies flat upon the object-stage, determines the axial position of the ellipse of elasticity in the diametral plane of the cylinder. For since the marginal portions of such a lamella act approximately as a plate of crystal, we shall immediately learn whether the two axes of the longitudinal and latitudinal directions are parallel, or whether they cut these directions obliquely. We may provisionally disregard this latter case wholly, since an intersection of this kind never occurs, so far as we can judge from our observations.

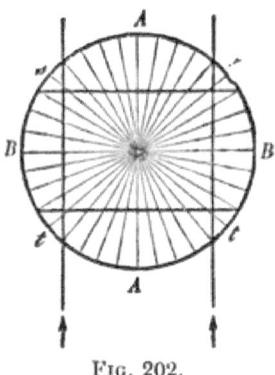

Fig. 202.

We may hence assume that the ellipses of elasticity effective in the latitudinal and longitudinal sections have their radial axis in common. The plane through the other two axes is therefore perpendicular to the radius. From this we can draw the further conclusion, if we consider the properties of the ellipsoid—that the common radial axis may be an axis of the ellipsoid, and that, consequently, the other two axes lie in a tangential plane. It is, therefore, only necessary to determine the ellipse of elasticity of one tangential section—as, for instance, of the segments $A\ A$ of the cylinder (Fig. 202)—in order to fix not only the directions of the axes in question, but also their relative magnitudes.

We have not, however, yet arrived at the ratio of the two tangential axes to the radial axes—except only in the case where the former correspond to the longitudinal and latitudinal directions of the cylinder. In every other case the observer is compelled to discover this ratio in sections which are made through the axes to be compared, or, if they do not lead to a result, by inclination of the cylinder, to which we shall refer later on.

If it should appear that sections, which have been made perpendicular to one of the three axes, act as single-refracting media, then the two other axes are equal—that is, the double-refracting elements are *optically uniaxial*. If, on the other hand, each section made through two axes produces colour, the elements of the cylinder are *optically biaxial*. Whether an object belongs to the one or the other category, cannot, as a rule, be determined until the direction of one optic axis is discovered experimentally, by sections or by rotation. It would indicate a very superficial knowledge of the phenomena if one imagined that the optical properties of an object can be discovered in every given section.[1]

---

[1] The differences observed in such sections have been denoted by some authors by expressions which have a totally different significance in crystallographic Optics. Mohl and **Valentin**, for instance, speak of *negative* and *positive* colour or composition of an object, according as the ellipse of elasticity, which comes into play in transverse sections, is situated radially or tangentially. In other places the major or the minor axis of the ellipse is briefly called the *optic* axis, and it is therefore tacitly assumed that it lies in the plane of the field of view, and so on. It is hardly necessary for us to point out the inaccuracy of these and similar explanations, such as are found in the literature of the subject; the reader who has followed our explanations will be able to exercise the needful criticism.

As an example we may, however, mention briefly a few results of Valentin on the action of cylindrical structures ("Die Untersuchung der Pflanzen- und Thiergewebe im polar. Licht," p. 161). Valentin assumes the uniaxial com-

Secondly, we take the case where the cylinder cannot be split up on account of its minuteness, or from other causes, and that hence we can only obtain *longitudinal views of the whole cylinder*. It may be foreseen that such views, in most cases, replace the diametral longitudinal sections (*B B*, Fig. 202), the effect of light which the two marginal zones produce being essentially the same. Their agreement will be the more complete, the more we diminish the refraction of the light which takes place on entrance into the substance of the cylinder. If we are successful in completely eliminating this refraction by proper selection of the surrounding medium (glycerine, oil, &c.), the rays incident from below act just as in a very thin plate—*i.e.*, they attain the same differences of path next to the margin, as if the middle piece *B B* (Fig. 203) alone were active. Further inwards, the gradually increasing effects of the triangular pieces *s* and *t* are added; though it is clear that they in no case preponderate in the peripheral layers of the cylinder. The two marginal zones must therefore,

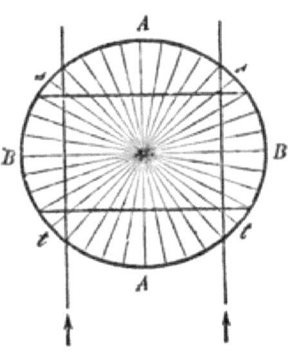

Fig. 203.

position of the elements of the cylinder, and then explains the appearance of the dark cross in the section as the "three principal directions of the optic axis:" the perpendicular, the tangential, and the radial. To the perpendicular position of the optic axis the "cross of the first order" corresponds; to the tangential position the "cross of the second order;" and to the radial position the "cross of the third order." And it is remarked that there are preparations from the vegetable kingdom (albumen of *Phytelephas*) in which crosses of the first, and others of the third order, appear side by side in one and the same section. Against this statement we must first of all observe that the assumptions themselves, so far at least as vegetable preparations are concerned, are at variance with reality, inasmuch as all the membranes of cells that are accurately known (it is true there are but few) have been found to be optically biaxial. But even disregarding this, the "cross of the first order" is a physical impossibility under the polarising Microscope (especially with amplifications such as are necessary with most of the objects of this class), for with a perpendicular axis the whole section appears dark. We must emphatically remark that by a *polarising Microscope* we understand an ordinary Microscope fitted with polarising prisms—the sole instrument, moreover, by which the microscopically small objects (such as bast-cells, vegetable albumen, &c.), mentioned by Valentin himself, can be observed. The image which an instru-

under all circumstances, exhibit the colours of a diametral section, if the surrounding medium is of approximately equal density.

The middle zone, however, acts in the longitudinal view, as before remarked, like two superposed plates of crystal. Observation can therefore, at most, decide whether the axes of elasticity of the longitudinal and latitudinal directions correspond, or intersect obliquely; and, in the latter case, whether the middle line of the acute angle, which the major axes together form, lies longitudinally or latitudinally, and which of the two ellipses is the upper one. After what has been noted above (pp. 348—349) on this point, no further explanation is needed to show that these data are in general insufficient for constructing the ellipsoid of elasticity. It is nevertheless important to test, one after another, the different cases which observation may afford, and to determine for each individual case the results obtained.

ment of this kind gives of transverse sections of cylinders, is produced by cones of light whose aperture corresponds with that of the diaphragm—that is, of the polarising Nicol. The total effect produced by such a cone, as regards double refraction, is approximately the same for the eye as if the differently inclined rays had traversed the object in the direction of the axis of the Microscope. A plate of crystal, whose optic axis is perpendicular, thus in reality exhibits no trace of double refraction. The action is quite different of the special *polarising apparatus* of Nœrrenberg, Dove, &c., which are sometimes designated as Microscopes, though incorrectly so. The real image which is here formed, and viewed through the eye-lens, is not the image of the object under examination; the rays which intersect in the plane of the image represent on the contrary, when produced backwards, a pencil of parallel rays, which traverses the object at an inclination to the axis of the Microscope the greater in proportion to the distance of the point of intersection from this axis. In that point of the image, therefore, incident rays of definite inclination produce interference, and upon this circumstance is based the formation of rings as they are observed in plates of calc-spar cut perpendicularly to the axis.

The "cross of the first order" occurs only in the polarising apparatus, but not in the polarising Microscope (very low amplifications excepted). Valentin commits the mistake in this as in other points (*e.g.*, combination with a selenite plate, determination of the axial direction, &c.) of basing his theoretical explanations upon the action of a special polarising apparatus, and then applying the results he has obtained without further consideration to observations with the polarising Microscope. It is manifest that such a method, combined with the arbitrary assumptions mentioned at the outset, must lead to error.

THE ACTION OF CYLINDRICAL OBJECTS. 355

Let A (Fig. 204) be the cylinder, placed horizontally, and B the transverse section of it; and let the ellipses of elasticity, which in the former are effective in the middle and at the margin, and similarly those of the section (here supposed to lie upon the line $C\,C$) be determined by observation. We consequently know of each of them, whether it is produced in the direction of the axis or at right angles to it, or whether it acts neutrally and therefore represents a circular section of the ellipsoid. We may also remark that these data for the middle of the horizontal cylinder, where the ellipses of elasticity of the side, turned towards and from, intersect, are to be referred to the imaginary ellipse which represents, as above pointed out, the action of the two superposed ones. But since each of these three ellipses may be placed longitudinally or latitudinally to the axis of the cylinder, or may even be a circle, which we will denote by l (longitudinal), q (latitudinal), and o, altogether twenty-seven different combinations result, of which, however, only the following thirteen are possible:—

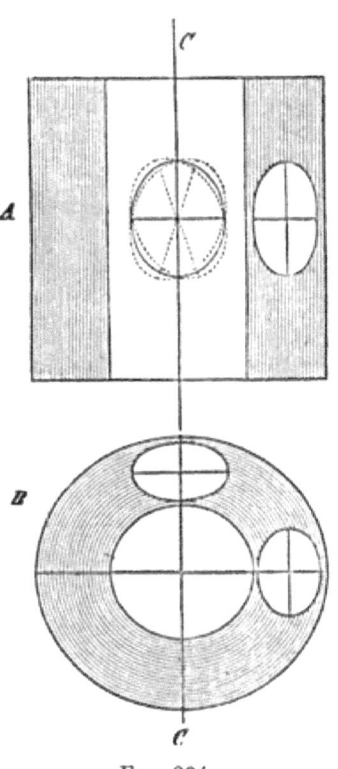

Fig. 204.

	1	2	3	4	5	6	7	8	9	10	11	12	13
Middle	l	q	l	q	l	q	l	q	l	q	o	o	o
Margin	l	q	l	q	q	l	l	q	o	o	l	q	o
Section	l	q	q	l	l	q	o	o	l	q	q	l	o

The other fourteen combinations are:—

	14	15	16	17	18	19	20	21	22	23	24	25	26	27
Middle	l	q	o	o	l	q	l	q	o	o	o	o	l	q
Margin	q	l	l	q	o	o	q	l	o	o	l	q	o	o
Section	q	l	l	q	q	l	o	o	l	q	o	o	o	o

The latter are impossible, because one always of the three positions taken up is not capable of being combined with the other two, as may easily be proved. For instance, from combination 14, L denoting the longitudinally situated axis of the ellipse (of the middle and marginal views), R the radial axis, and T the tangential axis, we obtain the impossible inequalities: $R < T, R > L, L > T$. The proof for the other combinations is just as easily supplied.

The conclusions which may be drawn from the thirteen possible combinations naturally arrange themselves in two series. The first series contains those cases where one of the three axes of the ellipsoid is parallel to the axis of the cylinder; the second embraces all the remainder having any oblique position of the tangential axes. Both series are placed side by side with the combination series in the following table, for more convenient comparison. The signs we have used therein are to be interpreted as follows: $\pm$ positive and negative—*i.e.*, undecided whether the one or the other; plane of the axes = plane of the optic axes; tangential = parallel to a tangential plane situated in the surface of the cylinder; radial = in the plane drawn through the radius and the longitudinal axis (L); transversal = in a plane perpendicular to the axis L; L = longitudinal axis—*i.e.*, axis of the ellipsoid of elasticity, which in the first series is parallel to the axis of the cylinder, and in the second series is inclined less than $45°$ to it; T = tangential axis—*i.e.*, the axis of the ellipsoid, which is perpendicular to the plane drawn through L and the radius of the cylinder; R = radial axis.

Combination.				Optical Character of the Elements of the Cylinder.	
No.	Middle.	Margin.	Section.	First Series: L parallel to the axis of the cylinder.	Second Series: L oblique, inclination $< 45°$.
1	l	l	l	Biaxial $\pm$, plane of axes tangential; L major, T minor axis of elasticity.	Biaxial $\pm$, plane of axes tangential; L major, T minor axis of elasticity.
2	q	q	q	Biaxial $\pm$, plane of axes tan.; T major, L minor axis.	Biaxial $\pm$, plane of axes tan.; T major, L minor axis.
3	l	l	q	Biaxial $\pm$, plane of axes radial; L major, R minor axis.	Biaxial $\pm$, plane of axes radial; L major, R minor axis. Or, biaxial $+$, plane of axes tan L major, T minor axis Or, uniaxial $+$ L = optic axis.

THE ACTION OF CYLINDRICAL OBJECTS.

No.	Combination.			Optical Character of the Elements of the Cylinder.	
	Middle.	Margin.	Section.	First Series: L parallel to the axis of the cylinder.	Second Series: L oblique, inclination < 45°.
4	q	q	l	Biaxial ±, plane of axes radial ; R major, L minor axis.	Biaxial ±, plane of axes radial; R major, L minor axis. Or, biaxial −, plane of axes tan. ; T major, L minor axis. Or, uniaxial − L = optic axis.
5	l	q	l	Biaxial ±, plane of axes transv. ; R major, T minor axis.	Biaxial ±, plane of axes transv. ; R major, T minor axis. Or, biaxial −, plane of axes tan. ; L major, T minor axis. Or, uniaxial − T = optic axis.
6	q	l	q	Biaxial ±, plane of axes transv. ; T major, R minor axis.	Biaxial ±, plane of axes transv. ; T major, R minor axis. Or, biaxial +, plane of axes tan. ; T major, L minor axis. Or, uniaxial + T = optic axis.
7	l	l	o	Uniaxial + L = optic axis.	Biaxial +, plane of axes tan. ; L major, T minor axis.
8	q	q	o	Uniaxial − L = optic axis.	Biaxial −, plane of axes tan. ; T major, L minor axis.
9	l	o	l	Uniaxial − T = optic axis.	Biaxial −, plane of axes tan. ; L major, T minor axis.
10	q	o	q	Uniaxial + T = optic axis.	Biaxial + T major, L minor axis.
11	o	l	q	Uniaxial − R = optic axis.	Biaxial +, plane of axes tan. ; L and T inclined 45°. Or, biaxial ±, plane of axes radial ; R minor axis. Or, uniaxial +, optic axis tan. and inclined 45°.
12	q	o	l	Uniaxial + R = optic axis.	Biaxial −, plane of axes tan. ; L and T inclined 45°. Or, biaxial ±, plane of axes radial ; R major axis. Or, uniaxial −, optic axis tan. and inclined 45°
13	o	o	o	Single-refracting.	Biaxial neutral, plane of axes tan. ; L and T inclined 45°.

It is a simple task to deduce the conclusions shown in the above table from the positions determined by observation, or to render them perceptible through models or figures, and no further explanation will be required by readers who clearly grasp the given ratios in their minds. It will amply suffice if we show by an example the course that has to be pursued.

Let the observed combination be *l l l* (the first in our series), and let the longitudinal axis *L* of the ellipsoid be placed obliquely to the axis of the cylinder. Then the limiting value of the angle which the axis *L* forms with the axis of the cylinder is determined

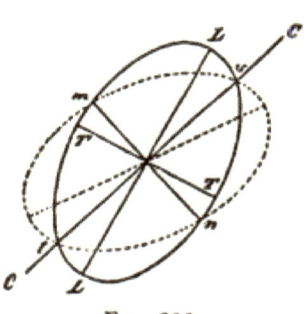

Fig. 205.

by the intersecting ellipses, which affect the middle of the horizontal cylinder —*i.e.*, it is in this case less than 45°. These ellipses are consequently situated as represented in Fig. 205—that is, if *C C* is parallel to the axis of the cylinder. We get, therefore, *s t* > *m n*, and *L L* greater than *T T*. The diameters *m n* and *s t*, however, which correspond to the latitudinal and longitudinal directions of the cylinder, represent respectively an axis of the ellipses which affect the transverse section and the margin of the horizontal cylinder; *m n* is the tangential axis of the ellipse of the section, *s t* the longitudinally situated one of the marginal view. But, from the form of these ellipses, as it is fixed by the combination *l l l*, it is clear that the radial axis *R* of the ellipsoid is greater than *m n*, and less than *s t*; it must therefore be the mean axis of the ellipsoid. Hence the elements are biaxial, and the optic axes lie in a tangential plane. The ratio of the axis *R* to the two others, and the situation of the segments dependent thereon, still remain undetermined, and the positive or negative character of the elements is therefore doubtful.

A few simple combinations, of the thirteen above-mentioned, deserve additional special consideration on account of the optical effect they produce. We will, first, emphasize the case given in the combinations 5 and 6 of our series, where the margin and the middle of the horizontal cylinder produce colours of opposite character. Its characteristic feature is that on observing with selenite plates, neutral longitudinal stripes at once appear, which form the transition between the opposite colours of the marginal and surface views, and which therefore appear without selenite as dark lines (Fig. 206). These lines represent the places where the resulting ellipse of elasticity of the surface view (which, of course, gradually passes over on both sides into the one of the opposite margin) has attained a circular form — that is, where the superposed ellipses of the half of the cylinder, turned towards and from, intersect at right angles. A rectangular intersection, however, presupposes, on account of symmetrical position, that the two ellipses are inclined less than 45° to the axis of the cylinder.

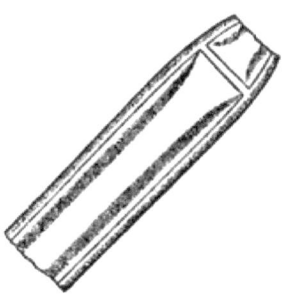

Fig. 206.

If the question were considered from a purely geometrical point of view, we might suppose that, from the known distance of the neutral lines from the margin, we could determine the unknown angle ϵ, which the superposed ellipses form in the median zone; for it is obvious that the distance from the middle line, at which the above-mentioned inclination of 45° is reached, is the greater the more the inclination to the middle line deviates from it. But since each layer in the cylinder has its own neutral line, which appears laterally displaced with regard to the adjoining layers, and more or less illuminated by their interference colour, the combined effect does not appear sufficiently decided to be of practical value in the way we have indicated, on account of the minuteness of the objects. The same is also true of the determination of the optic axes, where $\epsilon = 0$. It will suffice then to call attention in general to the rather complicated relations which in this case arise.

The neutral longitudinal stripes appear most perfectly in thick-walled, hollow, cylindrical cells, which refract the light strongly; yet with increased thickness also they are tolerably distinctly visible. We have observed this especially in vegetable hairs (*e.g.*, of *Stachys*), but here and there the tissue-cells also furnish good examples.

A second case, which to a certain extent forms the counterpart of the previous one, is given by combination 13. The margin and centre in the horizontal cylinder here appear neutral; the former because an optic axis is perpendicular, the latter because the two superposed ellipses intersect at right angles. With the distance from the middle line, however, these ellipses take up a more upright position; they now intersect acutely, and therefore give rise to an effect which corresponds to the position l upon the middle line. Towards the margin, however, this action becomes gradually eliminated again, because the difference between the major and the minor axis of the ellipses becomes ever smaller, and at the margin is *nil*. We consequently get two stripes illuminated by interference colours on either side of the middle line, while the latter and the two margins act neutrally.

If we now endeavour to reduce, where possible, the alternatives which figure in the above table to a smaller number, the first question is whether the *direction of the optic axes* may perhaps be experimentally determined. Considered theoretically, the matter is unusually simple, as is readily seen. It is only necessary to find out, by inclination of the section to different sides, the two positions in which the object acts as a single-refracting medium— that is, in which the optic axes are exactly perpendicular. Similarly, on inclination of the horizontal cylinder, we shall discover whether the optic axes lie in a tangential plane or not; for, where the former is the case, the margins, with a definite inclination which corresponds to the perpendicular position of an optic axis, must necessarily act neutrally.

Practically, however, these rules have only a very subordinate value, for the simple reason that observation is not possible with the accuracy that we might *à priori* expect. Transverse sections cannot in most cases be employed, and even marginal views frequently enough give a very doubtful effect. Beyond this, the latter afford only under the most favourable relations the data which are requisite for the determination of the position of the axes. For since the inclination of the ellipsoid of elasticity—that is to say, of its longitudinal axis L—to the axis of the cylinder is in general unknown, the positive or negative character of the elements of the cylinder remains doubtful, inasmuch as the angle which the optic axes together form cannot be approximately measured, nor the presence of a single optic axis testified. The few rules which may be of practical value, with regard to the action of the margins on inclination, are contained in the following paragraphs:—

(1.) If it is possible to discover the direction of the optic axes in the tangential plane, the angle which they make with the axes of elasticity, and consequently the inclination of the latter to the axis of the cylinder, are likewise determined. *Example*:—The middle of the horizontal cylinder gives a colour which corresponds to the position l; on inclination to the one side, whereby the end furthest from the observer falls and the near one rises, the right margin becomes neutral after a rotation of 40°, on rotation to the opposite side, however, at 20°; the left margin acts conversely. The elements of the cylinder are therefore negative, the axial angle is 60°, and the longitudinal axis L of the ellipsoid is inclined in the direction of a left-handed spiral (in botanical terminology) 10° to the axis of the cylinder.

(2.) If the middle of the horizontal cylinder gives the same colour as in the preceding example, and then neither margin becomes neutral on inclination to 45°, the elements of the cylinder are optically positive.

(3.) If the middle of the horizontal cylinder gives the colour of the position q, and then neither margin becomes neutral on inclination to 45°, the elements of the cylinder are optically negative.

In two special cases the middle also of the horizontal cylinder must act neutrally on a definite inclination. The one case occurs when the horizontal position produces a colour corresponding to the position l, and when, moreover, the radial axis of the ellipsoid is the minor one. Then the inclination of the cylinder implies a gradual approximation of the effective ellipses of elasticity to the crossed position which they would also attain after a revolution of 90°. These ellipses must therefore, at a definite inclination, deviate exactly 45° from the longitudinal direction, and hence intersect at right angles. Their effects consequently counteract each other; the centre behaves neutrally.—The second case is analogous to this first one; it occurs when the colour of the horizontal position corresponds to the position q, and when, moreover, the radial axis of the ellipsoid is the major one. Then the two ellipses approach each other on inclination of the cylinder to the longitudinal position, and hence attain also a position in which they intersect at right angles and counteract each other. The conditions of neutrality are, in general, given in both cases, if the radial axis of the middle elements is taken as equal, and hence their uniaxial constitution assumed.

The application of these relations to combinations 3 and 4 in the above table, which alone suffices for the prescribed conditions, affords us a means of at once striking out either the first and third, or the second of the possible cases there adduced.

2.—Spherical and Oval Objects.

We must at once remark that these terms refer only to objects whose double-refracting elements are grouped round one centre in more or less concentric layers, where therefore a certain congruity exists between the outer form and the inner structure. Spheres cut from a crystal or from a cylinder would of course act just as these objects themselves, and therefore do not require a special treatment.

The possibilities which are imaginable in stratified structures with regard to the optical action of the objects fall, on nearer examination of the relations, into two categories, which may always be readily distinguished on revolution of the object during observation. In all possible views that we get on rotation, either the polarisation colours retain the same character and the same order—i.e., we always observe the same position of the neutral lines of the acceleration and retardation quadrants—or this agreement of the different views does not take place,—the image exhibits, on rotation of the object, a very marked change of phenomena, which fails to take place only when perchance the axis of rotation coincides with a certain diameter of the object which is distinguished from the other.

It is apparent that with objects of the first class the double-refracting elements must be situated symmetrically with regard to the position of their axes of elasticity round each radius, and thence round each point of the surface; while in the latter case such symmetry, just as with a cylinder, is indicated only with regard to that particular direction which may be denoted as the axis. Rotation of the objects under the Microscope therefore discloses to us whether they are to be considered as spheres with equal diameters, or as bodies with a particular axial direction.

A. *Objects with Equal Diameters.*

The position of the neutral cross to the polarising prisms at once decides the question,—whether one of the three axes of elasticity of the double-refracting elements is situated radially or stands in a definite relation to the radius—or not; for in the former case the neutral lines correspond to the polarising planes of the prisms, whereas in the latter they cut them obliquely. Strictly speaking, indeed, this latter case cannot be combined with the equality of the diameters and with the constant position of the neutral cross thereby conditioned. For as soon as the axes of elasticity are situated obliquely to the layers or to the radius, the equality of the diameters demands all possible positions in the tangential plane—*i.e.*, the double-refracting elements must assume all the positions which a single element takes up, one after another, by rotation upon a radial diameter; the neutral line therefore lies in the same vertical plane with the axis of the cone which the element describes during rotation. If the axes of elasticity were situated in every direction in space, and if therefore the different positions were equally often represented, then the sphere would act as a single-refracting substance, and any further investigation would be purposeless.

A definite relation of the axes of elasticity to the radius is therefore already indicated by the mere presence of interference colours; the one of the axes having the preponderant number of elements must coincide with the radius, or, at any rate, form an angle of less than 45° with it. For our further consideration we will completely exclude this latter case, and assume here, just as we did with the cylinder and the single membrane-lamellæ, that the double-refracting elements are regularly situated throughout, and that consequently the axes of elasticity are so situated at all points of the object that one of them coincides with the radius, whilst the other two lie in a tangential plane—*i.e.*, in the plane of the surface of the layers.

On this supposition the sphere acts in any given position, at least in its marginal parts, as a middle lamella BB (Fig. 207) which we may imagine to be cut from it—or as a transverse section of the cylinder. For it may readily be shown that the spherical calotte turned to the observer and the one turned from him always act in the same way as the imaginary middle piece, and that they consequently increase its effect, but do not change it. This is at once apparent with uniaxial elements whose optic axis is situated radially. We arrive at the same result with biaxial elements if we consider that the tangential axes of elasticity are situated in every position, and therefore act upon the transmitted ray of light with a value which is mid-way between the minor and the major axes. The effect of the superposed elements is clearly the same as if, with similar position of the tangential axes, they were turned rapidly round their radial axis, and therefore also the same as with uniaxial elements with radially situated optic axis.

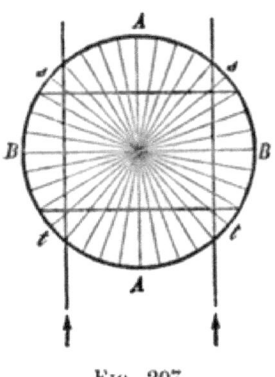

Fig. 207.

Hence the image which a spherical object gives in polarised light decides for us the question, whether the radial axis of the ellipsoid of elasticity is greater or less than the mean value of the two tangential ones; but it leaves undecided whether the elements are uniaxial or biaxial. Assuming the latter case, it further remains doubtful whether the radial axis is the mean one, or, in accordance with the character of the colour, the major or the minor—*i.e.*, it is doubtful whether the optic axes lie in a tangential plane or in a diametral one.

The neutral cross, which separates the acceleration and retardation quadrants from each other, acts exactly as in transverse sections of cylinders. It appears rectangular if the polarising planes of the Nicols carried through the centre of the layers intersect the layers themselves at right angles; and in the opposite case it appears oblique-angled. In circular layers the former is, of course, always the case; but in elliptical ones only when the axes of the ellipses lie in the imaginary polarising planes. In every other position the neutral lines form oblique angles, and if the ratio of the axes does not remain constant, they appear, in general, more or less curved. If, for instance, $P\,P$ and $N\,N$ (Fig. 208) are the planes of vibration of the Nicols, and $a\,b$ and $c\,d$ the axes of the elliptical layers, the neutral lines must appear as represented in the figure; they connect the points of contact of the tangents, leaving $P\,P$ and $N\,N$ at right angles.[1]

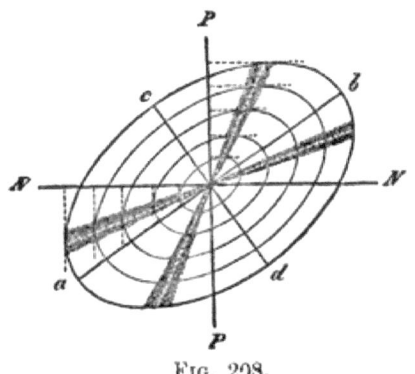

FIG. 208.

B. *Objects with One Axis.*

The double-refracting elements are here grouped round one definite diameter, instead of round a centre. Assuming that the grouping is uniform throughout, so that, for instance, the axes of the ellipsoid of elasticity always form the same angles with the radius and the meridian circle, which intersect in its centre, then such an object acts essentially like a cylinder. With horizontal position of the axis, at least in a middle latitudinal zone (which may be denoted as an equatorial zone), it produces the colours of the horizontal cylinder; and with perpendicular position of the axis, even if only in the peripheral portion, it produces the colour of the cylindrical transverse section. The effect may hence be

---

[1] The fact that the neutral lines in a system of such ellipses are *straight* lines—where consequently the ratio of the minor axis to the major remains the same—follows immediately from the relatively equal magnitude of the ordinates and abscissæ for the points of contact of the tangents. It is, in general, evident that the neutral lines in our figure—with the diameters lying in the planes of polarisation $P\,P$ and $N\,N$—represent two pairs of conjugate diameters, and consequently intersect at angles which vary within certain limits. The construction of the neutral lines for any system of oval layers is, from what has been already shown, a matter of no difficulty; it is only necessary to draw the tangents, and to join their points of contact.

interpreted in the same manner as with the cylinder. We will only remark that no object has hitherto been found which comes under this category.

V.

ON SOME STRUCTURAL PECULIARITIES OF ORGANISED SUBSTANCES.

HITHERTO we have strictly adhered to the supposition that the position of the ellipses of elasticity corresponds, in all points of the substance, to the effect which the superposed elements produce. We have assumed that the double-refracting elements, of which the layers of a membrane—a cylindrical or spherical object—consist, always exhibit the same regular arrangement, and, based upon this hypothesis, that one axis of elasticity is always situated radially, and the other two tangentially. We have then determined the position of the latter according to the effect which the surface views produce, or left them undetermined if these views acted neutrally.

This course would be self-evident with crystalline media with suitable arrangement of the smallest particles; with organised media it is always plausible; but it should be emphatically stated that no absolute necessity exists for the assumption of so regular a structure. Organised substances act in all essential points differently from non-organised ones; their optical character is not, as in the latter, dependent upon the changes of distance which the smallest particles undergo by pressure or tension, or even by swelling; it remains constant, even when the changes amount to a multiple of the original distances. We can stretch or bend a hair, a bast-fibre, &c., at will, without altering the character of its colours; whilst, for example, a fine glass tube, even on very slight curvature, produces the colour which corresponds to the change of distance of its atoms thereby occasioned. Just as little does the swelling of a piece of membrane cause an essential change of the optical properties when it is soaked in sulphuric acid or ammonio-oxide of copper, whereby possibly the thickness is increased fivefold, though the length and breadth but slightly. Hence it follows, however, that the optical properties are situated in those molecular groups—which we shall henceforth term micellæ—though standing in no connection with their distance from each other, or with the tensions which they at all times develope between them. Each separate micella acts as a small crystal, and if the effects become added, the resulting interference colour rises the higher the greater the number which the resulting ray of light has to pass, and thus becomes the more intense the greater the number brought to the unit-area of the microscopical image.

But since organised substances—as we shall further on prove—consist of a mass of different combinations which mutually commingle, and of which each perhaps forms its special micellæ, and which also are possibly differently situated; and since, further, the ratio in which these combinations commingle is, as may be shown, unequal in the different layers; and since, moreover, the necessity of a constantly equal position does not exist (or at least cannot always be proved) in the lodging of new micellæ between those already present though of the like constitution: the transmitted ray is in all probability subjected to the action of differently situated elements, and the ellipse which represents the observed effect can, strictly speaking, be explained only as an imaginary or resulting one, which in regard to form and position is between the real ellipses that act upon the ray. It is, for instance, quite immaterial whether one axis of the ellipse of elasticity is situated radially throughout, or whether in the superposed elements (assuming their number infinitely great) it presents all possible deviations between $0°$ and $30°$ to either side,—the effect is the same in both cases. Similarly, it is immaterial whether the successive micellæ of the same radial series are situated equally or unequally, in as far as the superposed series only exhibit the deviations just mentioned. We may in general assume any arrangement of the double-refracting elements, provided certain positions, which together produce the observed effect, are preponderantly represented.

A further peculiarity, which is connected with the inner structure and the method of growth of organised substances, is the unequal fluid-contents of somewhat dense and loose layers, and the inequality of double-refracting power resulting therefrom. Hence it is not permissible to compare transverse sections and surface views, with regard to their double-refracting power, with the intention of discovering the relative excentricity of the effective ellipses of elasticity from the colours which they produce. If, for instance, a membrane 6 mic. thick gives in the surface view, with a selenite plate red I., the acceleration colour blue II., whilst a transverse section produces the same effect with 4 mic., it does not follow that the ellipse of elasticity of the section possesses a relatively greater, or in general a greater, excentricity than that of the surface view; for with the latter the more effective dense layers form only a fraction of the total thickness, whereas in the section they run continuously from one sectional surface to the other. A comparison in the given case is therefore only admissible between the different tangential directions, as they become effective in transverse, longitudinal, or any oblique sections, if they pass through a radius.

It is manifest that surface views of membranes, whose layers produce colours of opposite character in transverse section, do not admit of interpretation according to the rules we have laid down, when more than two such layers are present.

VI.

COLLECTION OF EXAMPLES.

ALL structures hitherto investigated agree in that one of the three axes of elasticity is situated radially. The two axes lying in the tangential plane exhibit the most variable positions; occasionally they correspond to the longitudinal and latitudinal directions of the cells or fibres, or they may cut these directions at different angles. We will hereafter denote by *left-handed rotation* and *right-handed rotation* the positions in which the longer of the two tangential axes lies with reference to the direction of a left-handed or right-handed spiral (in accordance with botanical terminology).[1] The angle which the spiral coils form with the longitudinal axis we will call ϕ. As regards the position of the ellipses in the transverse

---

[1] A common screw has, according to botanical terminology, a left-handed thread, whilst in mechanics it is termed right-handed.

section, diametral longitudinal sections or marginal views, and surface views of cylindrical structures, we will employ a terminology agreeing in principle with that applied above (p. 356), though in so far differing from it that for transverse and longitudinal sections (and with spherical objects for any sectional views whatever) it is based upon the relation to the course of the layers, not to the diameter. We shall hence denote by p (parallel) and by t (transversal) the positions of the ellipses of elasticity, in which their major axes lie respectively in the direction of the layers, or cut them at right angles. This relation to the course of the layers is, moreover, not arbitrarily chosen, but derived directly from nature, inasmuch as the ellipses of the known objects in longitudinal and transverse sections, almost without exception, show the same position to the stratification. The position of the ellipses represented in Fig. 209, for example, would consequently be denoted as follows: transverse section

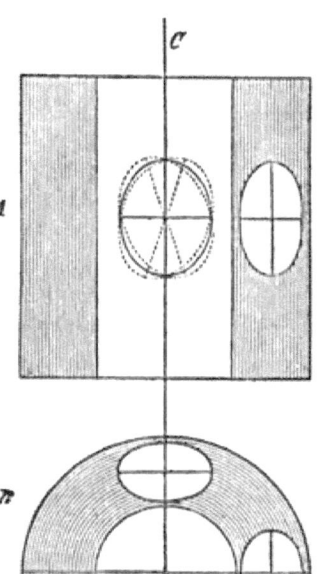

FIG. 209.

p, margin p, surface p. When diametral longitudinal sections were examined in place of the marginal view, the term "longitudinal section," instead of "margin," has been employed. The two orthogonal positions of the object, which serve to distinguish the consecutive and the alternative positions, according to what has been stated above, are represented by horizontal ($=$) or upright (∥) strokes, in which, of course, the same position of the selenite plates, and the usual position of the polarising prisms corresponding to these strokes, are throughout assumed. Since the optic axes always lie in the planes of the major and minor axes of elasticity, it has been thought superfluous to specially name the position of the axial plane. On the other hand, the positive or negative constitution is given with the few objects which have supplied the necessary data.

B B

The ellipsoid of elasticity to which our results refer has a different position from that usually assumed, when the axes are similarly directed. With reference to this, it is an ellipsoid rotated to 90°, or a reciprocal ellipsoid, since its minor axis takes the position of the major according to the usual representation, and conversely (cf. on this point pp. 322-3).

A. Cylindrical Objects.

* A tangential axis is parallel to the axis of the cylinder.

Caulerpa prolifera Lamourx.—Stem. Transv. sect. t, long. sect. t, surface doubtful, in the same piece of membrane agreeing with position t, in other places p. Hence, major axis radial. According to action on rotation, biaxially positive.

Chamædoris annulata Mont.—Transv. sect. p, long. sect. p, surface p. Hence, major axis longitudinal, minor radial, mean cross-tangential. Probably optically negative; on account of unevenness of surface of membranes not observed with certainty.

Acetabularia mediterranea Lamourx.—Stalk. Action similar to *Chamædoris annulata*.

Valonia Ægagropila Ag.—Transv. sect. p, long. sect. p, surface t. Hence, major axis cross-tangential, minor radial, mean longitudinal. Optically positive; angle of optic axes about 80°.

** The tangential axes stand acutely to the axis of the cylinder.

Stachys spec.—Thin-walled hairs of the stalk. Produce neutral longitudinal stripes near the margin. The orthogonal position = appears to be consecutive, the position at right angles to it alternative; consequently left-handed rotation. Angle ϕ, after measurements in split hairs, about 17°; left-handed rotation confirmed. In agreement with the above we get transv. sect. t, margin t, surface p. Hence, minor axis transverse to the left-handed spiral; the ratio of two other axes unknown.

Trifolium rubens L.—Thick-walled hairs of calyx of fruit. Action similar to hairs of *Stachys*; with neutral longitudinal stripes and left-handed rotation; angle ϕ unknown.

Abies excelsa DC.—Old wood, macerated in nitric acid and chlorate of potash. (a.) *Thin-walled wood-cells.* Transv. sect. p, margin p, surface p. Position = appears to be consecutive, position ∥ alternative; consequently left-handed rotation. Angle φ according to observations on bisected cells corresponding to the inclination of the pores and (stronger) spiral stripes—which latter, however, vary between 0° and about 45°; left-handed rotation confirmed. Hence, major axis of elasticity parallel to the stripes · the ratio of the two other axes in general unknown, though in longitudinally-marked cells the radial one always the minor. (b.) *Thick-walled cells.* Transv. sect. of the so-called primary membrane *p*, of the thickened layers *p* or neutral, seldom *t*; margin *p*; surface in combined effect *p*. The orthogonal positions neutral. Hence for the primary membrane, major axis longitudinal, minor radial, mean cross-tangential. The comparison of the heights of the layers which produce definite colours in transverse and longitudinal sections (marginal views) further yields the fact that the two tangential axes can only differ slightly, or are possibly equal (since the surface view of the primary membrane alone is not observed). The elements are therefore negatively uniaxial or negatively biaxial with small axial angle. For the thickened layers we get from the above: major axis longitudinal, the two others equal or only slightly different, judging from the feeble action of the transverse section; the elements consequently positively uniaxial or positively biaxial, with small axial angle.

Cedrus libanotica Lk.—(a.) *Porous wood-cells* with oblique pores corresponding to a left-handed spiral, inclination of which to the axis of the cells amounts to 45° to 80°. Transv. sect. *p*, long. sect. *p*, surface *t*, left-handed rotation. Hence, major axis of elasticity in the direction of a left-handed spiral, the two others unknown. (b.) *Thick-walled wood-cells*, with delicate left-handed spiral fibres. Transv. sect.? Margin in the outer layers neutral, in the inner *p*; surface *p*. Hence, major axis in a tangential plane, the rest unknown.

Taxus baccata L.—Thick-walled wood-cells with left-handed spiral fibres and similarly inclined pores; inclination to axis of cells about 30° to 40°. Transv. sect. *p*, margin *p*, surface *t*. Hence, major axis in a tangential plane, the others unknown. By far the greatest number of wood-cells with right-handed, left-handed, steeply rising or compressed fibres, as also those with circular ones, act similarly; the pores have nevertheless about the same inclination corresponding to a steep left-handed spiral.

Gunnera scabra Ruiz et Pav.—Spiral vessels with right-handed compressed threads, whose inclination to the axis > 45°. Transv. sect. p, margin p, surface t. Consecutive position with regard to the axis of the cells $\parallel$, alternative position $=$, consequently right-handed rotation. The single spiral fibres act tolerably neutrally in transv. sect.; their major axis of elasticity is longitudinal. Hence, major axis of elasticity of the vessels in the direction of a compressed right-handed spiral; the two others unknown, though probably differing but slightly (judging from the action of the isolated spiral fibres).

Periderm and Cuticle.—Any section t. Hence, major axis radial.

Fibres of Muscle.—According to Brücke, and our own observations with muscle of *Coleoptera* agreeing therewith. Long-view p (longitudinal), transv. sect. neutral, orthogonal positions neutral. Hence, major axis of the longitudinal direction parallel, the two others equal or differing but slightly; optically positive. (The opposite result of Mohl is due to an error.)

B. Spherical Objects.

Starch-Grains.—Any sectional view t, surface neutral. Hence, radial axis the major or a mean one situated nearer to the major than to the minor. Possibly uniaxial and then positive.

Inulin granules from the cells of the discs of *Acetabularia*, arising from their being placed in alcohol. Sectional view p, surface neutral. Hence, radial axis the minor or a mean one situated nearer to the minor than to the major. Possibly uniaxial and then negative.

C. Disc-like Crystals of Amylodextrine.

The minute discs of amylodextrine have a similar structure to that of sphero-crystals; the crystal needles are not, however, arranged round a centre, but radially round an axis. They are distinguishable from all the above-named objects by the fact that with vertical position of the axis the black cross is situated diagonally, not orthogonally. The position of the effective ellipses of elasticity is demonstrated by Fig. 210. In accordance with this we get with selenite plate red I., for example, yellow above and below, blue to the right and left, and, if we invert the minute discs, blue above and below, and yellow to the right and left. With horizontal position of the axis we do not observe any double-refraction.

From this action it follows, in the first place, that all crystal needles are situated exactly as if we had led the same needle round in a circle in a plane at right angles to the axis; and further, that no axis of the ellipsoid of elasticity coincides with the radius. Various possible cases are, of course, imaginable as to form and position of the ellipsoid of elasticity.

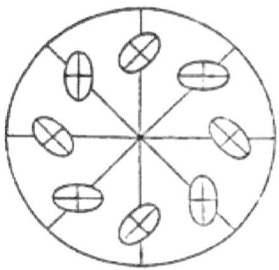

FIG. 210.

VII.

ON THE EMPLOYMENT OF NÖRRENBERG'S "POLARISING MICROSCOPE."

We devote a few remarks to Nörrenberg's "Polarising Microscope," the optical arrangement of which we must here assume to be essentially known. Valentin says of this instrument that in many cases it helps us in researches in polarised light where the ordinary Microscope would no longer be sufficient. We cannot agree with this opinion. Our conviction is that even the best constructed polarising apparatus is of no exceptional importance for microscopic purposes. Such apparatus is, according to its construction, only applicable in cases where the object to be investigated acts as a plate of crystal, and where it, moreover, fills the whole field of view. The powerful lenses with which the Nörrenberg instrument is provided certainly make it possible to reduce the field of view to ·1 mm.; but the condition always remains that only those rays reach the eye which have undergone the double-refracting influence of the object. The lens serving as eye-piece is indeed adjusted to a plane which coincides with the posterior focal plane of the object, or is only slightly distant from it. We do not therefore see the real image of the object under examination through the eye-piece, but merely the system of the interference curves which the incident rays produce in the plane of intersection.

Hence the applicability of the instrument is confined to crystalloid objects or parts of objects, which are large enough for us to be able to observe them separately. As such, for instance, we may name somewhat larger sections of horn and chitine, sections of muscle, cell-membranes of considerable extent, &c. To keep off all stray light, the preparation should be placed upon a small aperture in an otherwise opaque surface, for which purpose a thin leaf of tin-foil, for instance, fastened upon the object-stage is convenient. The interference curves which then appear naturally agree with those which very thin plates of crystal produce in the polarising apparatus. As, however, most text-books of Physics treat this point in sufficient detail, it is unnecessary for us to enter into it more fully.

The great majority of microscopic objects remain inaccessible for the so-called polarising Microscope. Take, for example, any cellular tissue, say a section of fir, where each cell-wall acts as a correspondingly situated plate of crystal,—we should in vain try to cover up all the cell-walls to the very last one, in order to observe their interference curves. It would therefore be futile to think of succeeding with cylindrical and spherical structures, whose crystalloid elements are much smaller than those in prismatic wood-cells!

The observation of cellular tissue in the polarising Microscope can lead to a successful result only if certain walls are so preponderantly represented, that in sections made parallel to them they influence the optical effect—*i.e.*, give rise to phenomena which are not essentially disturbed by the other walls.

INDEX.

ABBE'S method of determining Angle of Aperture, 176.
,, process for testing, 157, 160.
,, theorem of Inclination of Rays, 25
Abbe, E., on Angle of Aperture, 95.
,, on Illuminating Apparatus, 111.
Aberration, Chromatic, 50-53.
,, ,, testing the, 158-167.
,, ,, of air-bubbles in water, 201-202.
,, Spherical, 53-65.
,, ,, testing the, 149-158.
,, ,, in the Eye-piece, 62.
Abies excelsa, D.C., example of polarisation, 370.
Achromatism only approximate, 53.
,, and aplanatism, 53-56.
See "Eye-piece," "Lenses."
Adjustment, Plane of, Interferences in, 240-247.
Air, as medium for dry substances, 281.
,, in prepared sections, how remedied, 279.
Air-bubbles in water, 191-202.
Algæ, study of Division in lower, 280.
Amylodextrine, crystals of, example of polarisation, 372-373.
Analyser, the, 311-315.
Angle of Aperture, determination of, 172-177.
,, ,, effect of Illuminating Apparatus upon, 103-111.
,, ,, importance of, 93-97.
,, ,, relation between Focal Length and, 99-101.
Aperture, Angle of, *see* Angle.
,, of Lenses, diffractional action of, 102-103.
Aplanatism, 53-56, 64, 65.
See "Eye-piece," "Lenses."

Base, the (microscope-stand), 120.
Bénèche, No. 9 objective, determination of cardinal points of, 186.
Binocular Microscopes, 48-50.
,, stereoscopic, 222-226.
Bodies. *See* "Liquid," and "Solid."
Brightness of Field, 84-90.
,, in simple Microscope, 259.
Browning's Spectral Eye-piece, 47-48.
Bubbles, Air-, in water, 191-202.

CAMERA LUCIDA, for drawing, 298.
Campani ("Huyghenian") Eye-piece, 35-42.
Canada balsam, preserving medium, 281.
,, ,, refractive Index of, 305-307.
Cardinal points, Optical, 9.
,, ,, determination of, 23, 184-188.
Caulerpa prolifera Lamourx, example of polarisation, 370.
Cedrus libanotica Lk., example of polarisation, 371.
Cell, Culture-, 280.
Cells, gutta-percha, india-rubber, and glass, 285-287.
,, wood, examples of polarisation, 379.
Cements for sealing, 286.
Centering of systems of lenses, 79-84.
,, testing the, 170-171.
Chamædoris annulata Mont., example o polarisation, 370.
Chevalier's lenses, 263, 264.
Chloride of calcium, preserving medium, 282.
Chromatic Aberration, 50-53; 158-167.
Coddington's lenses, 262.
Colours, Interference, of thin plates, 245-247.
Cover-glass, employment of, 271-272.
,, influence of, 65-68.

INDEX.

Crystals, action of anisotropic crystalloid bodies, 319-341.
,, action of two superposed crystalloid bodies, whose planes of vibration intersect obliquely, 341-349.
,, of *amylodextrine*, example of polarisation, 372-373.
,, Herapathite, 310.
Curvature of the Image-surface, 169.
,, ,, Image-surface, influence of Curvature of refracting surfaces upon, 75-78.
,, ,, Field-of-view, 64-65, 260.
,, ,, Field-of-view, Image-distortion, so-called, 69-70.
,, ,, Field-of-view, distinguished from Image-distortion, 71.
,, ,, Field-of-view, elimination of, 72.
,, ,, Field-of-view, in connexion with determination of Magnifying-power, 178.
Cylinder-diaphragms, 120.
Cylinders, Hollow, 206-217, 349.

DEFINITION, Power of, 148, 149.
Determination of Angle of Aperture, 172-177.
,, Magnifying power and Focal length, 178-183.
,, Cardinal points, 184-188.
Diameters, equal, objects with, 363-365.
Diaphragms, 120.
Dicatopter, Hagenow's, for drawing, 293.
Diffraction in connexion with delineation of fine structures, 226-236.
Diffractional action of Aperture of lenses, 102-103.
Discrimination, testing absolute power of, 126-127.
,, testing relative power of, 140-147.
Distance. *See* "Focal," "Object."

Distortion, Image, 64; 68-78.
,, ,, distinguished from Curvature, 71.
,, ,, elimination of, 72.
See "Aberration," "Curvature."
Drawing of microscopic objects, 296-303.
,, photomicrographic, 303.

ELASTICITY, axes of, determination of, 330-341.
,, ellipsoid or surface of, 319-323.
,, ,, phenomena of polarisation in relation to, 323-330.
Elevations and Depressions, as opposed to dense and soft layers, 221-222.
Eye, Chromatic Aberration of, 166-167.
,, 34-49.
,, fluctnations in magnifying power of, 37.
,, Spherical Aberration in, 62.
,, correction of Aberration in, 72-75.
Eye-piece, aplanatic and orthoscopic, 43-45.
,, Campani's (or Huyghens'), 35-42; 74; 78.
,, Erecting, 45-46.
,, error of supposing that combined action of Field-glass and Eye-glass compensate defective Achromatism of Objective, 161-167.
,, for elimination of distortion and curvature, 72.
,, Holosteric or solid, 44.
,, Ramsden's, 42-43; 74; 78.
,, Spectral, 47-48.
,, micrometer, 179-181; 182.
Eye-point, the, 40-42.
Eyes, care of the, 273-274.

FARRANTS, preparation of glycerine recommended by, 282.
Field of view, Brightness of, 84-90.
,, ,, Curvature of, 64-65; 260.
,, ,, Curvature of, in connexion with determination of Magnifying-power, 178.
,, ,, determination of diameter of, 180.

Field of view, extent of, 261.
,, ,, Flatness of, 64 ; 68-78.
,, ,, ,, testing the, 168-169.
Flatness of the field of view, 64 ; 68-78.
,, testing the, 168-169.
Focal length Definition, 1-2.
,, and Angle of Aperture, 99-101.
,, determination of, 181-183.
,, optical power dependent upon, 271.
,, conjugate Focal lengths, 8-9 ; 18.
,, of a system of lenses, 8.
,, of Objective-system, 27, 34.
,, of refracting systems, 18, 33.
Focusing, 116-117.
Frey (H.), improved Revolving-stage, and method of sealing, 285.

Gauss (C.F.), his exposition of the laws of Refraction, 3 *et seq.*, 24.
Geissler's moist chamber, 280.
Globules, Oil, in water, 202-206.
Glycerine as preserving medium, 282.
Goniometer, the, 294-295.
Goring, application to Microscope of Herschel's distinction between Defining and Penetrating power, 91.
,, on Illumination, 104 (*note*).
Grammatophora marina, 136, 138.
,, *subtillissima*, 137, 138.
Granules, Reflexion of light by, 236-239.
Gunnera scabra Ruiz et Pav., example of polarisation, 371-372.

HARTING, his erroneous statement that irregularities of Centering are magnified as much as the object, 84.
,, his measurements of wire-gauze, 129 (*note*).
,, method of drawing, 299.
,, method of testing Chromatic Aberration, 159.
,, proposed standard of unity in micrometric dimensions, 293.
,, on illumination, 104 (*note*).
,, on Image distortion, 69.

Harting, on Limit of Discrimination, 142 and *note*.
Hartnack's Analyser, 315.
,, Eye-piece, 45, 50.
,, and Prazmowski's Polariser, 309.
,, ,, Stand, 122.
Helmholtz on Limit of Discrimination, 142, and *note*.
,, on Stereoscopic effect of binocular Microscopes, 225 (*note*).
Herapathite crystals in polarisation, 310.
Hermetic sealing of objects, 283.
Herschel's distinction between Defining and Penetrating power, 91.
Hipparchia janira, 133, 138.
Holosteric or solid Eye-piece, 44-45.
Huyghens' Eye-piece, 35-42.

Illumination, 103-115, 268-271.
,, apparatus for, 118, 119.
,, oblique, 110-111 ; 247-251.
,, with reflected light, 111-115.
,, with transmitted light, 104-111.
,, erroneous ideas of influence of, 104 (*note*).
Image-distortion, 64 ; 68-78.
,, distinguished from curvature, 71.
,, elimination of, 72.
Images of wire-gauze as test-objects, 127-132.
Interference phenomena, 226-247.
Inulin granules, example of polarisation, 372.

KELLNER's Orthoscopic Eye-piece, 44.
Krafft's Periscopic Eye-piece, 44.

LABELS for objects, 287.
Lamps used in determining Angle of Aperture, 173-177.
Lantern Microscope, the, 267.
Lateral displacement of coloured objective images, 52.
Layers, alternate, solid, and aqueous, 221.
,, dense and soft, elevations and depressions as opposed to, 221-222.

Lepisma saccharinum, 133, 138.
„   „   for demonstration of diffraction phenomena, 231.
Lenses—Centering of the systems of, 79-84.
„ cleaning and preservation of, 273.
„ construction of, for elimination of Distortion and Curvature, 72.
„ difference and analogy between an infinitely thin lens and given systems of, 4-9.
„ differential action of Aperture of, 102-103.
„ in connexion with Illumination, 107, 110.
„ testing the Centering of, 170-171.
„ Achromatic and Aplanatic, 54.
„ Coddington's, 262.
„ combination of, for determination of Cardinal points, 57-62, 184-188.
„ condensing lens in Polarising Microscope, 304, 308.
„ Cylinder, 262-263.
„ of Eye-pieces, 35-48, 63-65.
„ of Lantern Microscope, 267.
„ of Objective, 27-34.
„ of simple Microscope, 256-266.
  See "Aberration," "Angle of Aperture," "Eye-piece," "Focal length," "Illumination."
Level, differences of, 254-255.
Lieberkühn Mirror, contrast to Illumination, 114.
Light, Interference of direct with reflected, 240-243.
„ Interference of refracted with reflected, 244.
„ Interference of refracted and direct, 245.
„ Interference lines caused by withdrawal of source of, 239-240.
„ technical directions with regard to Illumination, 268-271.
  See "Diffraction," "Rays," "Reflexion," "Refraction."
Lister's method of determining Angle of Aperture, 173-174.
Lycæna argus, 133 ; 138.

MAGNIFYING POWER, determination of, 36, 37, 178-181, 260.
„   „   for different ranges of sight, 37.
„   „   selection of, 271.
Measurement of microscopic objects, 288-296.
„ Units of, compared, 294.
  See "Determination."
Membranes, bounded by one plane and one undulating surface, 219-220.
„ bounded by parallel undulated surfaces, 220-221.
„ with small depressions or holes, 217, 219.
„ drawing of, 300, 302.
Merz's Spectral Eye-piece, 48.
Micrometer, determination of diameter of Field-of-view by, 181.
„ determination of Focal length by, 181-183.
„ determination of Magnifying power by, 179-181.
„ determination of Objective amplification by, 181.
„ Glass-, 288-291.
„ Screw-, 291.
„ Eye-piece Screw-, 291.
Micromillimetre, proposed standard of unity, 293.
Microscope, optical power of, 90-101.
„ testing the, 126-188.
„ component parts of, 27-50.
„ base, 120.
„ diaphragms, 119.
„ focusing, 116.
„ illuminating apparatus, 118.
„ length and position of the tubes, 120-121.
„ object stage, 117.
„ placing the, 269.
„ stands of modern opticians, 122-125.
„ treatment, cleaning, &c., 273.
„ Lantern, 267.
„ Multocular, 48-50.
„ Polarising, 304-319 ; 354 (*note*).

Microscope, Polarising, Nörrenberg's, employment of, 373-374.
,, Simple, 256-266.
  See "Eye-piece," "Lenses," "Objective."
Microtomes, 278.
Mirror, general conditions and directions, 269-270.
,, Sœmmerring's Single Reflexion, used for drawing, 298.
,, Illuminating, used with Polariser, 310.
Mohl on the influence of Cover-glasses on Objectives of medium power, 67.
,, his disapproval of movement of stage in focusing, 117.
,, his method of testing Chromatic Aberration, 159.
,, on the position of the Nicol prism to the source of light, 310-311.
,, on cylindrical structures, 352 (*note*).
,, screw-micrometer, 291-292.
Mœller Test-plates of, 138.
Motion, phenomena of, 252-253.
Mueller, (Otto), Limit of Discrimination, 142.
Multocular Microscopes, 48-50.
Muscle, fibres of, example of polarisation, 372.

NACHET's Erecting Eye-piece, 46.
,, stereoscopic binocular Microscope, 50 ; 224.
Navicula rhomboides, 137, 138.
Needle for preparing specimens, 275.
Nicol's prism, 305-315.
Nitzschia sigmoidea 137, 138.
Nobert's Test-plate, 139-140.
Nodal-points, 8.
Nörrenberg's Polarising Microscope, employment of, 373-374.

OBERHÆUSER's camera lucida, for drawing, 298.
,, Cylinder-diaphragms, 120.
,, Erecting Eye-piece, 45.
,, Horse-shoe stand and Drum stand, 122.

Object-distance increased by approximation of Refracting surfaces, 33.
,, of objective-system, 34.
Object-slide, Bourgogne's, 287.
,, stage, 117.
Objective, the, 27-34.
,, magnifying power of, not constant, 36-37.
,, of multocular Microscopes, 50.
,, combination of lenses in, 61.
Objects, Air-bubbles in water, 191-202.
,, alternate solid and aqueous Layers, 221.
,, cylindrical, 349-362.
,, Elevations and Depressions as, opposed to dense and soft layers 221-222.
,, globules of Oil in water, 202-206.
,, Hollow spheres and Hollow cylinders, 206-217.
,, Membranes with small depression or holes, 217-219.
,, ,, bounded by one plane and one undulating surface, 219-220.
,, ,, bounded by parallel undulated surfaces, 220-221.
,, sections opaque in conseq. of containing air, how remedied, 279.
,, spherical and oval, 362-366.
,, ,, ,, with equal diameters, 363-365.
,, ,, ,, with one axis, 365, 366.
,, action of anisotropic crystalloid bodies, viewed separately, 319-341.
,, action of two superposed crystalloid bodies, whose planes of vibration intersect obliquely, 341-349.
,, apparatus for turning the, 315-319.
,, delineation of the fine structure of, by interference, 226-236.
,, drawing of, 296-303.
,, measurement of, 288-296.
,, polarisation phenomena, collection of examples, 368-373.

Objects, preparation and treatment of specimens, 275-280.
„ preservation of specimens, 280-288. *See* "Crystals," "Liquid," "Membranes," "Solid," "Substances."
Observation, microscopic, theory of, 189-255.
Oil-globules in water, microscopic observation of, 202-206.
Optical Power, 90-92.
Organic test-objects, 132-139.
Orthoscopic Eye-piece, 44, 72.
Oscillarieæ, observation of motion of, 252.

PACINI's liquid for preserving specimens, 283.
Periderm and Cuticle, example of polarisation, 372.
Periscopic Eye-pieces, 44.
Photomicrographic representation of objects, 303.
Pinnularia viridis, 134, 138.
Pipette, the, 278.
Plane of Adjustment, interferences in, 240-247.
Plane, Focal planes, and principal plane of an infinitely thin lens, 5.
„ Principal, of compound Refracting systems, 6, 16.
„ Principal and Focal, of Objective systems, 185-188.
Plate, Rotating, 315-319.
Pleurosigma angulatum, 134, 138.
„ *attenuatum*, 135, 138.
„ „ for demonstration of diffraction phenomena, 234.
„ *balticum*, 136, 138.
Plœssl's Aplanatic Eye-piece, 43.
Points, Principal, of compound refracting systems, 6, 16.
„ Nodal, 8.
„ Optical Cardinal, 9.
„ determination of Optical Cardinal, 23; (in tabular form); 184-188.
Polarisation, phenomena of, 304-374.
„ „ in relation to the Ellipsoid of Elasticity, 323-330.

Polariscopes, use of, 305.
Polariser, the, 304-311.
Power, Defining and Penetrating, 91-92. *See* "Magnifying," "Optical."
Prisms, used for drawing, 298.
„ Foucault's, 308-309.
„ Hartnack and Prazmowski's, 309-310.
„ Nicol's, 305-315.
Pritchard's lenses, 263, 264.
Pseudoscopic effect in binocular Microscopes, 223-224.

RAMSDEN's Eye-piece, 42-43.
Rays of direction, 8, 16.
„ analytical determination of the path of, in refracting systems, 9-23.
„ determination of the distances of corresponding image-points from the axis in the case of any given inclination of the, 24-26.
„ refracted by Eye-piece, 34.
„ course of, within Eye-piece, subsequent to their emergence from Eye-glass, 39-42.
„ means of dividing Pencil of, 48-50; 223.
„ in relation to brightness of field, 84-90.
„ „ Illumination, 103-115; 247-251.
„ Aperture of effective cones of, 257-259.
„ devices for drawing which project the, 298.
„ Polarisation of, 304.
  See "Diffraction," "Illumination," "Light," "Reflexion," "Refraction."
Razors for preparation of sections, 276.
Reflector, Side-, 113 (*note*).
Reflexions by small spheres, granules, fine threads, &c., and Interference phenomena thereby produced, 236-239.
„ Interference due to, 240-244.
„ in drawing of microscopic objects, 297-303.

Reflexions in Polarisation, 305.
Refracting surfaces, distances between, 6, 33.
,, ,, property of re-uniting in a point rays emanating from a point, 6.
,, ,, determination of Optical Cardinal points of, 23; 184-188.
,, ,, Curvature of, influence upon Curvature of Image-surface, 75-78.
,, systems, analytical determination of path of rays in, 9-23.
,, ,, Convergent and Divergent systems, 17.
,, ,, determining Optical Cardinal-points of, 23; 184-188.
Refraction, as expounded by Gauss, 3 et seq., 24.
,, Chromatic aberration, 50-53.
,, Interference due to, 244-247.
,, Lateral Displacement of coloured Objective images, 52.
,, Spherical aberration, 53-65.
,, of Eye-piece, 34.
Riddell's observation of pseudoscopic effect in his catoptric binocular Microscope, 224.
Robinson's method of determining Angle of Aperture, 175-176.
Rotation of objects, 315-319.

S. HACHT's "Lehrbuch der Anatomie and Physiologie," 301.
Schieck's Aplanatic Eye-piece, 44.
Schleiden on Illumination, 104 (*note*).
Schmidt's (C.) Goniometer, 294-295.
Schroeder's Aplanatic Eye-piece, 44.
Scissors for preparing specimens, 278.
Sealing, hermetic, of objects, 283.
Sections—*See* "Objects."
Seibert's Periscopic Eye-piece, 44.
Side-reflector, silver or parabolic, 113 (*note*).

Sight, short, in persons, differences in magnifying power caused by, 37.
,, comparisons of magnifying powers for different ranges of, 37.
Solid glass Eye-piece, 45.
Sœmmering's Mirror, 298.
Specimens—*See* "Objects."
Spectral Eye-piece, 47-48.
Spermatozoa, observation of motion of, 252.
Spheres, Hollow, microscopic observation of, 206-217.
,, small, reflexion of light by, 236-239.
Spherical Aberration, 53-65.
,, ,, Testing the, 149-158.
,, ,, in Eye-piece, 62.
Spiral-threads, drawing of, 302.
Stachys spec., example of polarisation, **370.**
Stage, Object-, 117.
Stands, construction of, 116-121.
,, of modern opticians, 122-125.
Starch-grains, example of polarisation, 372.
Stilling's method of drawing, 299.
Substances, organised, on some structural peculiarities of, 366-368.
Sugar-water as preserving medium, 283.
Surirella gemma, 136, 138.
Swarm-spores, observation of motion of, 252.

Taxus baccata L., example of polarisation, 371.
Test-objects, images of wire-gauze as, 127-132.
,, organic, 132-139.
Test-plate, Nobert's, 139, 140.
Testing angle of aperture, 172-177.
,, Cardinal points, 184-188.
,, Centering, 170-171.
,, Chromatic Aberration, 158-167.
,, Flatness of Field-of-view, 168-169.
,, Focal length, 181-183.
,, Magnifying power, 178-181.
,, Optical Power, 126-149.
,, Power of Definition, 148-149.
,, Power of Discrimination, (absolute) 126-140; (relative) 140-147.

Testing, Spherical Aberration, 149-158.
Threads, fine, reflexion of light by, 236-239.
,, spiral, drawing of, 302.
Tolles's Solid Eye-piece, 45.
Tourmaline, employment of, in polarisation, 309.
Trifolium rubens L., example of polarisation, 370.
Tubes, Microscope, length and position of, 120-121.

VALENTIN, Object-stage with double Rotation, 319.
,, on Cylindrical Structures, 352; (*note*).
Valonia Ægagropila Ag., example of polarisation 370.

Varnishes, for sealing, 286.
Vegetable tissues, preservation of, 282-283.

WATER, air-bubbles in, 191-202.
,, oil-globules in, 202-206.
Welcker's Rotating-plate, 316.
Wenham's method of determining Angle of Aperture, 174-175.
,, observation of pseudoscopic effect in his dioptric binocular Microscope, 224.
Paraboloid, 114 and *note*.
Wire-gauze, images of, as test-objects, 127-132.
Wollaston on relative distances of lenses, 264-265.
Work-table, the, 274.

www.ingramcontent.com/pod-product-compliance
Lightning Source LLC
Chambersburg PA
CBHW031413230426
43668CB00007B/300